Gerhard Lutz

Semiconductor Radiation Detectors

Device Physics

With 167 Figures and 11 Tables

Springer

Dr. Gerhard Lutz
Semiconductor Laboratory of the Max-Planck-Institutes
for Physics and Extraterrestrial Physics
Otto-Hahn-Ring 6
81739 Munich, Germany
e-mail: Gerhard.Lutz@hll.mpg.de

Cover picture: A large area double sided silicon detector and several small detector and test structures fabricated on a silicon wafer of four inch diameter in the MPI Semiconductor Laboratory at Munich. (Foto Filser)

2nd printing of the 1st ed. 1999

Library of Congress Control Number: 2007925696

ISBN 978-3-540-71678-5 Springer Berlin Heidelberg New York

Springer is a part of Springer Science+Business Media

springer.com

© Springer-Verlag Berlin Heidelberg 1999, 2007

Typesetting: LE-TEX Jelonek, Schmidt & Vöckler GbR, Leipzig
Production: LE-TEX Jelonek, Schmidt & Vöckler GbR, Leipzig
Cover: WMXDesign GmbH, Heidelberg

Printed on acid-free paper   54/3180/YL - 5 4 3 2 1 0

Sen   conductor Radiation Detectors

Preface

Semiconductor detectors for nuclear radiation and particles have experienced a rather rapid development in the last few years. Although these developments have been documented in a large number of publications, it seemed useful to collect this information in the form of an introductory textbook that also includes the basic concepts behind the most recent developments.

The book is intended to serve as a basis for academic teaching as well as a guide and reference for all those active in the development or use of semiconductor detectors.

Semiconductor detectors are now used in a large variety of fields in science and technology, including nuclear physics, elementary particle physics, optical and x-ray astronomy, medicine, and materials testing – and the number of applications is growing continually. Closely related, and initiated by the application of semiconductors, is the development of low-noise low-power integrated electronics for signal readout.

The success of semiconductor detectors is due to several unique properties that are not available with other types of detectors. Examples of these properties are: the combination of extremely precise position measurement with high readout speed; direct availability of signals in electronic form; the simultaneous precise measurement of energy and position; and the possibility of integrating detector and readout electronics on a common substrate.

It is worth noting that all these developments have grown out of the need to provide investigative tools for basic research – in this case for elementary particle physics – and also that the fruits of these developments are now of benefit to other fields of science and technology.

In presenting the material, emphasis is given to the principles of physics in detection and device structures, while specific applications and detector systems are left to one side. A major part is devoted to readout electronics and considerations of noise in detector–amplifier systems.

Although detector systems per se are not covered, the demands are dealt with that are made on detector properties by the presently planned applications of tens of thousands of detectors in the harsh radiation environment of newly constructed particle colliders. The production of this large number of detectors requires a simple design that can be produced economically and that can nevertheless cope with the drastic radiation-induced changes to material properties. The field of radiation damage and device stability is therefore also given broad coverage.

There are also some aspects of semiconductor detectors that are barely covered or completely neglected. The emphasis of the book is on silicon detectors. Although other semiconductor materials (and in particular compound semiconductors) are given less prominence, a large number of the physical principles and device structures can also be applied to other detector materials. Completely missing is any reference to cryogenic detectors, which partially also use semiconductors and operate at very low temperatures.

The important subject of detector technology has been treated only in a rudimentary way because it is planned to follow up this book by several volumes from different authors treating other aspects of semiconductor detectors. The next two, already in preparation, will cover detector technology and the physical limits on the measuring precision of detectors.

I have tried to introduce subjects in an intuitive way before resorting to a more formal mathematical treatment. Understanding the physics of the devices is in my opinion not only of importance for people working in detector development but also necessary for selecting and making proper use of the detectors for specific applications. I have also tried to write in a self-contained way so that the book can be read without the need for frequent consultation of other standard literature. Therefore the book contains an introduction to basic semiconductor physics and an appreciable amount of information also available elsewhere in standard textbooks.

In writing this book I have profited from many discussions with colleagues, in particular from my coworkers at the MPI Semiconductor Laboratory. Several of them have also performed the tedious task of careful checking of part or all of the manuscript and have made valuable suggestions for improvements. I want to mention in particular E. Gatti, R. Wunstorf, R.H. Richter, J. Kemmer, L. Strüder, K. Kandiah, P.F. Manfredi, M. Doser, L. Andricek, D. Hauff, N. Hörnel, P. Holl, P. Klein, P. Lechner, H. Soltau, C.v. Zanthier, C. Fabjan, and D. Atkins.

Special thanks go to my wife Ette for her patience and support during the years of my writing this book.

Munich, March 1999   Gerhard Lutz

Contents

1 Introduction

The detection of nuclear charged particles and radiation is made possible by their interaction with matter. The most widely used materials for this purpose are gases – one of the earliest examples being the Geiger–Müller counter – but also liquids (e.g. liquid Argon) and even solid materials. Charged particles and electromagnetic radiation are capable of ionising atoms along their path of flight or in the vicinity of their interaction point, thus producing free electric charge carriers that may be collected and measured directly. One may also use secondary effects, for example the generation of light in scintillators due to a recombination of ions and electrons.

A very large variety of detectors for ionizing nuclear radiation based on ionization in gases has been developed in the last decades, and these detectors are capable of measuring both the position and energy of the radiation. Although semiconductor detectors have been used for nuclear spectroscopy for quite some time, the use of semiconductors for position measurement is of more recent date, but the development in this field has been very fast.

The development of position-sensitive semiconductors was initiated by experimental particle physics, which needed detectors capable of measuring particle tracks with approximately $10 \, \mu m$ precision that at the same time could operate at high rates. This was done for the purpose of investigating the rare 'charmed' particles that were discovered in 1976. Their relatively long lifetime (10^{-13} to 10^{-12} seconds) made a direct observation of their short decay length observable with these detectors.

The development of detectors with these properties was made possible by the adaptation of technologies used in microelectronics for the fabrication of silicon detectors, work that was pioneered by Josef Kemmer in the 1970s. The introduction of silicon strip detectors, produced with this planar technology into a charm search experiment in 1979, marked the start to a revolution in experimental techniques of particle physics that included the development of low-noise–low-power analog microelectronics for the readout of semiconductor detectors. Another important event was the invention of the semiconductor drift detector by Emilio Gatti and Pavel Rehak in 1983, in which the signal charge transport is parallel to the semiconductor wafer surface.

The drift chamber principle and the technological possibilities opened up by the planar process were the source of many new ideas and concepts: electronics was integrated with the detector and completely new semiconductor structures were invented.

Simultaneously, these new detectors spread to other fields of application. The improvement in the capability for energy measurement, which was possible with these new structures, made them interesting for purely spectroscopic applications, and these can now be performed at higher rates and higher temperature. Of particular interest is the possibility of simultaneous precision measurement of position and energy as is required, for example, in x-ray astronomy.

It is the purpose of this book to give a basic understanding of the principles and properties of semiconductor detectors while applications are only touched peripherally. Emphasis is on silicon detectors which have seen the fastest development for which the most advanced technology is available and for which the best knowledge of material properties is available due to the widespread use in electronics.

The basic properties of semiconductors and well known structures – as relevant to detectors – are reviewed in Chaps. 2 and 3. The special properties of semiconductors, which make them so useful for the detection of nuclear radiation, are discussed in Chap. 4. The basic mechanisms of charge generation, which are also the basis for the detection of nuclear radiation, are also discussed in Chap. 2, and further information is given in Chap. 4 for several semiconductor materials and in Chap. 11 with respect to radiation damage. Chapter 4 also gives information on metals and insulators that are used for building detector structures.

The most basic radiation detector, a rectifying junction with or without reverse bias, is discussed in Chap. 5, while more sophisticated detectors are described in Chap. 6. Common to these detectors is their additional ability to measure the position of the incident radiation, even though in some applications one only makes use of their excellent energy measurement capability. The detectors described in Chaps. 5 and 6 do not have intrinsic charge amplification properties, although in some cases the first amplification element of the electronics for the readout function has been integrated with the detector. The question of integration between detector and ancillary electronics (Chap. 8) and the description of detectors with intrinsic amplification (Chap. 9) follow the rather lengthy Chap. 7 on detector electronics in general. A basic explanation of the principles of transistors precedes in Chap. 7 the discussion on the measurement of charge. The emphasis is on the physical processes that limit the measurement precision of signal charge produced in the detectors. Because microelectronics plays an increasingly important role in semiconductor detector readout functions, electronics circuit elements suitable for these technologies are described in the same chapter, and the basic properties of some integrated circuit technologies are discussed. A short treatment of strip detector biasing methods and their noise in conjunction with electronic readout concludes this chapter.

Chapter 10 gives a very short review on device technology and will only be of interest for people not at all familar with detector production. This topic will be extensively covered in a separate volume of this series.

Chapter 11 combines the (to some degree) related subjects of device stability, radiation damage and radiation hardness. The physical mechanisms of electric breakdown of devices and methods to prevent their occurence apply not only to detectors before the introduction of radiation damage but also during and after irradiation. In designing detectors, one has to take into account the changes in materials' properties that are induced through radiation damage. A large part of Chap. 11 deals with the physical reasons for these changes and the way in which these changes are parameterized.

Throughout the book the results of numerical simulations are shown. The design of more sophisticated detectors is hardly possible without numerical simulations. An indication of how device simulation can be performed is presented in Chap. 12. For the sake of simplicity, the presentation has been restricted to one dimension; however, deep-level defects – which in most cases are neglected and which play a dominant part in radiation damaged detectors – have been included.

The final part of the book sets out useful reference material for the work. Appendix A gives a list of the common mathematical symbols used in the text. Appendix B compiles well known physical constants. The choice of units kept throughout the text does not wholly conform to the international standards but follows the tradition in semiconductor physics. The basic units in this system are centimeters for length, seconds for time, and electron-Volts for energy. The last portion of the reference material presented is the list of books and journal articles referred to in the text (using the Harvard system).

Part I

Semiconductor Physics

2 Semiconductors

Basic semiconductor physics is treated in many excellent textbooks (Spenke 1965; Smith 1979; Kittel 1976; Grove 1967; and Sze 1981 and 1985), to which the interested reader is referred. For a semiquantitative understanding of detectors a short treatment of the subject is included. It is based on the usual corpuscular descriptions of electrons and holes within the crystals, with parameters such as effective mass, mean free path etc. obtained from a quantum-mechanical treatment of electrons in the periodic potential of the crystal. The quantum-mechanical basis of semiconductor physics will not be dealt with in this text. It can be found in standard literature (see, for example, Spenke 1965 and Wang 1989).

The treatment of semiconductor physics will be restricted to crystalline material because amorphous semiconductor detectors will not be covered in this book.

2.1 Crystal Structure

Most commonly used semiconductors are single crystals with diamond (with Si and Ge) or zinc blende (with GaAs and other compound semiconductors) lattice type as shown in Fig. 2.1.

Both lattices may be viewed as being composed of two interpenetrating face-centered cubic (fcc) sublattices that are displaced by one quarter of the

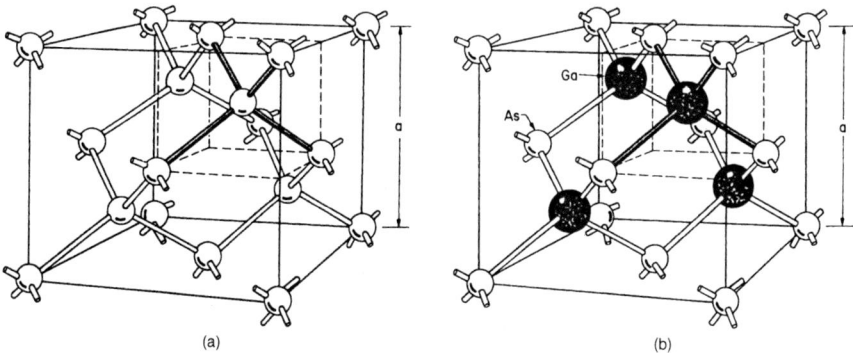

(a)   (b)

Fig. 2.1a,b. Diamond (a) and zinc blende (b) lattice. (After Sze 1985, p. 5 Fig. 3)

distance along the diagonal of the cube. All atoms in the diamond lattice are identical, while the two fcc sublattices are built of different atoms in the case of III–V compounds such as GaAs. Each atom is surrounded by four close neighbors belonging to the other fcc sublattice. They are arranged in a tetrahedron (Fig. 2.2a) and each atom shares its four outer (valence) electrons with those of the neighbors, thus forming covalent bonds. A schematic two-dimensional representation of this situation, which does not conserve the relative position of the atoms, is given in Fig. 2.2b.

Notice in Fig. 2.1 that the $\langle 1, 1, 1 \rangle$ plane (defined by lower-left front, lower-right back and upper-left back corners of the lattice cell) has the highest packing density of atoms. This fact will become important when considering the influence of cutting direction on surface and detector properties.

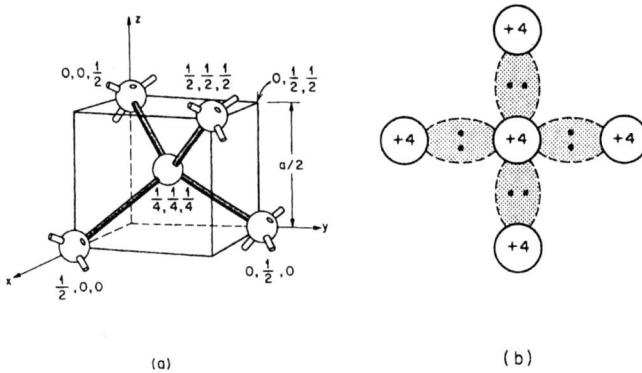

(a)   (b)

Fig. 2.2a,b. Tetrahedron bond (a) and schematic two-dimensional representation (b). (After Sze 1985, p. 8 Fig. 6)

2.2 Energy Bands

The schematic two-dimensional representation of the tetrahedron may be generalized to present a complete crystal (Fig. 2.3).

At low temperatures all valence electrons remain bound in their respective tetrahedral lattice. At higher temperatures thermal vibrations may break the covalent bond and a valence electron may become a free electron, leaving behind a free place or hole. Both the electron and the hole (to be filled by a neighboring electron) are available for conduction. This rather naive qualitative picture can be improved if we imagine the crystal to be assembled from single atoms originally very far apart, so that they do not influence each other and each of them shows the well known discrete energy levels for electrons. One may assume that the atoms are already on a lattice with very large lattice spacing and that this lattice spacing is gradually shrinking.

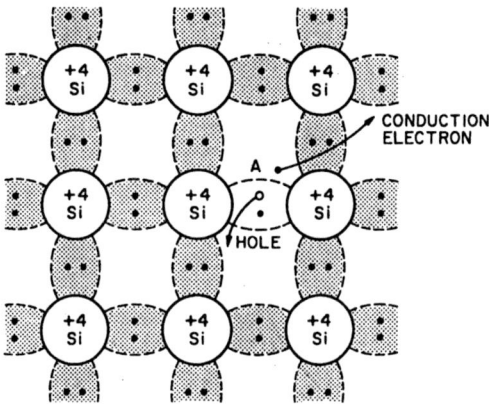

Fig. 2.3. Schematic bond representation of a single crystal with one broken bond in the center. (After Sze 1985, p. 9 Fig. 7 top)

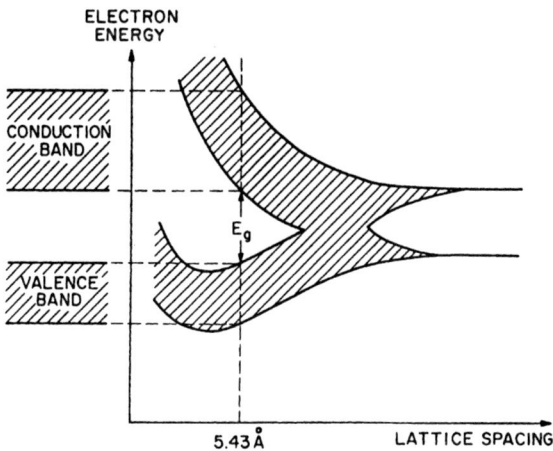

Fig. 2.4. Energy levels of silicon atoms arranged in a diamond structure, as a function of lattice spacing. (After Sze 1985, p. 10 Fig. 8)

The energy levels as a function of the lattice spacing have been calculated using quantum mechanics and are shown for two energy levels in silicon in Fig. 2.4.[1] At very large distances each atom has the same two energy levels; the energy levels are N-fold degenerate (N being the number of atoms), they split into N closely spaced levels when the atoms are brought closer together. For $N \to \infty$, one speaks of energy bands, rather than levels, and these bands broaden, merge and split again with even closer spacing. The spacing corresponding to silicon is indicated in Fig. 2.4 and corresponds to the minimum

---

[1] For a description of the quantum mechanical methods as applied to semiconductors, the reader is referred to standard textbooks, such as Spenke 1965 and Wang 1989.

total energy of the electrons and the lattice, not very far from the minimum energy of the electrons in the filled valence band. At low temperature one has a completely filled valence band and an empty conduction band; at room temperature the thermal energy is high enough to lift a few electrons to the conduction band, thus creating a weak conductivity due to free electrons and holes (Fig. 2.5b).

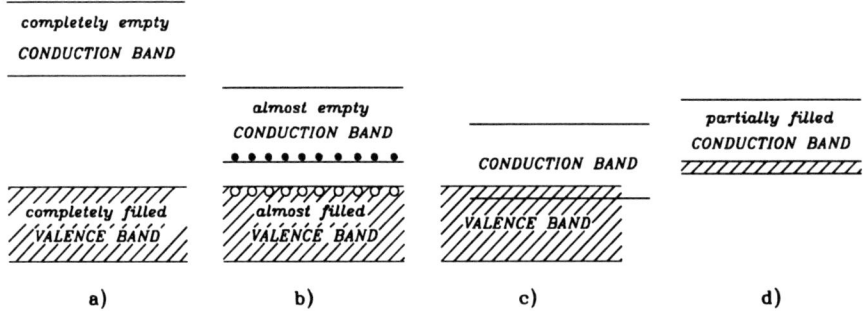

Fig. 2.5a–d. Energy band structure of insulators (a) semiconductors (b) and conductors (c,d)

The structure of an insulator (e.g. SiO_2) is similar (Fig. 2.5a), except that the band gap is much larger so that the occupation probability of states in the conduction band is zero. Conductors may either have overlapping valence and conduction bands (Fig. 2.5c) or a partially filled conduction band (Fig. 2.5d).

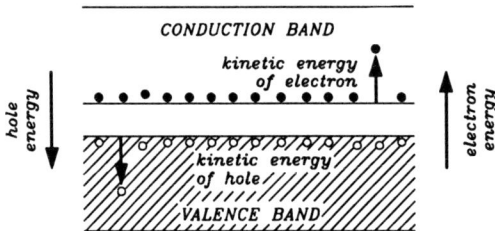

Fig. 2.6. Potential and kinetic energy in the band representation

Without giving any justification but only referring to results from quantum mechanical treatments as found in standard textbooks, we can state that it is possible for many purposes to treat electrons in the conduction band and holes in the valence band similar to free particles, but with an effective mass (m_n or m_p) different from elementary electrons not imbedded in the lattice. This mass is furthermore dependent on other parameters such as the direction of movement with respect to the crystal axis. The kinetic energy of electrons is measured from the lower edge of the conduction band upwards, that of the holes downward from the upper edge of the valence band (Fig. 2.6).

This simple picture has its limitations, however. It does not distinguish between direct (e.g. GaAs) and indirect (e.g. Si) semiconductors. In order to do so, one has to consider the relationship between momentum and energy as functions of the crystal direction (Spenke 1965, Sze 1985). For indirect semiconductors the momentum of holes at the top of the valence band is different from the momentum of electrons at the bottom of the conduction band.[2] Thus a transition between the two states requires a momentum transfer to the crystal lattice. This has consequences for the generation and recombination of charge carriers, as well as the possibility of use in optoelectronic devices.

2.3 Intrinsic Semiconductors

Intrinsic semiconductors contain no (in practice, very few) impurities compared with the number of thermally generated electrons and holes. For this condition we shall attempt to estimate the number of free charge carriers (electrons and holes) under equilibrium conditions. The occupation probability for an electronic state is given by the Fermi-Dirac function

$$F(E) = \frac{1}{1 + \exp\left(\frac{E - E_F}{k \cdot T}\right)} \tag{2.3.1}$$

where E_F, the Fermi energy, is the energy at which the occupation probability of a (possible) state is one half, k is the Boltzmann constant and T is the absolute temperature.

This expression can be approximated separately for electrons and holes if E_F is within the band gap at a distance of more than roughly $3kT$ from either edge, as is certainly the case for intrinsic semiconductors, where it is close to the middle of the band gap. Then we have

$$F_n(E) \simeq e^{-\frac{E - E_F}{k \cdot T}} \tag{2.3.2}$$

$$F_p(E) = 1 - F(E) \simeq e^{-\frac{E_F - E}{k \cdot T}} \quad .$$

The density of states in the conduction and valence bands is obtained in the standard way of considering standing waves in a unit volume of physical space, from which the number of states in a spherical layer of momentum space, corresponding to a range of kinetic energy is obtained. Multiplying by two for the two electron spin directions in addition leads to the number of states $N(E_{kin})$ in the unit volume in a small kinetic-energy interval dE_{kin} around E_{kin}:

$$N(E_{kin})\, dE_{kin} = 4\pi \cdot \left(\frac{2m}{h^2}\right)^{\frac{3}{2}} E_{kin}^{\frac{1}{2}}\, dE_{kin} \quad , \tag{2.3.3}$$

---

[2] See also Sect. 2.6.6 and Fig. 2.17.

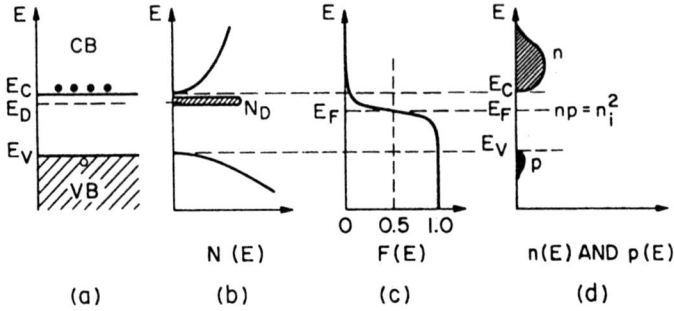

Fig. 2.7a–d. Energy band structure (a) density of states (b) occupation probability (c) and carrier concentration (d) for intrinsic semiconductors. (After Sze 1985, p. 18 Fig. 15)

where we have to take for m the effective mass of electrons m_n or holes m_p respectively, h is Planck's constant, and the kinetic energy measured from the band gap is as indicated in Fig. 2.6.[3]

The density of free electrons n is obtained by integrating the carrier concentration (Fig. 2.7d) given by the product of the density of states N ((2.3.3) and Fig. 2.7b) and the occupation probability $F_n(E)$ (Fig. 2.7c) over the conduction band:

$$n = 2 \left(\frac{2\pi m_n kT}{h^2} \right)^{\frac{3}{2}} e^{-\frac{E_C - E_F}{kT}} = N_C e^{-\frac{E_C - E_F}{kT}} \ . \tag{2.3.4a}$$

Similarily for holes, we have:

$$p = 2 \left(\frac{2\pi m_p kT}{h^2} \right)^{\frac{3}{2}} e^{-\frac{E_F - E_V}{kT}} = N_V e^{-\frac{E_F - E_V}{kT}} \ . \tag{2.3.4b}$$

N_C and N_V are the effective densities of states in the conduction and valence bands respectively. The product of electron and hole concentration, given by $n \cdot p = N_C N_V e^{-\frac{E_C - E_V}{kT}}$, depends on the band gap $E_G = E_C - E_V$ and is thus independent of the Fermi level E_F.

So far the value of the Fermi level E_F has not been specified and equations (2.3.1) to (2.3.4) will also be valid for extrinsic semiconductors (to be dealt with in the following chapter). The Fermi level for intrinsic semiconductors E_i

---

[3] As mentioned earlier, the simple picture of treating electrons and holes as particles with effective masses m_n and m_p has its limitations. The effective mass is given by the inverse of the second derivative of energy with respect to momentum in the minimum of the conduction band (electrons) and the maximum in the valence band (holes). It thus has a strong dependence on the momentum direction with respect to the crystal axes. This direction dependence has been neglected in the derivation of density of states. One may make allowance for it because at low electric fields the isotropic thermal motion of electrons allows us to define the "density of states' effective mass" by an average over all directions.

can be found from the requirement that the numbers of electrons and holes are equal: $n = p = n_i$. One thus finds that:

$$n_i = \sqrt{N_C N_V}\, e^{-\frac{E_C - E_V}{2kT}} = \sqrt{N_C N_V}\, e^{-\frac{E_G}{2kT}} \qquad (2.3.5)$$

and

$$E_i = \frac{E_C + E_V}{2} + \frac{3kT}{4} \ln\left(\frac{m_p}{m_n}\right) , \qquad (2.3.6)$$

where E_i is the Fermi level close to the middle of the band gap, the deviation being due to the unequal effective masses of electrons and holes.

The intrinsic level E_i, according to the definition of (2.3.6), will be of general use also in the case of extrinsic semiconductors, to be dealt with in Sect. 2.4.

Introducing the intrinsic carrier density n_i and the intrinsic level E_i, one may reformulate the more generally valid (2.3.4) in the useful form:

$$n = n_i e^{\frac{E_F - E_i}{kT}} \qquad p = n_i e^{\frac{E_i - E_F}{kT}} . \qquad (2.3.7)$$

Example 2.1
Problem: *Scale the intrinsic carrier concentrations for germanium, silicon and GaAs from room temperature ($T = 300\,\mathrm{K}$) to $0\,°\mathrm{C}$ and liquid nitrogen temperature ($T = 77\,\mathrm{K}$).*
Solution: *According to (2.3.4) and (2.3.5), the effective density of states scales with $T^{3/2}$, so that the intrinsic charge-carrier density varies with temperature as*

$$\frac{n_i(T)}{n_i(300\,\mathrm{K})} = \left(\frac{T}{300}\right)^{3/2} \cdot \frac{\exp(-\frac{E_G(T)}{2kT})}{\exp(-\frac{E_G(300\,\mathrm{K})}{2k \cdot 300})} .$$

Taking the values for intrinsic concentration n_i, band gap E_G and temperature dependence $\frac{\partial E_G}{\partial T}$ from Table 4.1, we obtain with the Boltzman constant $k = 8.62 \times 10^{-5}\,\mathrm{eV/K}$:

	Si	Ge	GaAs
$E_G\,[\mathrm{eV}]$	1.12	0.67	1.35
$dE_G/dT\,[\mathrm{eV/K}]$	-2.3×10^{-4}	-3.7×10^{-4}	-5.0×10^{-4}
$n_i(300\,\mathrm{K})\,[\mathrm{cm}^{-3}]$	1.45×10^{10}	2.4×10^{13}	1.79×10^{6}

One may then rewrite the above formula as

$$\frac{n_i(T)}{n_i(300\,\mathrm{K})} = \left(\frac{T}{300}\right)^{3/2} e^{-\frac{E_G(300\,\mathrm{K})}{2k \cdot 300}(300/T - 1)} e^{\frac{1}{2k}\frac{\partial E_G}{\partial T}(300/T - 1)} ,$$

where the first factor describes the phase-space variation of the available states in the bands, the second describes the dominating exponential temperature dependence and the third describes the effect of the band-gap variation with temperature. Numerical results are summarized in the following Table 2.1.

Table 2.1. Numerical results for Example 2.1

	Si	Ge	GaAs
$T = 300\,\mathrm{K}$			
$n_{\mathrm{i}}(300\,\mathrm{K})\,[\mathrm{cm}^{-3}]$	1.45×10^{10}	2.4×10^{13}	1.79×10^{6}
$T = 0°\mathrm{C} = 273\,\mathrm{K}$	$\left(\frac{T}{300}\right)^{3/2} =$	0.8681	
$\mathrm{e}^{-\frac{E_{\mathrm{G}}(300\,\mathrm{K})}{2k\cdot300}(300/T-1)}$	0.1175	0.278	0.076
$\mathrm{e}^{\frac{1}{2k}\frac{\partial E_{\mathrm{G}}}{\partial T}(300/T-1)}$	0.8764	0.809	0.751
$n_{\mathrm{i}}(273\,\mathrm{K})\,[\mathrm{cm}^{-3}]$	1.30×10^{9}	4.68×10^{12}	8.83×10^{4}
$T = 77\,\mathrm{K}$	$\left(\frac{T}{300}\right)^{3/2} =$	0.130	
$\mathrm{e}^{-\frac{E_{\mathrm{G}}(300\,\mathrm{K})}{2k\cdot300}(300/T-1)}$	5.80×10^{-28}	5.09×10^{-17}	1.48×10^{-33}
$\mathrm{e}^{\frac{1}{2k}\frac{\partial E_{\mathrm{G}}}{\partial T}(300/T-1)}$	0.021	0.002	2.25×10^{-4}
$n_{\mathrm{i}}(77\,\mathrm{K})\,[\mathrm{cm}^{-3}]$	2.30×10^{-20}	3.17×10^{-7}	7.74×10^{-32}

2.4 Extrinsic Semiconductors

Intrinsic semiconductors are rarely used in semiconductor devices since it is extremely difficult to obtain sufficient purity in the material. Moreover, in most cases one intentionally alters the property of the material by adding small fractions of specific impurities. This procedure, which can be performed either during crystal growth or later in selected regions of the crystal, is called doping. Depending on the type of added material, one obtains n-type semiconductors with an excess of electrons in the conduction band or p-types with additional holes in the valence band. We will look at extrinsic semiconductors through the simple bond representation and also through a band model.

Figure 2.8a shows a two-dimensional schematic bond representation of a silicon crystal with one silicon atom replaced by an arsenic atom with five valence electrons. Only four are used for the formation of covalent bonds with neighboring atoms, while the fifth is not bound to a specific atom but is free for conduction. It should be stressed that the crystal as a whole remains uncharged, since the charge of the free electron is compensated for by the excess charge of the arsenic nucleus bound in the crystal lattice.

If a silicon atom is replaced by an atom with only three valence electrons (Fig. 2.8b) one electron is missing in the covalent bonds and a hole is thus created. This hole may be filled by an electron from a neighboring atom, this being equivalent to a movement of the hole. The hole is free for conduction. (That the moving hole is more than a missing electron whose place is filled by a neighboring electron follows from quantum mechanical considerations and is experimentally verified in the Hall experiment.)

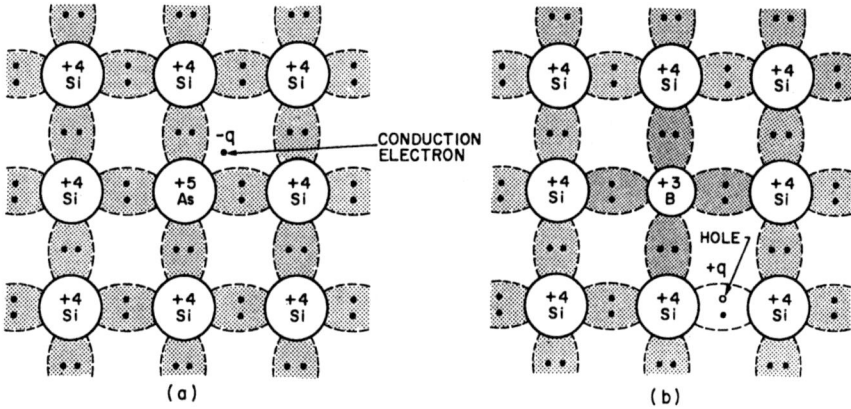

Fig. 2.8a,b. Bond representation of n-type (a) and p-type (b) semiconductors. (After Sze 1985, p. 21 Fig. 21)

Looking at the situation in the band model, we can state that the replacement of a proper atom of the lattice by a different atom is accompanied by the creation of localized energy levels in the band gap. These energy levels may be of the donor (E_D) or acceptor (E_A) type. If donor levels E_D are close to the conduction band, as is the case for phosphorous $(E_C - E_D = 0.045\,\text{eV})$ or arsenic $(E_C - E_D = 0.054\,\text{eV})$ atoms in silicon, for instance, these states will be almost completely ionized at room temperature and the electrons will be transported to the conduction band (Fig. 2.9a). This is due to the many states with similar energy level nearby in the conduction band, with which the donor states have to share their electrons.

Equivalent considerations hold for acceptor-type states, e.g. boron in silicon $(E_A - E_V = 0.045\,\text{eV})$. These states will be filled almost completely and holes will be created in the valence band (Fig. 2.9b).

The situation that the donor levels are almost completely ionized can be described by a movement of the Fermi level E_F from the intrinsic level E_i

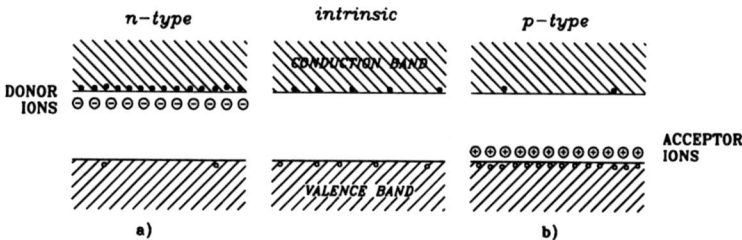

Fig. 2.9a,b. Schematic energy band representation of extrinsic n-type (a) and p-type (b) semiconductors

towards the conduction band. Up to fairly high doping concentrations, the value of E_F follows from (2.3.4) by setting the electron concentration in the conduction band n equal to the donor concentration N_D, thus:

$$E_C - E_F = kT \ln \frac{N_C}{N_D} \ . \tag{2.4.1}$$

Similarily one obtains for p-type material and acceptor concentration N_A:

$$E_F - E_V = kT \ln \frac{N_V}{N_A} \ .$$

A somewhat more complicated treatment is required for the simultaneous presence of donors and acceptors and for very high doping concentrations. It may also be mentioned in this context that the number of donor or acceptor states does not necessarily equal the number of corresponding impurity atoms, since in order to become electrically active doping atoms they have to be properly built into the crystal lattice.[4]

Using the results of the previous section, one may rewrite equations (2.3.4) with the help of (2.3.5) and (2.3.6) in the following rather instructive form already given in (2.3.7):

$$n = n_i \, e^{\frac{E_F - E_i}{kT}} \tag{2.4.2}$$

$$p = n_i \, e^{\frac{E_i - E_F}{kT}} \ .$$

The increase of majority carriers (electrons in the case of n-type material) is accompanied by a decrease of minority carriers according to the mass-action law

$$n \cdot p = n_i^2 \ , \tag{2.4.3}$$

in agreement with (2.4.2).

2.5 Carrier Transport in Semiconductors

So far we have considered semiconductors in equilibrium. This means semiconductors as electrically neutral bodies without application of external voltages and after waiting a sufficiently long time to reach thermal equilibrium. In this chapter we will consider phenomena that occur either through the application of an external electric field (drift) or because of an inhomogeneous distribution of movable charge carriers (diffusion). The process of creation or destruction of free charge carriers will also be considered here.

Movable charge carriers (electrons in the conduction band and holes in the valence band) are essentially free particles, since they are not associated

---

[4] For very high doping concentrations (as used for example in tunnel diodes), the Fermi level may move into the conduction or valence band. In this case the approximations of (2.3.2) introduced for simplifying analytic integration over the state densities become invalid.

with a particular lattice site. Their mean kinetic energy is $\frac{3}{2}kT$, so that the mean velocity at room temperature is of the order of 10^7 cm/s. They scatter on imperfections within the lattice, which are due to thermal vibrations, impurity atoms, and defects. A typical mean free path is 10^{-5} cm and a mean free time $\tau_c \approx 10^{-12}$ s.

2.5.1 Drift

In the field-free case, the average displacement of a movable charge-carrier due to random motion will be zero. However, if an electric field is present, the charge carriers will be accelerated in between random collisions in a direction determined by the electric field and a net average drift velocity will be obtained of

$$\boldsymbol{\nu}_n = -\frac{q \cdot \tau_c}{m_n}\mathcal{E} = -\mu_n \mathcal{E} \qquad (2.5.1)$$

$$\boldsymbol{\nu}_p = \frac{q \cdot \tau_c}{m_p}\mathcal{E} = \mu_p \mathcal{E} \ .$$

This relationship holds for fields small enough that the velocity change due to acceleration by the electric field is small with respect to the thermal velocity and the mean collision time is independent of the electric field. Alternatively, if the field is high enough, such that the electron and/or hole energies become appreciably larger than the thermal energies, then strong deviations from linearity are observed and the drift velocities finally become independent of the electric field at their saturation values $\nu_{s,n}$ and $\nu_{s,p}$.

Scattering occurs on imperfections of the crystal lattice that are due to thermal vibrations and other sources, such as crystal defects and doping atoms. Thus mobilities μ_n, μ_p are dependent on temperature and doping concentration. For a closer description of scattering mechanisms and the resulting dependence of mobility, the reader is referred to the standard literature (see, for example, Sze 1985).

2.5.2 Diffusion

Consider now the situation of an inhomogeneous distribution of free charge carriers in a semiconductor crystal and neglect all effects that are due to electric fields, i.e. the electric field due to inhomogeneous charge-carrier distribution and/or doping concentration. We are thus treating (for the moment) electrons and holes as if they were electrically neutral and we choose a boundary such that on one side the carrier concentration is higher than on the other. Although the net average displacement of an individual charge-carrier is zero in the absence of forces due to an electric field, the probability of carriers crossing from the side showing the higher concentration to that showing the lower is larger than the probability for crossing in the opposite direction, as there are more particles having a chance to do so. This effect, which is called diffusion, will result in a smoothening of the charge distribution. It is mathematically described by the diffusion equation

$$\boldsymbol{F}_n = -D_n \nabla n \tag{2.5.2}$$

$$\boldsymbol{F}_p = -D_p \nabla p \ .$$

Here $\boldsymbol{F}_n$ is the flux of electrons, D_n the diffusion constant and ∇n the gradient of carrier concentration. The corresponding symbols for holes are $\boldsymbol{F}_p$, D_p and ∇p.

Combining the effects of drift and diffusion, one obtains the current densities:

$$\boldsymbol{J}_n = q\mu_n n \mathcal{E} + q D_n \nabla n \tag{2.5.3}$$

$$\boldsymbol{J}_p = q\mu_p p \mathcal{E} - q D_p \nabla p \ .$$

Mobility and diffusion are related to each other by the Einstein equation

$$D_n = \frac{kT}{q} \mu_n \tag{2.5.4}$$

$$D_p = \frac{kT}{q} \mu_p \ .$$

The Einstein equation can be derived by considering on a microscopic scale the drift and diffusion processes taking into account the scattering of charge carriers on crystal imperfections. Alternatively it can be derived from (2.5.3) when considering a system in thermal equilibrium. There the current densities given by the sum of diffusion and drift at any point of the system has to be zero.

Example 2.2

Problem: *Derive Einstein's relation from the requirement of zero current density and constant Fermi level in the case of thermal equilibrium. Consider the case of two semiconductors with different n-type doping densities being joined with each other. Restrict the treatment to one dimension only.*

Solution: *Writing (2.5.3) in one dimension and setting the current density* $\boldsymbol{J}_n = 0$, *we obtain*

$$\mu_n n \frac{\partial V}{\partial x} = -D_n \frac{\partial n}{\partial x}$$

and thus

$$\frac{1}{n} \frac{\partial n}{\partial x} = -\frac{\mu_n}{D_n} \frac{\partial V}{\partial x} \ .$$

Integrating this equation between two points x_1 and x_2 deep within the respective doping regions, we obtain for the ratio between the electron densities

$$\ln\left(\frac{n_2}{n_1}\right) = -\frac{\mu_n}{D_n}(V_2 - V_1)$$

and thus

$$\frac{n_2}{n_1} = e^{-\frac{\mu_n}{D_n}(V_2 - V_1)} \ .$$

From Boltzmann statistics given by (2.4.2), we have

$$\frac{n_2}{n_1} = \frac{e^{\frac{E_F - E_{i,2}}{kT}}}{e^{\frac{E_F - E_{i,1}}{kT}}} = e^{-\frac{E_{i,2} - E_{i,1}}{kT}} = e^{-q\frac{V_2 - V_1}{kT}}$$

and from the two expressions for n_2/n_1 we derive Einstein's equation

$$\frac{\mu_n}{D_n} = \frac{q}{kT} \ .$$

2.5.3 Magnetic Field Effects

One of the advantages of semiconductor detectors, compared with (for example) detector readout by photomultipliers, is their easy operation inside magnetic fields. However, certain effects have to be understood and taken care of. Here we will only consider the movement of single electrons or holes under the Hall effect and refer to later sections for specific devices.

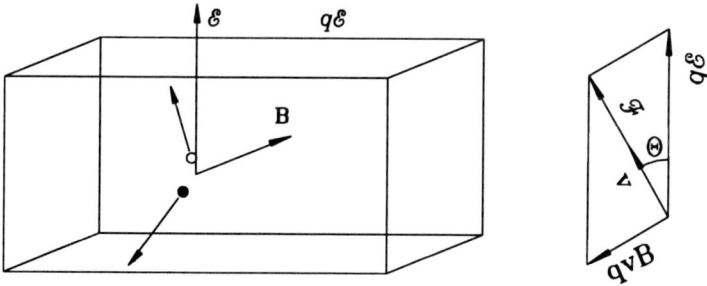

Fig. 2.10. Movement of electrons and holes in the presence of a magnetic field

The movement of electrons and holes in the simultaneous presence of an electric and magnetic field is shown diagramatically in Fig. 2.10 for the special case of orthogonal field directions. Without magnetic field an electron–hole pair would just separate in the electric field, the hole moving along the field due to the electrostatic force $\mathcal{F} = q \cdot \mathcal{E}$ with a drift velocity of $\nu_p = \mu_p \mathcal{E}$, and the electron in the opposite direction with velocity $\nu_n = -\mu_n \mathcal{E}$. In the presence of the magnetic field, however, one has to add to the electrostatic force the Lorentz force $\mathcal{F} = q(\mathcal{E} + \boldsymbol{v} \times \mathcal{B})$.

This leads to a sideways displacement of the charge movement from the lines of electric field in such a way that in Fig. 2.10 electrons and holes both move to the left. The angles of inclination θ between velocity and field vector are read from the diagram of force as

$$\tan \theta_p = \mu_p^{\text{H}} \mathcal{B} \tag{2.5.5}$$

$$\tan \theta_n = \mu_n^{\text{H}} \mathcal{B} \ ,$$

where the superscript H stands for "Hall", indicating that the Hall mobility differs somewhat from the drift mobility.[5]

For silicon at room temperature, $\mu_n^{\text{H}} = 1670 \, \text{cm}^2/\text{Vs}$, $\mu_p^{\text{H}} = 370 \, \text{cm}^2/\text{Vs}$. Thus the inclination angle for electrons is a factor $\mu_n^{\text{H}}/\mu_p^{\text{H}} = 4.5$ larger than that for holes. For a $1\,T$ field, one has $\tan(\theta_n) = 1670 \times 10^{-4} \, \frac{\text{m}^2}{\text{Vs}} \cdot 1 \, \frac{\text{Vs}}{\text{m}^2} = 0.167$, yielding $\theta_n = 9.5°$. Similarily, θ_p is found to be $2.1°$.

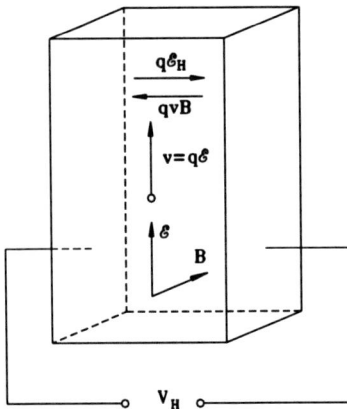

Fig. 2.11. The Hall effect

The Hall effect is based on the same physical process, although a different experimental setup is used (Fig 2.11). Here the Lorentz force $q\boldsymbol{v} \times \mathcal{B}$ is counterbalanced by a naturally built-up electric field $q\mathcal{E}_{\text{H}}$ in such a way that the (single-type) charge carriers keep the flow direction that was present without magnetic field. Thus the Hall voltage can be used to determine the carrier velocity, and consequently mobility as well as carrier concentration.

The Hall effect also proves that the concept of moving holes, rather than just being an artifact for holes being filled by electrons, is a more fundamental

---

[5] The standard simple model on diffusion assumes that the scattered electron or hole "forgets" its history (and direction) and has random thermal velocity distribution immediately after the scattering process. Furthermore, independence of the effective mass on the direction of motion is assumed. With these assumptions one expects the Hall mobility to be exactly equal to the drift mobility. A more sophisticated analysis (Wang 1989, (6.6.12); Shockley 1950 pp 270–277) yields $\frac{\mu_{\text{H}}}{\mu} = \frac{\langle v^2 \tau^2 \rangle \langle v^2 \rangle}{\langle v^2 \tau \rangle^2}$.

concept, since these two interpretations lead to the opposite sign of the Hall effect in the case of holes. This is perhaps seen more directly when considering again the movement of electrons and holes in the presence of a magnetic field (Fig. 2.10). In the simple picture, the electron filling a hole would take the same direction as a real electron; thus the motion of the hole would be exactly opposite to that of an electron, in apparent contradiction to experiment. This contradiction points out the weakness of describing an essentially quantum mechanical system in a corpuscular representation. This contradiction is resolved to a certain degree by interpreting missing electrons in the valence band as positively charged particles (holes), their effective mass being defined by the second derivative of energy with respect to momentum, taken at the maximum of the valence band.

2.6 Carrier Generation and Recombination in Semiconductors

Free electrons and holes may be generated by the lifting of electrons from the valence band into the conduction band, thus creating simultaneously equal numbers of electrons and holes. This can be accomplished by various mechanisms that have to supply the necessary energy, such as thermal agitation, optical excitation and ionization by penetrating charged particles. Carrier generation by radiation is more fully described in standard literature (see for example Knoll and Glenn 1989 or Leo 1994).

It is also possible to inject free carriers of a single type only, e.g. through a forward-biased diode, or to deplete the semiconductor of its free carriers by application of a reverse-bias voltage.

Here we will consider various methods for carrier generation as well as the mechanisms for returning from a non-equilibrium situation thus created back to an equilibrium condition.

2.6.1 Thermal Generation of Charge Carriers

Thermal generation of charge carriers usually has a detrimental effect in semiconductor radiation detectors because it leads to noise superimposed onto the signals. In some direct semiconductors, the band gap is small enough compared with the thermal voltage at room temperature $\left(\frac{kT}{q} = 0.0259 \text{ V} \right)$ so that electrons may be excited directly from the valence to the conduction band. Therefore these detectors (e.g. Ge) have to be operated at low temperature. In others (e.g. Si and GaAs) the probability of direct excitation at room temperature is extremely low. Here the thermal excitation occurs in two steps through intermediate local states in the band gap (Fig. 2.12). These intermediate states are created by imperfections within the crystal and by impurities.

Note that for indirect semiconductors (Si, Ge) the minimum energy needed for a band-to-band transition (electron–hole pair generation) is not simply given

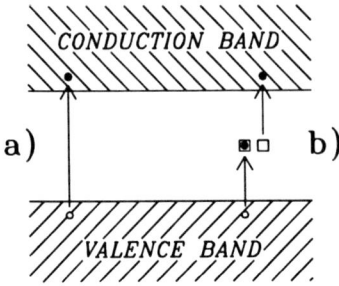

Fig. 2.12a,b. Direct (a) and indirect (b) excitation of electrons

by the width of the band gap. As the maximum of the valence band and the minimum of the conduction band are located at different momenta (Fig. 2.17), additional momentum has to be transferred in the process.

2.6.2 Generation of Charge Carriers by Electromagnetic Radiation

This effect is the basis of photo detectors and solar cells. The schematics of the basic process is shown in Fig. 2.13. A photon is absorbed and its energy is used to lift the electron from the valence band into the conduction band. If the photon energy is above the band gap E_G, the electron will be lifted into one of the empty states of the conduction band, leaving behind a hole in the valence band.[6] Electron and hole will subsequently move towards the band-gap edges, thereby emitting energy in the form of phonons (lattice vibrations) or lower-energy photons. Absorption of photons with energies below E_G is in principle also possible if local states in the band gap due to lattice imperfections exist. In the example shown, a hole is created together with the ionization of the local state.

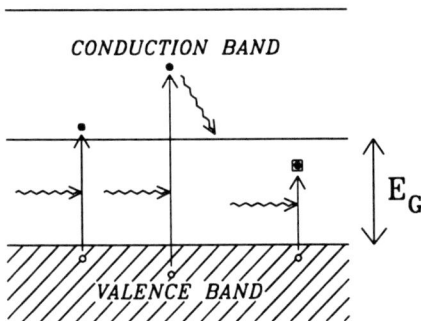

Fig. 2.13. Generation of electrons and holes by absorption of photons of energies $E = E_G$, $E > E_G$ and $E < E_G$

---

[6] For indirect semiconductors (see Sect. 2.6.6) additional energy is needed to supply momentum transfer to the electron or to make a direct transition without change of momentum but with larger energy requirement.

2.6.3 Generation by Charged Particles

Charged particles traversing material lose part of their energy through elastic collisions with electrons. This process has been investigated very thoroughly both experimentally and theoretically. The basic theory has been developed first by Bohr using classical arguments, and later in a quantum mechanical way by Bethe (1930), Bloch (1933) and Landau (1944).

Bohr was considering the momentum transfer given to a free electron at rest, when a charged particle was passing at its closest distance b (the impact parameter). The transferred energy, which is proportional to the square of the momentum transfer, is then integrated over the possible impact parameters. Integration limits to the range of impact parameters come from the maximum kinematically allowed momentum transfer and a comparison of the orbital period with the time the passing particle exerts a force. As the time that the electrostatic force acts on the electron is inversely proportional to the velocity, the energy loss is inversely proportional to the square of the particle velocity.

The Bethe-Bloch formula gives the rate of ionization loss of a charged particle in matter. We write it in a form containing corrections for density and shell effects, as for example presented in Leo 1994:

$$\frac{dE}{dx} = 2\pi N_0 r_e^2 m_e c^2 \rho \frac{Z}{A} \frac{z^2}{\beta^2} \left[\ln \left(\frac{2m_e \gamma^2 v^2 W_{max}}{I^2} \right) - 2\beta^2 - \delta - 2\frac{C}{Z} \right] , \quad (2.6.1)$$

where

$2\pi N_0 r_e^2 m_e c^2 = 0.1535 \,\mathrm{MeVc^2/g}$
x is the path length in g/cm^2;
$r_e = \frac{e^2}{4\pi m_e c^2} = 2.817 \times 10^{-13}$ cm and is the classical electron radius;
m_e is the electron mass;
$N_0 = 6.022 \times 10^{23} \,\mathrm{mol^{-1}}$ and is Avogadro's number;
I is the effective ionization potential averaged over all electrons;
Z is the atomic number of the medium;
A is the atomic weight of the medium;
ρ is the density of medium;
z is the charge of a traversing particle;
$\beta = v/c$, the velocity of a traversing particle in units of speed of light;
$\gamma = \frac{1}{\sqrt{1-\beta^2}}$;
δ is a density correction;
C is a shell correction; and
W_{max} is the maximum energy transfer in a single collision.

The energy loss rate as a function of particle energy is shown in Fig. 2.14. Energy and velocity of the incident particle with mass M are related by relativistic kinematics as

$$E = \gamma M c^2 = \frac{Mc^2}{\sqrt{1-\beta^2}} , \quad \text{with} \quad \beta = \frac{v}{c} ,$$

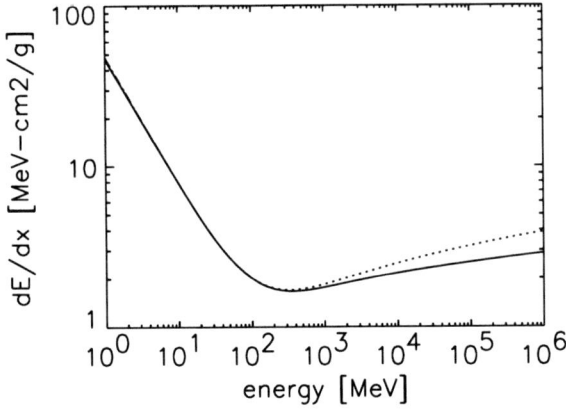

Fig. 2.14. Rate of energy loss due to ionization as a function of kinetic energy of a charged pion traversing Silicon (effective ionization potential $I = 173\,\text{eV}$) with (*continuous line*) and without (*dotted line*) density and shell corrections

which for small velocities reduces to $E_{\text{kin}} = E - Mc^2 = Mv^2/2$. The maximum energy transfer produced by a head-on collision is given by

$$W_{\max} = \frac{2m_e c^2 \beta^2 \gamma^2}{1 + 2s\sqrt{1 + \beta^2\gamma^2} + s^2} \,, \quad \text{with} \quad s = \frac{m_e}{M} \,,$$

which for $M \gg m_e$ reduces to

$$W_{\max} \approx 2m_e c^2 \beta^2 \gamma^2 \,.$$

For a parameterization of the mean excitation potential I, the shell correction C that is important at very low velocity and the density correction δ that flattens the relativistic rise the reader is referred to Leo 1994.

For low (nonrelativistic) energies, one observes the energy loss rate to be inversely proportional to the energy (or velocity squared). With rising energy a minimum is reached followed by the relativistic rise showing logarithmic characteristics, and finally a saturation due to polarization of the medium.

The average energy loss in a sample of finite thickness can be calculated from the Bethe-Bloch formula by integration. There are in addition, however, statistical fluctuations about this value (see Fig. 2.15), a subject treated in depth by Landau. A review of this subject, including refinements of Landau's original treatment and a comparison with experimental data, can be found in Bichsel 1988 and references quoted therein.

In semiconductors, only part of the energy loss is used for the creation of electron–hole pairs. In silicon the average energy used for the creation of a pair is 3.6 eV, three times larger than the band gap of 1.12 eV. This is true for radiation energies that are large with respect to the band gap.

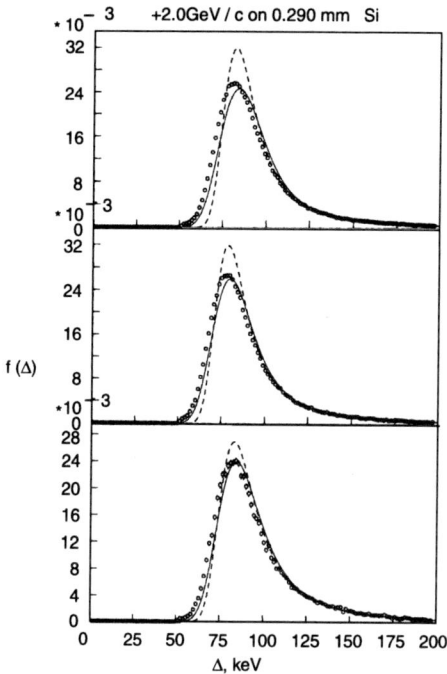

Fig. 2.15. Experimental energy loss distribution (*points*) for (from *top* to *bottom*) 2 GeV/c positrons, pions and protons traversing a 290 μm-thick silicon detector. (After Bak 1987, Fig. 12). Theoretical expectations from Landau (*dashed*) and a more refined theory (Bak et al. 1987, *solid line*) are given for comparison

2.6.4 Shape of a Radiation-Generated Charge Cloud

Depending on the type of radiation, the generation of the charge cloud may involve rather complicated processes, sometimes including interactions in which a primary particle produces several secondary particles. Here we will restrict ourselves to a qualitative description for radiation types for which semiconductor detectors are well suited:

- *Visible and ultraviolet light*: in general, a single electron–hole pair will be produced by a photon. The photon will be absorbed close to the surface (typically a fraction of a micrometer in silicon).
- *X-rays*: a "point" interaction with the production of many electron–hole pairs in a small spatial region is expected. The number of pairs can be estimated from the average energy necessary to create an electron–hole pair (3.6 eV in silicon).
- *α particles*: because of the high, strongly velocity-dependent ionization, the penetration depth is rather short (a few micrometers). The density of electron–hole pairs increases with path length as the velocity decreases (compare 2.6.1) and has a pronounced maximum at the stopping point of the particle. The ionization density plotted as a function of penetration depth is known as a Bragg curve.

- *β radiation*: due to the much lower mass of electrons with respect to a helium nucleus and the factor-of-two lower charge, β radiation ionizes much more feebly than α radiation. β radiation will therefore penetrate deeply into the semiconductor, or even pass through it, producing roughly uniform electron–hole pair densities along its path so long as the velocity remains relativistic and producing increased density at the end of its path. A relativistic singly charged particle is often referred to as minimum-ionizing particle or m.i.p.
- *High-energy charged particles*: This type of radiation will penetrate the detector with nearly constant (relativistic) velocity, producing uniform electron–hole density along its path. The density is nearly independent of the particle energy and is proportional to the square of the charge of the ionizing particle.
- *Nonrelativistic charged particles* such as protons and nuclei will produce an ionization density inversely proportional to their energy (which decreases with path length) and proportional to the square of their charge. Simultaneous measurement of the energy loss in a thin detector and of the total energy can therefore be used for particle identification.

Other types of radiation, such as neutrons and very high-energy photons, may also produce signals in semiconductors, for example by recoiling a silicon nucleus or creating electron–positron pairs that are capable of producing electron–hole pairs by ionization. The probability for this to happen is so small, however, that semiconductors by themselves are inappropriate for neutron and high-energy photon detection. However, interleaving semiconductors with other materials, which converts this radiation into something detectable, is possible.

2.6.5 Multiplication Processes

If an electron or hole is created in, or moved into, a high-field region inside a semiconductor, it may be accelerated strongly enough in between collisions to obtain sufficient energy for the creation of an electron–hole pair. An avalanche may thereafter develop.

Although charge multiplication may cause problems in badly designed semiconductor devices, leading to electrical breakdown, one can also make use of this effect in a controlled way for signal amplification, for example in avalanche diodes. A schematic description of the process in the band model is shown in Fig. 2.16. Here for the first time the band is represented as a function of position, in contrast to momentum as was done before. The electric field strength $\mathcal{E}$ is proportional to the slope of the bands, respectively the intrinsic energy E_i according to $\mathcal{E}_x = \frac{1}{q}\frac{\partial E_i}{\partial x}$. Acceleration of an electron in the electric field is represented by a horizontal arrow, and the vertical distance of this line from the edge of the conduction band gives the kinetic energy at the respective position.

A single primary electron–hole pair is assumed to be generated at position x_0 by (e.g.) a photon. At low field strengths (Fig. 2.16a), corresponding to a small inclination of the bands, the gain in kinetic energy of the charge carriers in between collisions is too small to create a secondary electron–hole pair. At

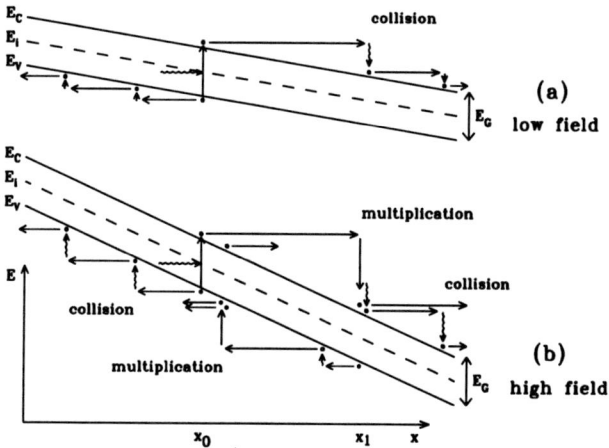

Fig. 2.16a,b. Charge multiplication by the development of avalanches in high-field regions of a semiconductor. A slowdown of charge carriers without the creation of secondary electron–hole pairs in the low-field region (a). An avalanche breakdown at high field (b)

collision the kinetic energy of electrons and holes will simply be transferred to the crystal lattice before they are accelerated in the electric field again, and so no multiplication occurs.

In contrast, at high field strengths (Fig. 2.16b), the energy gain in between collisions may be high enough to allow the creation of an electron–hole pair. This is shown in the figure at position x_1 for the accelerated primary electron. Part of the kinetic energy is used for the creation of the additional electron–hole pair (full arrow downwards), and part goes into lattice vibration. One thus arrives after collision with two electrons and one hole, all at position x_1. Each of these will be accelerated by the field again, creating with a certain probability further electron–hole pairs in subsequent collisions. The probability for creating secondary pairs will be different for electrons and for holes. Raising the electric field by only a moderate amount one may find a condition whereby essentially only one type of charge-carrier (electrons in silicon) will produce secondary pairs. In such a case the charge generated by multiplication processes will be proportional to the primary generated charge. At very high fields, the gain in kinetic energy in between collisions for both types of carriers is high enough to create additional electron–hole pairs. In this condition an avalanche breakdown occurs.

Example 2.3

Problem: *Estimate the minimum energy required for a primary electron in order to be able to create an electron–hole pair. Require energy and momentum conservation and assume equal effective masses m for electrons and holes.*

Solution: *The minimum energy is required when the three particles (scattered primary electron, created electron, and hole) have equal momentum and are* *collinear with the primary electron. One then has the conservation relationships:*

$$E_{e,\,min} = 3E' + E_G$$
$$P_{e,\,min} = \sqrt{2mE_{e,\,min}} = 3\sqrt{2mE'}$$
$$E_{e,\,min} = \tfrac{1}{3}E_{e,\,min} + E_G \;,$$

from which one finds

$$E_{e,\,min} = \tfrac{3}{2}E_G \;.$$

2.6.6 Recombination

Once an excess of minority charge carriers (e.g. electrons in p-type material) is created, it will take some time for the system to come back to thermal equilibrium. Excess carriers could, for example, be created by a pulse of light shining onto the semiconductor. The transition back to equilibrium is due to recombination of the excess minority carriers (electrons) with the majority carriers (holes).

This process of recombination is significantly different for "direct" and "indirect" semiconductors, a classification that has already been mentioned in Sect. 2.2. While for direct semiconductors (e.g. GaAs) electrons located at the minimum of the conduction band and holes concentrated at the maximum of the valence band have the same crystal momentum, this is not the case for indirect semiconductors (e.g. Si), as shown in Fig. 2.17, where the energy band structure is plotted as a function of the crystal momentum in two specific (111 and 100) crystal directions. For indirect semiconductors, therefore, the direct band-to-band recombination is suppressed, as it requires a large momentum

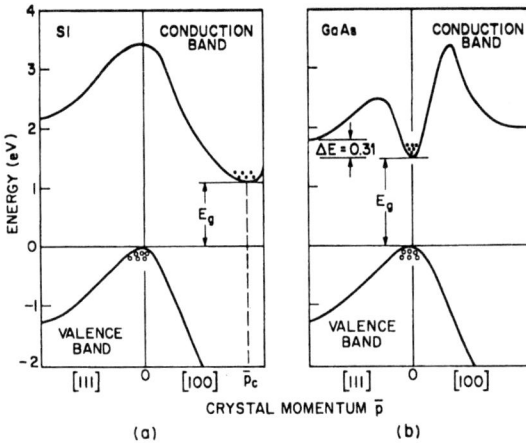

Fig. 2.17a,b. Energy band structure of silicon (a) and gallium-arsenide (b). (After Sze 1985, p. 14 Fig. 12)

transfer to the crystal lattice. Recombination occurs instead in two step processes, involving the capture and emission of electrons and holes into and out of the intergap generation/recombination centers. For a more complete treatment of this subject, the reader is referred to Shockley and Read (1952), Hall (1952), Spenke (1965) or Sze (1985).

2.6.7 Charge-Carrier Lifetime

A very important parameter of detector-grade semiconductor material is the charge-carrier lifetime, and there needs to be a distinction between recombination and generation lifetime. These terms describe the transient behavior from a nonequilibrium charge distribution obtained either by injection of additional carriers or by their removal, back to an equilibrium condition.

Consider for instance the case where a direct p-type semiconductor is in thermal equilibrium. Although the concentration of electrons and holes does (on average) not change, thermal generation and recombination of electron–hole pairs is continuously occurring. The recombination rate, expected to be proportional to the product of electron and hole concentrations via

$$R = \beta np \ , \tag{2.6.2}$$

in thermal equilibrium equals the generation rate, so that

$$R_{\text{th}} = G_{\text{th}} = \beta n_0 p_0 = \beta n_i^2 \ . \tag{2.6.3}$$

We can thus define the excess recombination rate U as

$$U = R - R_{\text{th}} = \beta(np - n_i^2) \ . \tag{2.6.4}$$

Under illumination with light there is an additional generation rate G_{L}, and both minority and majority carrier concentrations will be increased from their thermal equilibrium values by the same amount Δn as electrons and holes are created in pairs. Thus the recombination rate will increase to

$$R = \beta(n_0 + \Delta n)(p_0 + \Delta n)$$

and, for constant illumination, a relationship between the additional generation rate G_{L} – equaling the excess recombination rate U – and the amount of excess minority carriers Δn_{L} generated by light can be derived as

$$G_{\text{L}} = U = R - R_{\text{th}} = \beta \Delta n_{\text{L}} (p_0 + n_0 + \Delta n_{\text{L}}) \ .$$

For low injection levels ($\Delta n \ll p_0$), this simplifies to

$$G_{\text{L}} = \frac{\Delta n_{\text{L}}}{\tau_{\text{r}}} \quad \text{with} \quad \tau_{\text{r}} = \frac{1}{\beta p_0} \ . \tag{2.6.5}$$

The significance of the lifetime τ_{r} can be illustrated by considering a sudden turnoff of the light source. Then the excess recombination rate U will be proportional to the excess minority carrier density Δn, where

$$U = \frac{n - n_0}{\tau_r} \quad , \tag{2.6.6}$$

and the minority carrier density will decrease from its value $n_0 + \Delta n_L = n_0 + G_L \tau_r$ and will approach the thermal equilibrium value n_0 at a time-constant τ_r, the recombination lifetime.

Consider now another situation in which all charge carriers have been removed from the semiconductor, for example by applying an external voltage. One may again address the question of any possible return to equilibrium. Here the generation rate will be the thermal generation rate

$$G_{th} = \beta n_0 p_0 = \beta n_i^2 \quad ,$$

while the initial recombination rate is zero. As electrons and holes are created in pairs, the equilibrium condition will be (if the sample is small enough that the unequal distribution due to the electric field of the donor charges can be neglected)

$$p = n = n_i \quad .$$

The time constant for return to this equilibrium is defined as the generation lifetime, given by

$$\tau_g = \frac{n_i}{G_{th}} = \frac{1}{\beta n_i} \quad , \tag{2.6.7}$$

a value significantly different from the recombination lifetime.

The generation lifetime is closely related to the current generated in space-charge regions of electronic devices, such as a reversely biased diode which, as we will see later on, is the most common semiconductor detector.

Above derivations were for direct semiconductors. In the case of indirect semiconductors, the relationship between generation and recombination lifetimes is not as straightforward, because the much more complicated mechanism of recombination involves transitions through intermediate states in the band gap. However, the definition of generation and recombination lifetime remains unchanged and device performance can be estimated equivalently using these parameters.

Example 2.4

Problem: *Find the dark current per unit area of a fully depleted silicon detector of $d = 300\,\mu m$ thickness if the generation lifetime is 1 ms.*
Solution: *Inverting (2.6.7), we find the number of charge carriers generated per unit volume and unit time to be*

$$G_{th} = \frac{n_i}{\tau_g} = \frac{1.45 \times 10^{10}\,\mathrm{cm}^{-3}}{10^{-3}\,\mathrm{s}} = 1.45 \times 10^{13}\,\mathrm{cm}^{-3}\mathrm{s}^{-1}$$

and a current per area in a wafer of 300 μm thickness of

$$J_{th} = q d G_{th} = 1.6 \times 10^{-19}\,\mathrm{As} \cdot 0.03\,\mathrm{cm} \cdot 1.45 \times 10^{13}\,\mathrm{cm}^{-3}\mathrm{s}^{-1}$$
$$= 70\,\mathrm{nA/cm}^2 \quad .$$

2.6.8 Carrier Lifetime in Indirect Semiconductors

In indirect semiconductors such as silicon or germanium, a band-to-band re-combination process is very unlikely because electrons at the bottom of the conduction band and holes at the top of the valence band have different crystal momentum, contrary to direct semiconductors (e.g. GaAs) as shown in Fig. 2.16. Recombination occurs instead through localized energy states in the for-bidden gap. These localized states are due to the presence of impurities and crystal defects leading to lattice distortion and irregular charge distributions. Depending on the type, defects can assume two or more charge states. Changes between the charge states of the defect occur by four processes: electron and hole emission, and electron and hole capture, as indicated in Fig. 2.18.

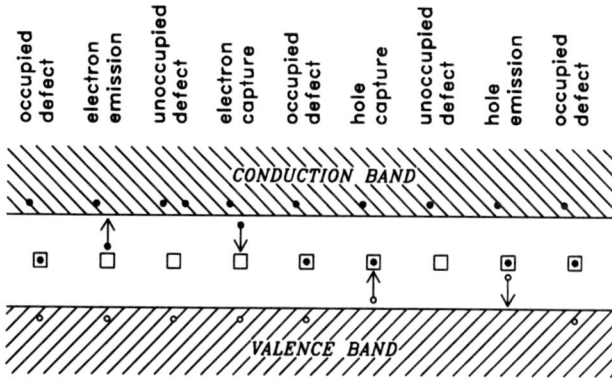

Fig. 2.18. Charge-changing processes of a simple defect with one energy level only

The single-energy-level defect shown in the figure is capable of assuming two charge states only (say neutral and positive, as is the case for donors). If the state is positive, it may change to neutral by for instance capturing an electron from the conduction band or emitting a hole into the valence band (which may also be viewed as lifting an electron from the valence band to the defect). Changing the charge in the other direction is accomplished by hole capture or electron emission. Electron–hole recombination can thus be accomplished by two capture processes in sequence, such as electron capture followed by hole capture.

Electron–hole pair generation occurs by emission of an electron and a hole in sequence. In some circumstances only one type of charge-carrier, e.g. elec-trons, plays a role. An electron is captured by the defect and after some time emitted to the conduction band again. Such a process is called "trapping" and is normally followed by detrapping, and the crystal defects are often referred to as trapping centers. Note, however, that there is no difference in principle between generation/recombination and trapping centers.

Knowing electron and hole capture cross-sections and emission probabilities and the initial value of the electron and hole densities, one can find the time development of the charge-carrier densities and thus their lifetime. Thermal equilibrium considerations allow the derivation of a relationship that connects capture and emission processes of electrons and of holes.

Let n, p be the electron and hole concentration, N_t the defect concentration, and P_t the occupation probability of the defect (the defect is occupied if it is in the more negative state, i.e. after capturing an electron or emitting a hole). Then the rates of the four processes of electron capture $R_{c,n}$, electron emission $R_{\epsilon,n}$, hole capture $R_{c,p}$ and hole emission $R_{\epsilon,p}$ are given as

$$R_{c,n} = \nu_{\mathrm{th},n}\sigma_n n N_t (1 - P_t) \tag{2.6.8}$$

$$R_{\epsilon,n} = \epsilon_n N_t P_t \tag{2.6.9}$$

$$R_{c,p} = \nu_{\mathrm{th},p}\sigma_p p N_t P_t \tag{2.6.10}$$

$$R_{\epsilon,p} = \epsilon_p N_t (1 - P_t) \tag{2.6.11}$$

with ν_{th} the thermal velocity, σ the capture cross-section and ϵ the emission probability for electrons and holes respectively.

In thermal equilibrium the rates of capture and emission have to be equal, separately for both electrons and for holes. This follows from the requirement that electron and hole concentration, as well as the average defect occupation probability, do not change. Furthermore, the average defect occupation probability has to be given by the Fermi function, evaluated at the energy of the defect level E_t, thus:

$$R_{c,n} = \nu_{\mathrm{th},n}\sigma_n n N_t (1 - P_t) = \epsilon_n N_t P_t = R_{\epsilon,n} \tag{2.6.12}$$

$$R_{c,p} = \nu_{\mathrm{th},p}\sigma_p p N_t P_t = \epsilon_p N_t (1 - P_t) = R_{\epsilon,p} \tag{2.6.13}$$

$$P_t = F(E_t) = \cfrac{1}{1 + e^{\frac{E_t - E_F}{kT}}} \ , \tag{2.6.14}$$

from which one arrives at

$$\epsilon_n = \nu_{\mathrm{th},n}\sigma_n n \, \frac{1 - F(E_t)}{F(E_t)} = \nu_{\mathrm{th},n}\sigma_n n \, e^{\frac{E_t - E_F}{kT}} = \nu_{\mathrm{th},n}\sigma_n n_{\mathrm{i}} \, e^{\frac{E_t - E_{\mathrm{i}}}{kT}} \tag{2.6.15}$$

$$\epsilon_p = \nu_{\mathrm{th},p}\sigma_p p \, \frac{F(E_t)}{1 - F(E_t)} = \nu_{\mathrm{th},p}\sigma_p p \, e^{\frac{E_F - E_t}{kT}} = \nu_{\mathrm{th},p}\sigma_p n_{\mathrm{i}} \, e^{\frac{E_{\mathrm{i}} - E_t}{kT}} \ . \tag{2.6.16}$$

Here we have made use of (2.4.2) to express the charge-carrier densities by the intrinsic densities and the Fermi level.

Turning now back to the stationary nonequilibrium case, the net recombination rate can be calculated either for electrons or for holes by taking the difference between capture and emission rates. One expects the same answer for electrons and holes because the average occupation rate of the defects has to be constant. One thus has

$$U = R_{c,n} - R_{\epsilon,n} = N_t \nu_{\mathrm{th},n}\sigma_n \left[n(1 - P_t) - n_{\mathrm{i}} \, e^{\frac{E_t - E_{\mathrm{i}}}{kT}} P_t \right]$$

$$= R_{c,p} - R_{\epsilon,p} = N_t \nu_{\mathrm{th},p}\sigma_p \left[p P_t - n_{\mathrm{i}} \, e^{\frac{E_{\mathrm{i}} - E_t}{kT}} (1 - P_t) \right] \ , \tag{2.6.17}$$

from which one finds the defect-occupation probability P_t as follows:

$$\frac{1 - P_t}{P_t} = \frac{\nu_{\text{th},p}\sigma_p p + \nu_{\text{th},n}\sigma_n n_i e^{\frac{E_t - E_i}{kT}}}{\nu_{\text{th},n}\sigma_n n + \nu_{\text{th},p}\sigma_p n_i e^{\frac{E_i - E_t}{kT}}} \tag{2.6.18}$$

$$P_t = \frac{\nu_{\text{th},n}\sigma_n n + \nu_{\text{th},p}\sigma_p n_i e^{\frac{E_i - E_t}{kT}}}{\nu_{\text{th},p}\sigma_p p + \nu_{\text{th},n}\sigma_n n_i e^{\frac{E_t - E_i}{kT}} + \nu_{\text{th},n}\sigma_n n + \nu_{\text{th},p}\sigma_p n_i e^{\frac{E_i - E_t}{kT}}} \tag{2.6.19}$$

$$1 - P_t = \frac{\nu_{\text{th},p}\sigma_p p + \nu_{\text{th},n}\sigma_n n_i e^{\frac{E_t - E_i}{kT}}}{\nu_{\text{th},p}\sigma_p p + \nu_{\text{th},n}\sigma_n n_i e^{\frac{E_t - E_i}{kT}} + \nu_{\text{th},n}\sigma_n n + \nu_{\text{th},p}\sigma_p n_i e^{\frac{E_i - E_t}{kT}}} . \tag{2.6.20}$$

Inserting (2.6.19) and (2.6.20) into (2.6.17), one obtains for the excess recombination rate as follows:

$$U = \frac{N_t \nu_{\text{th},n}\sigma_n \nu_{\text{th},p}\sigma_p (np - n_i^2)}{\nu_{\text{th},n}\sigma_n \left[n + n_i e^{\frac{E_t - E_i}{kT}} \right] + \nu_{\text{th},p}\sigma_p \left[p + n_i e^{\frac{E_i - E_t}{kT}} \right]} . \tag{2.6.21}$$

Comparison with (2.6.4) yields

$$\beta = \frac{N_t \nu_{\text{th},n}\sigma_n \nu_{\text{th},p}\sigma_p}{\nu_{\text{th},n}\sigma_n \left[n + n_i e^{\frac{E_t - E_i}{kT}} \right] + \nu_{\text{th},p}\sigma_p \left[p + n_i e^{\frac{E_i - E_t}{kT}} \right]} . \tag{2.6.22}$$

As was done before for the case of direct semiconductors, we can derive from this carrier-concentration-dependent recombination factor β the low-level injection recombination time constant τ_{r} and the generation lifetime τ_{g}. With (2.6.5), one gets, after ignoring the small terms in the denominator of (2.6.22), for n- and p-type semiconductors respectively:

$$\tau_{\text{r},n} = \frac{1}{\beta n_0} \approx \frac{1 + \frac{n_i}{n_0} \left[\frac{\nu_{\text{th},p}\sigma_p}{\nu_{\text{th},n}\sigma_n} e^{\frac{E_i - E_t}{kT}} + e^{\frac{E_t - E_i}{kT}} \right]}{\nu_{\text{th},p}\sigma_p N_t} \tag{2.6.23}$$

$$\tau_{\text{r},p} = \frac{1}{\beta p_0} \approx \frac{1 + \frac{n_i}{p_0} \left[\frac{\nu_{\text{th},n}\sigma_n}{\nu_{\text{th},p}\sigma_p} e^{\frac{E_t - E_i}{kT}} + e^{\frac{E_i - E_t}{kT}} \right]}{\nu_{\text{th},n}\sigma_n N_t} . \tag{2.6.24}$$

In these (last two) equations, n_0 and p_0 are the majority carrier concentrations, usually large with respect to the intrinsic concentration n_i. For defect levels E_t very close to the intrinsic level E_i, the second term in the numerators of (2.6.23) and (2.6.24) can therefore be neglected.

The generation lifetime defined in (2.6.7) as the ratio of intrinsic charge density and initial generation rate of the fully depleted semiconductor is found, from (2.6.21) with $n = p = 0$, as

$$G_{\text{th}} = -U = \frac{N_t \nu_{\text{th},n}\sigma_n \nu_{\text{th},p}\sigma_p n_i}{\nu_{\text{th},n}\sigma_n e^{\frac{E_t - E_i}{kT}} + \nu_{\text{th},p}\sigma_p e^{\frac{E_i - E_t}{kT}}} \tag{2.6.25}$$

$$\tau_{\text{g}} = \frac{n_i}{G_{\text{th}}} = \frac{1}{N_t} \left[\frac{1}{\nu_{\text{th},p}\sigma_p} e^{\frac{E_t - E_i}{kT}} + \frac{1}{\nu_{\text{th},n}\sigma_n} e^{\frac{E_i - E_t}{kT}} \right] . \tag{2.6.26}$$

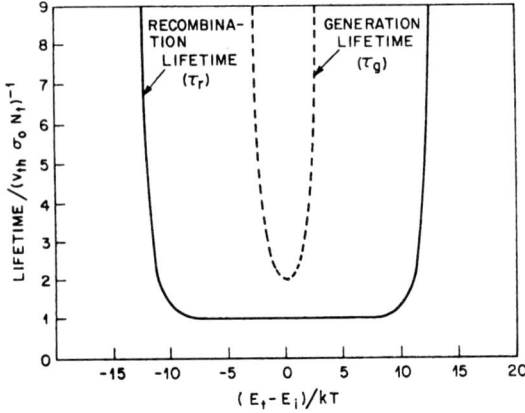

Fig. 2.19. Dependence of the recombination and generation lifetimes on the energy level of the defect, for the special case $\nu_{\text{th},n}\sigma_n = \nu_{\text{th},p}\sigma_p = \nu_{\text{th}}\sigma_0$. (After Sze 1985, p. 53 Fig. 15)

Generation and recombination time constants are shown as function of the trap energy level in Fig. 2.19 for the special case of equal capture cross-section of electrons and holes ($\nu_{\text{th},p}\sigma_p = \nu_{\text{th},n}\sigma_n$). One easily reads from (2.6.23–2.6.26) that recombination time constants in this case have an energy dependence of $a + \cosh(\frac{E_t - E_i}{kT})$ and $\cosh(\frac{E_t - E_i}{kT})$ respectively.

2.7 Simultaneous Treatment of Carrier Generation and Transport

In the previous sections of this chapter, charge generation and recombination has been treated separately from charge transfer phenomena (drift and diffusion). Here their interplay – described by the continuity equation and the electric field configuration – will be considered.

The continuity equation states that the increase of the number of charge carriers of a given type in an arbitrary part of the semiconductor is given by the difference of generation and recombination in the volume and the inward flux through the surface. Restricting ourselves for simplicity to the one-dimensional case, we have for the small volume $A\,\mathrm{d}x$ in Fig. 2.20:

$$\frac{\partial n}{\partial t} A\,\mathrm{d}x = \Phi_n(x)A - \Phi_n(x + \mathrm{d}x)A + (G_n - R_n)A\,\mathrm{d}x \ ,$$

where Φ_n, G_n and R_n are flux, generation and recombination rates of electrons. Replacing the flux by current density ($J_n = -q\Phi_n$), one obtains

$$\frac{\partial n}{\partial t} = \frac{1}{q}\frac{\partial J_n}{\partial x} + G_n - R_n$$

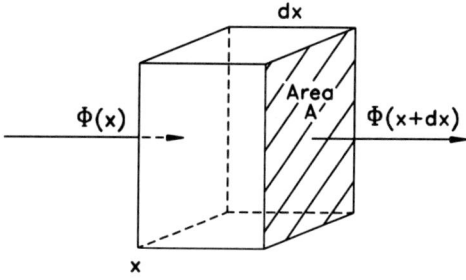

Fig. 2.20. Charge-carrier flow and generation-recombination in a small volume of thickness dx and area A

for electrons, and similarily for holes

$$\frac{\partial p}{\partial t} = -\frac{1}{q}\frac{\partial J_p}{\partial x} + G_p - R_p \ .$$

This may be generalized for the three-dimensional case to

$$\frac{\partial n}{\partial t} = \frac{1}{q}\nabla \boldsymbol{J}_n + G_n - R_n$$

$$\frac{\partial p}{\partial t} = -\frac{1}{q}\nabla \boldsymbol{J}_p + G_p - R_p \ .$$

Using (2.5.3) one arrives at the continuity equations

$$\frac{\partial n}{\partial t} = \mu_n n \nabla \mathcal{E} + D_n \nabla^2 n + G_n - R_n \tag{2.7.1}$$

$$\frac{\partial p}{\partial t} = -\mu_p p \nabla \mathcal{E} + D_p \nabla^2 p + G_p - R_p \ ,$$

where the current density has been separated into a drift term and a diffusion term.

The electric field $\mathcal{E}$ is determined by the charge distribution through Poisson's equation

$$\nabla \mathcal{E} = \frac{\rho}{\epsilon \epsilon_0} \ , \quad \text{with } \rho = q(p - n + N_\mathrm{D} - N_\mathrm{A}) \ . \tag{2.7.2}$$

The space-charge density ρ has been expressed by carrier and doping concentrations.

An exact analytical solution of this set of simultaneous differential equations with given boundary conditions is in most cases not possible. However numerical solutions can be found using computer techniques. This is done extensively in device simulations, which are described in Chap. 12. We will also attempt to give approximate solutions for specific cases through the text.

Example 2.5

Problem: *Find the width (due to diffusion only) after $1\,\mu s$ of an concentrated but expanding electron cloud produced in silicon.*

Solution: *As the problem is spherically symmetrical, it is sufficient for the linear differential equation (2.7.1) to look at a single projection only. We then solve the continuity equation in the absence of an electric field and neglecting recombination. Thus:*

$$\frac{\partial n(x,t)}{\partial t} = D_n \frac{\partial^2 n(x,t)}{\partial x^2} \ .$$

Assuming a Gaussian distribution with time-dependent width, we have

$$n(x,t) = a(t)e^{-b(t)x^2}$$

$$\frac{\partial n(x,t)}{\partial t} = (a'(t) - a(t)b'(t)x^2)e^{-b(t)x^2}$$

$$\frac{\partial^2 n(x,t)}{\partial x^2} = (-2ab + 4ab^2 x^2)e^{-b(t)x^2}$$

and one then obtains the following:

$$a' - ab'x^2 = D_n[-2ab + 4ab^2 x^2]$$

$$a' = -2abD_n$$

$$- ab' = 4ab^2 D_n$$

$$- \frac{b'}{b^2} = 4D_n$$

$$\frac{1}{b} = 4D_n(t + t_0)$$

$$\frac{a'}{a} = -2bD_n = -\frac{1}{2(t + t_0)}$$

$$\ln(a) = -\frac{1}{2}\ln(t + t_0) + \ln(\text{const})$$

$$a = \frac{\text{const}}{\sqrt{t + t_0}}$$

leading to

$$n(x,t) = \frac{C}{\sigma}e^{-\frac{x^2}{2\sigma^2}} \quad \text{where} \quad \sigma = \sqrt{2D_n(t + t_0)} \ .$$

The width of the charge cloud thus increases with the square root of time. For $1\,\mu s$, we have the following result for silicon:

$$\sigma = \sqrt{2\frac{kT}{q}\mu_n t} = \sqrt{2 \cdot 0.0259\,\text{V} \cdot 1350\,\text{cm}^2/\text{Vs} \cdot 10^{-6}\,\text{s}} = 84\,\mu\text{m} \ .$$

Generalizing the result from the previous example from electrons to holes, we find that for an intrinsically concentrated charge cloud of either electrons or holes, we can show that after time t, from diffusion only, a Gaussian distribution with width σ is such that

$$\sigma = \sqrt{2Dt} = \sqrt{2\frac{kT}{q}\mu t} \quad . \tag{2.7.3}$$

$D = \frac{kT}{q}\mu$ is the diffusion constant of either electrons or holes.

2.8 Summary and Discussion

Commonly used semiconductors are single crystals with diamond or zincblende lattice. Their properties can be described in a simple bond representation or in a more sophisticated band model that incorporates some quantum mechanical results in a semiclassical picture. Electron energies are constrained to lie in bands. Valence and conduction bands are separated by the band gap in which no energy levels exist. At low temperatures, the valence band is completely filled and the conduction band completely empty; the semiconductor is a perfect insulator. At elevated temperatures, thermal excitation succeeds in lifting a small fraction of the electrons from the valence band to the conduction band.

Semiconductors can be doped by replacing a small fraction of proper atoms by foreign atoms with a higher or lower number of valence electrons. The excess or missing electrons in the crystal structure act as conduction electrons or holes. In the band model this situation is described by local energy states in the band gap.

In thermal equilibrium the occupation probability of energy states is described by the Fermi function shown in (2.3.1). An energy state at the Fermi energy has the occupation probability $\frac{1}{2}$. The value of the Fermi level E_F is obtained in a large uniformly doped semiconductor, far from boundaries with other materials, from the requirement of charge neutrality. Electron and hole densities are related to the intrinsic carrier concentration n_i and the deviation of the Fermi level from the intrinsic level E_i thus:

$$n = n_i\, e^{\frac{E_F - E_i}{kT}} \qquad p = n_i\, e^{\frac{E_i - E_F}{kT}} \quad . \tag{2.4.2}$$

The product of electron and hole concentration in thermal equilibrium, given by

$$n \cdot p = n_i^2 \quad , \tag{2.4.3}$$

is independent of position and doping.

Carrier transport is due to drift and diffusion. At moderate electric fields the drift velocity ν is proportional to the electric field $\mathcal{E}$, where

$$\nu_n = -\mu_n \mathcal{E} \qquad \nu_p = \mu_p \mathcal{E} \tag{2.5.1}$$

while at high field strengths the drift velocity approaches a saturation value. Diffusion is due to random thermal motion that tends to smooth a nonuniform carrier distribution. The diffusion equation relates the carrier flux $\boldsymbol{F}$ to the gradient of the carrier density:

$$\boldsymbol{F}_n = -D_n \nabla n \qquad \boldsymbol{F}_p = -D_p \nabla p \ . \tag{2.5.2}$$

Values for diffusion D and mobility μ are related by the Einstein relation

$$D = \frac{kT}{q} \mu \ . \tag{2.5.4}$$

The presence of a magnetic field changes the direction of motion of electrons and holes due to the Lorentz force.

Charge carrier generation occurs by thermal and optical excitation and by ionizing radiation. Electrons and holes can recombine, thus reducing the number of free charge carriers. In thermal equilibrium the rate of carrier creation by thermal excitation equals the rate of recombination.

The mechanisms for generation and recombination are quite different for direct and indirect semiconductors. In indirect semiconductors the mechanisms predominantly involve the presence of crystal defects (with local energy states close to the center of the band gap). In direct semiconductors recombination may lead to radiative transitions with photon emission.

The return from a nonequilibrium state to the equilibrium state is described by the minority carrier lifetime. A distinction has to be drawn between recombination and generation lifetimes, which can be numerically very different.

3 Basic Semiconductor Structures

The structures described in this chapter form the basis for semiconductor electronics as well as for detectors. A thorough knowledge of their functioning and properties is a prerequisite for an understanding of semiconductor detectors. Although excellent treatments of this subject may be found in many textbooks (Grove 1967, Sze 1985), a short description with emphasis on later applications to detectors will be given.

3.1 The p–n Diode Junction

Arguably, the most important electronic structure is the p–n junction, which is obtained by joining together extrinsic semiconductors of opposite doping (Fig. 3.1). Such a structure shows diode characteristics, and that means it will conduct current mainly in one direction. To understand this phenomenon, we will first consider the structure in thermal equilibrium without application of an external voltage, then with application of voltages of either polarity, and finally under irradiation with light and other ionizing radiation.

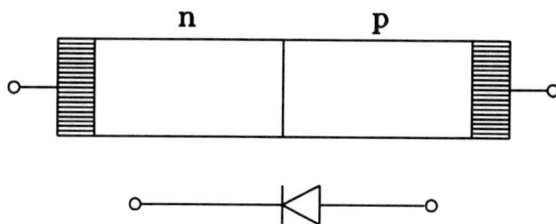

Fig. 3.1. A p–n diode junction: structure and device schematic

3.1.1 A p–n Diode in Thermal Equilibrium

We start from the hypothetical condition that the homogeneously doped p and n regions are initially separated, electrically neutral and in thermal equilibrium, with electrons and holes homogeneously distributed in their respective volumes (Fig. 3.2a). Once the bodies are brought into contact (Fig. 3.2b), electrons will diffuse into the p region and holes into the n region and a surplus will be created of negative electric charge in the p region and of positive charge in the n region. This creates an electric field that counteracts the diffusion. The electric field

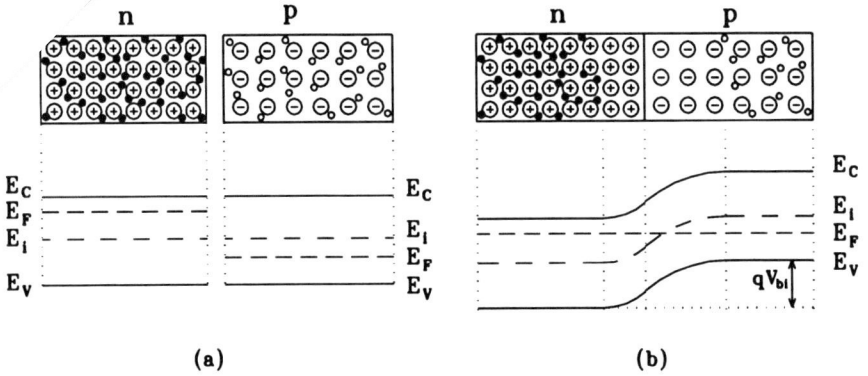

Fig. 3.2a,b. A p–n diode junction in thermal equilibrium, with its parts separated (a) and brought together (b)

also sweeps away any mobile charge carriers (electrons and holes) in the region around the boundary, so that a space-charge region is obtained in which the excess nuclear charge from the doping atoms is not neutralized by the movable carriers.

The situation may also be considered in the band model shown in the same figure. The built-in voltage V_{bi} is obtained from the requirement that the Fermi levels have to line up in thermal equilibrium. It may be worthwhile to mention that this built-in voltage, which is also referred to as a diffusion voltage, will not appear on the same metal electrical connections attached to the device, since it will be compensated for by the built-in voltages in the metal–semiconductor contacts. This point will be taken up again after discussing the metal–semiconductor contact.

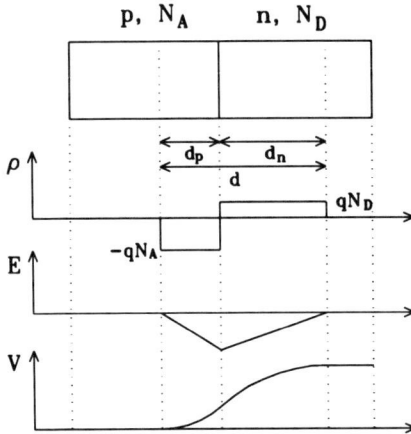

Fig. 3.3. A p–n diode junction in the "abrupt change" approximation

Example 3.1

Problem: Find the built-in voltage for a silicon p–n junction at room temperature with doping concentrations $N_A = 10^{16}\,\mathrm{cm}^{-3}$ (acceptor) and $N_D = 10^{12}\,\mathrm{cm}^{-3}$ (donor).

Solution: The built-in voltage can be calculated from the difference in the intrinsic levels on the neutral p and n regions. Using (2.4.2) and setting the majority carrier concentrations equal to, respectively, the acceptor and donor concentrations, we have

$$n_n = N_D = n_i\, e^{\frac{E_F - E_i^n}{kT}} \qquad p_p = N_A = n_i\, e^{\frac{E_i^p - E_F}{kT}}$$

$$N_A \cdot N_D = n_i^2\, e^{\frac{E_i^p - E_i^n}{kT}}$$

$$V_{\mathrm{bi}} = \frac{1}{q}(E_i^p - E_i^n) = \frac{kT}{q}\ln\frac{N_A N_D}{n_i^2} \tag{3.1.1}$$

$$= 0.0259\,\ln\frac{10^{16}\cdot 10^{12}}{(1.45\times 10^{10})^2} = 0.458\,\mathrm{V}\ .$$

As mentioned earlier, the space-charge region around the *p–n* junction boundary will be almost free of movable charge carriers, while the majority carrier concentration in the neutral regions will be large. The transition regions for most practical cases will be very thin compared with the space-charge region, so that for calculation of the potential, one may in good approximation assume an abrupt change between a neutral and a completely depleted space-charge region (see Fig. 3.3).

Example 3.2

Problem: Derive the width d of the natural space-charge region and the maximum electric field $\mathcal{E}_{\max}$ for an abrupt p–n junction as a function of the doping concentrations N_A and N_D. Calculate the width for a very highly doped $(N_A = 10^{16}\,\mathrm{cm}^{-3})$ p-type junction on n-type detector-grade material with $N_D = 10^{12}\,\mathrm{cm}^{-3}$.

Solution: Using the abrupt change approximation, we may find the extension of the space-charge region from the junction into the n and p regions from the requirement that the electric field at the boundaries of the space-charge region is zero and that the potential difference equals the built-in voltage. Requiring zero electric field at the boundaries corresponds to zero total charge in the space-charge region, expressed algebraically

$$N_D d_n = N_A d_p\ .$$

The electric field at the junction is then

$$\mathcal{E}_{\max} = \frac{1}{\epsilon\epsilon_0}q N_D d_n = \frac{1}{\epsilon\epsilon_0}q N_A d_p$$

and the voltage steps are

$$V_n = \frac{\mathcal{E}_{\max} d_n}{2} = \frac{1}{\epsilon\epsilon_0} \frac{q N_D d_n^2}{2} = \frac{1}{\epsilon\epsilon_0} \frac{q N_A d_p d_n}{2}$$

$$V_p = \frac{\mathcal{E}_{\max} d_p}{2} = \frac{1}{\epsilon\epsilon_0} \frac{q N_A d_p^2}{2} = \frac{1}{\epsilon\epsilon_0} \frac{q N_D d_p d_n}{2} .$$

Setting the voltage V across the junction, where

$$V = V_n + V_p = \frac{q}{2\epsilon\epsilon_0}(N_A + N_D)d_n d_p$$

$$= \frac{q}{2\epsilon\epsilon_0}\frac{(N_A + N_D)N_D}{N_A}d_n^2 = \frac{q}{2\epsilon\epsilon_0}\frac{(N_A + N_D)N_A}{N_D}d_p^2 ,$$

equal to the built-in voltage V_{bi}, we obtain the following for the depletion depths d_n, d_p, d and the maximum electric field $\mathcal{E}_{\max}$:

$$d_n = \sqrt{\frac{2\epsilon\epsilon_0}{q}\frac{N_A}{N_D(N_A + N_D)}V_{bi}}$$

$$d_p = \sqrt{\frac{2\epsilon\epsilon_0}{q}\frac{N_D}{N_A(N_A + N_D)}V_{bi}}$$

$$d = d_n + d_p = \sqrt{\frac{2\epsilon\epsilon_0 V_{bi}}{q(N_A + N_D)}}\left[\sqrt{\frac{N_A}{N_D}} + \sqrt{\frac{N_D}{N_A}}\right]$$

$$= \sqrt{\frac{2\epsilon\epsilon_0(N_A + N_D)}{q N_A N_D}V_{bi}} \tag{3.1.2}$$

$$\mathcal{E}_{\max} = \frac{1}{\epsilon\epsilon_0}q N_D d_n = \sqrt{\frac{2q}{\epsilon\epsilon_0}\frac{N_A N_D}{N_A + N_D}V_{bi}}$$

with V_{bi} given by (3.1.1).

For the case of very asymmetric doping, where $N_A \gg N_D$, we obtain

$$d \approx \sqrt{\frac{2\epsilon\epsilon_0}{q N_D}V_{bi}} = \sqrt{\frac{2\cdot 11.9\cdot 8.854\times 10^{-14}\,\mathrm{F/cm}}{1.6\times 10^{-19}\,\mathrm{As}\cdot 10^{12}\,\mathrm{cm}^{-3}}\times 0.485\,\mathrm{V}} = 25.3\,\mathrm{\mu m}$$

$$\mathcal{E}_{\max} \approx \sqrt{\frac{2q}{\epsilon\epsilon_0}N_D V_{bi}}$$

$$= \sqrt{\frac{2\cdot 1.6\times 10^{-19}\,\mathrm{As}}{11.9\cdot 8.854\times 10^{-14}\,\mathrm{F/cm}}\times 10^{12}\,\mathrm{cm}^{-3}\cdot 0.485\,\mathrm{V}} = 384\,\mathrm{V/cm} .$$

Although the net current in an unbiased diode has to be zero, it may still be instructive to consider the mechanisms leading to this balance. Considering electrons, for example, we notice that their concentration is very high in the neutral n region ($n_{n_0} \approx N_D$), while it is very low in the neutral p region ($n_{p_0} \approx \frac{n_i^2}{N_A}$). Thus a diffusion electron current is expected to flow from the n region to the p region. It is counterbalanced by the drift current flowing in the opposite direction. This balance holds at any point in the junction and for electrons and holes separately.

The carrier concentration at an arbitrary position within the junction can be expressed by the potential, or the intrinsic energy level E_i, at that position using (2.4.2):

$$n = n_i\, e^{\frac{E_F - E_i}{kT}} \qquad\qquad n_n = n_i\, e^{\frac{E_F - E_i^n}{kT}}$$

$$\frac{n}{n_n} = e^{-\frac{E_i - E_i^n}{kT}} \qquad\qquad \frac{p}{p_p} = e^{-\frac{E_i^p - E_i}{kT}} \,. \tag{3.1.3}$$

In these equations n_n and E_i^n (p_p and E_i^p) are the electron (hole) concentration and intrinsic energy level in the neutral n region (p region). For the ratio of carriers of the same type in the two neutral regions, one obtains similarly:

$$\frac{n_p}{n_n} = \frac{p_n}{p_p} = e^{-\frac{E_i^p - E_i^n}{kT}} = e^{-q\frac{V_{bi}}{kT}} \,. \tag{3.1.4}$$

3.1.2 A p–n Diode with Application of an External Voltage

If an external voltage is applied to the diode, the system is not in thermal equilibrium and the previous equilibrium considerations can only be applied in an approximate way. In order to find the current–voltage characteristics of a diode, we will estimate the minority carrier densities (e.g. electrons in the p region) at the edges of the neutral regions. From this value we then will derive the minority carrier diffusion currents at the edges of the neutral regions, which to good approximation represent the total currents of the respective carrier types both in the space-charge region and at the boundary of the other neutral regions. Minority carrier drift currents in the neutral regions can be ignored because of the large ratio between majority and minority carrier concentrations.

For a forward bias, the voltage across the junction will decrease from the equilibrium value V_{bi} by the externally applied voltage $V > 0$ to $V_{bi} - V$. The width of the space-charge region (see (3.1.2)) will shrink to

$$d = \sqrt{\frac{2\epsilon\epsilon_0 (N_A + N_D)}{q N_A N_D}(V_{bi} - V)} \tag{3.1.5}$$

and the minority carrier concentration at the edge of the space-charge region will increase. Using for the minority charge carrier concentrations at the edges of the space-charge region the relationship given in (3.1.3), which has been developed for the thermal equilibrium case[7], one expects

---

[7] This assumption is equivalent to using the Fermi level for majority carriers not only in the thermal equilibrium condition but also when extending the validity range throughout the space-charge region. In that region one then has different "Fermi levels" for electrons and holes, which one calls "Quasi-Fermi" levels and which are strictly speaking only a convenient parameterization of carrier densities.

$$n_p = n_n e^{-q\frac{V_{bi}-V}{kT}} = n_{p_0} e^{q\frac{V}{kT}} \;, \qquad p_n = p_{n_0} e^{q\frac{V}{kT}} \;. \tag{3.1.6}$$

Here, n_{p_0} is the electron density at the edge of the neutral p region in the thermal equilibrium case, which equals the electron density in the nonequilibrium case inside the neutral p region far away from the edge. As the minority carrier diffusion currents will be proportional to the deviation of the minority carrier concentrations from their equilibrium values, exponential behavior of electron and hole, as well as the total diode current, is expected. This can be expressed as

$$J = (J_{s_n} + J_{s_p})\left(e^{\frac{qV}{kT}} - 1\right) = J_s \left(e^{\frac{qV}{kT}} - 1\right) \tag{3.1.7}$$

and J_s, the reverse bias saturation current, can be calculated by solving the time-independent continuity equation for minority carriers with the boundary condition of zero minority carrier concentration at the edge towards the space-charge region.

Writing down (2.7.1) for one dimension and neglecting the drift term, one obtains for the stationary case ($\frac{\partial n}{\partial t} = 0$)

$$D_n \frac{\partial^2 n_p}{\partial x^2} + G_n - R_n = 0 \;. \tag{3.1.8}$$

With use of the excess recombination rate $U \equiv R_n - G_n$ from (2.6.6), the solution to the differential equation

$$D_n \frac{\partial^2 n_p}{\partial x^2} - \frac{n_p - n_{p_0}}{\tau_r} = 0 \tag{3.1.9}$$

with the boundary condition $n_p(x = 0) = 0$ is given by

$$n_p = n_{p_0}\left(1 - e^{-\frac{x}{\sqrt{D_n \tau_r}}}\right) \;. \tag{3.1.10}$$

Here we have chosen $x = 0$ for the space-charge region boundary in the p substrate, with x pointing in the direction from n to p side, so that for a position far inside the neutral p region, $x \to \infty$ the minority carrier concentration n_p equals the thermal equilibrium value n_{p_0}.

Using (2.6.6) and (2.5.3) respectively, we can find the excess generation rate $G = G_n - R_n$ and the diffusion minority carrier current J_n as a function of x, the distance from the edge of the depletion region. Then we have

$$G = \frac{n_{p_0} - n_p}{\tau_r} = \frac{n_{p_0}}{\tau_r} e^{-\frac{x}{\sqrt{D_n \tau_r}}} \tag{3.1.11}$$

$$J_n = qD_n \frac{\partial n_p}{\partial x} = \frac{qn_{p_0} D_n}{\sqrt{D_n \tau_r}} e^{-\frac{x}{\sqrt{D_n \tau_r}}} \;. \tag{3.1.12}$$

The value at $x = 0$ gives the diffusion current emitted from the p side into the space-charge region, yielding

$$J_{s_n} = \frac{qn_{p_0} D_n}{L_n} \quad \text{with } L_n = \sqrt{D_n \tau_r} \tag{3.1.13}$$

L_n, called diffusion length, is a measure for the depth of the region from which charge will diffuse into the space-charge region.[8]

Equivalent considerations for the hole current in the n region lead to a total reverse bias saturation current of

$$J_s = q \left(\frac{n_{p_0} D_n}{\sqrt{D_n \tau_{r_n}}} + \frac{p_{n_0} D_p}{\sqrt{D_p \tau_{r_p}}} \right) . \tag{3.1.14}$$

Thus the diode current can be seen to be inversely proportional to the doping concentration and to the square root of the lifetime.

The above expression has been derived under the assumption that no charge is being generated in the space-charge region. This is certainly a oversimplification for reversely biased diodes built on detector-grade material.

In order to discuss the role of the depletion region in the current–voltage characteristics, it is again advantageous to first consider an unbiased diode in thermal equilibrium. Then the electron and hole concentrations will vary in such a way that the product $n \cdot p = n_i^2$ remains constant, while generation and recombination rate balance each other in any point in the junction. The electron and hole concentrations can be obtained from (3.1.3), which we used as an approximation for the minority carrier concentration also in the case of applied bias. If a forward bias is applied to the junction, the product $n \cdot p$ will increase above n_i^2 and recombination will exceed generation in the space-charge region. This leads to an increase of the forward current above the value obtained from (3.1.7) and (3.1.14).[9] For reverse biasing, $n \cdot p$ will fall below n_i^2 and generation will dominate. In the extreme case of $n \cdot p \ll n_i^2$ – the standard situation for biased detectors – one may simply calculate the generation current originating in the depletion region by multiplying the depletion volume with the generation rate $G_{th} = \frac{n_i}{\tau_g}$ from (2.6.7), as has been done already in the example given in Sect. 2.6.7.

A slightly better estimation for the additional current generated in the space-charge volume is obtained by multiplying only the change in the depletion volume due to the applied voltage with the thermal generation rate:[10]

$$J_v = - q \frac{n_i}{\tau_g} (d - d_0)$$

$$= - q \frac{n_i}{\tau_g} \sqrt{\frac{2 \epsilon \epsilon_0 (N_A + N_D)}{q N_A N_D}} (\sqrt{V_{bi} - V} - \sqrt{V_{bi}}) . \tag{3.1.15}$$

---

[8] The derivation (3.1.10) is valid for the situation in which the metal contacts are at large distances compared with the diffusion lengths L_n and L_p from the junction. The more general case will be considered in Chap. 5, in Example 5.1.

[9] The forward recombination current has less steep exponential behavior with voltage $\left(J_r \approx e^{\frac{qV}{2kT}} \right)$ than the diffusion current (see (3.1.7)) and therefore dominates at low applied voltage (below $\approx 0.5V$ for Si-diodes).

[10] At zero applied voltage, the natural depletion region has formed already, but the current has to be exactly zero.

This expression follows from (3.1.5), with the sign convention of taking currents in forward direction as positive.

3.1.3 A p–n Diode Under Irradiation with Light

Irradiation with light will generate electron–hole pairs in a semiconductor. Pairs generated in the space-charge region of a p–n junction will be separated by the electric field. If the voltage across the junction is kept at the built-in potential by, for instance, shortening the external leads of the device, a photocurrent will be generated. If alternatively the leads are open, the voltage across the junction will drop to such a value that the photocurrent is compensated for by an increase of the diffusion current as given in (3.1.7). This voltage change appears as an output voltage at the external leads of the device.

Example 3.3
Problem: *Not only electron–hole pairs created in the space-charge region will contribute to the photocurrent, but also those generated in the neutral region – although with smaller efficiency. Find this efficiency as a function of the distance from the space-charge region and the diode bias voltage, using similar approximations to those used in deriving the current–voltage characteristics of a diode. See Fig. 3.4.*

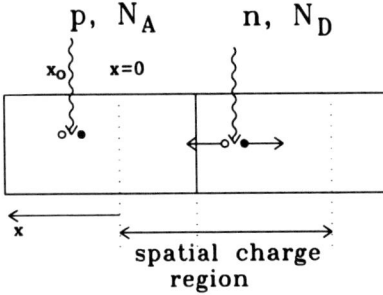

Fig. 3.4. Creation of a photocurrent in a unbiased p–n diode junction. Photons may be absorbed in the space-charge region or in the neutral n and p regions

Solution: *We consider, as in Sect. 3.1.2, minority carriers (electrons) in the p region. The origin of our one-dimensional coordinate system ($x = 0$) is put at the edge of the space-charge region on the p side, x pointing towards the neutral p region, and light is supposed to generate charge at $x = x_0$ so that the generation rate of electrons by light is described by $G_L(x) = G_L \cdot \delta(x - x_0)$. So the problem reduces to solving the continuity equation (3.1.8) and (3.1.9) separately for the regions $0 < x < x_0$ and $x_0 < x < \infty$ with boundary conditions at $x = 0$, $x = x_0$ and $x \to \infty$.*

At $x = 0$ the electron density is assumed to obey the exponential behavior of (3.1.6), and thus

$$n(x = 0) = n_{p_0} e^{\frac{qV}{kT}} \ .$$

At $x = x_0$ we require continuity of the electron density, and so

$$n(x_0 + \epsilon) = n(x_0 - \epsilon) \qquad\qquad \epsilon \to 0$$

and a discontinuity of the electron flux equaling the generation rate at $x = x_0$:

$$F(x = x_0 + \epsilon) - F(x = x_0 - \epsilon) = G_L \ .$$

At $x \to \infty$ the flux has to be zero:

$$F(x \to \infty) \to 0$$

The general solution to the linear differential equation

$$D_n \frac{\partial^2 n(x)}{\partial x^2} - \frac{n(x)}{\tau_r} = -\frac{n_{p_0}}{\tau_r} \tag{3.1.16}$$

is given by

$$n(x) = A e^{-\frac{x}{L}} + B e^{\frac{x}{L}} + n_{p_0} \tag{3.1.17}$$

$$F(x) = -D_n \frac{\partial n(x)}{\partial x} = \frac{D_n}{L} \left[A e^{-\frac{x}{L}} - B e^{\frac{x}{L}} \right]$$

with $L = \sqrt{D_n \tau_r}$, the diffusion length.

Using subscripts 1 and 2 for the two regions, we find from the boundary conditions

at $x \to \infty : B_2 = 0$;
at $x = x_0 : A_2 e^{-\frac{x_0}{L}} = A_1 e^{-\frac{x_0}{L}} + B_1 e^{\frac{x_0}{L}}$

$$\frac{D_n}{L} \left(A_2 e^{-\frac{x_0}{L}} - A_1 e^{-\frac{x_0}{L}} + B_1 e^{\frac{x_0}{L}} \right) = G_L$$

leading to
$$B_1 = \frac{L}{2D_n} G_L e^{-\frac{x_0}{L}} ;$$

at $x = 0 :$ $A_1 + B_1 + n_{p_0} = n_{p_0} e^{\frac{qV}{kT}}$

leading to
$$A_1 = -\frac{L}{2D_n} G_L e^{-\frac{x_0}{L}} + n_{p_0} \left(e^{\frac{qV}{kT}} - 1 \right) \ .$$

Putting A_1 and B_1 into (3.1.17) we find the region 1 electron concentration and flux after combining terms containing the photon generation G_L as

$$n_1(x) = n_{p_0} + \frac{L}{D_n} G_L \sinh \frac{x}{L} e^{-\frac{x_0}{L}} + n_{p_0} \left(e^{\frac{V}{kT}} - 1 \right) e^{-\frac{x}{L}}$$

$$F_1(x) = -G_L \cosh \frac{x}{L} e^{-\frac{x_0}{L}} + \frac{D_n}{L} n_{p_0} \left(e^{\frac{V}{kT}} - 1 \right) e^{-\frac{x}{L}} \ .$$

From the flux at $x = 0$ we find the electron current density in the diode:

$$J_n = qF(x = 0) = q\frac{D_n}{L}n_{po}\left(e^{\frac{qV}{kT}} - 1\right) - qG_Le^{-\frac{x_0}{L}} \ . \tag{3.1.18}$$

The first term represents the electron current in the unilluminated diode derived already in Sect. 3.1.2, the second term represents the contribution due to illumination. Thus in our model, minority carriers generated by light contribute with 100% efficiency to the reverse-bias diode current when generated in the space-charge region and with efficiency

$$\epsilon = e^{-\frac{x_0}{L}} \tag{3.1.19}$$

when produced in the neutral region.[11] $L = \sqrt{D_n\tau_r}$ is the diffusion length and x_0 the distance from the space-charge region. The dependence of the illumination-generated current on the bias voltage is only indirect, through the variation of the extent of the space-charge region.

Example 3.4
Problem: Find the voltage appearing on the open terminals of a diode illuminated homogeneously, such that the generation rate per volume is G_L. Find the current in the device when the terminals are short-circuited.
Solution: Using the result of the previous example, we may integrate the contribution of light to the current density in the neutral n and p regions, as well as in the space-charge region, and we obtain

$$J_L = -q\,G_L\,(L_n + d + L_p) \ , \tag{3.1.20}$$

where L_n and L_p are the diffusion lengths in the respective regions and d is the width of the space-charge region, given by (3.1.5). We then set the total current density

$$J = J_s(e^{-\frac{qV}{kT}} - 1) + J_v - qG_L(L_n + L_p + d) \tag{3.1.21}$$

to zero, to obtain the voltage V appearing on the terminals. The diffusion current J_s and the volume current J_v are given in (3.1.14) and (3.1.15), the depletion depth d by (3.1.2). If we ignore the change of depletion depth due to illumination, we have $J_v = 0$ and $d = d_0$ and obtain

$$\begin{aligned}
V &= \frac{kT}{q}\ln\left[1 + \frac{qG_L(L_n + L_p + d_0)}{J_s}\right] \\
&= \frac{kT}{q}\ln\left[1 + \frac{G_L(\sqrt{D_n\tau_n} + \sqrt{D_p\tau_p} + d_0)}{n_{po}\sqrt{\frac{D_n}{\tau_n}} + p_{no}\sqrt{\frac{D_p}{\tau_p}}}\right] \ .
\end{aligned} \tag{3.1.22}$$

---

[11] In the derivation, we have assumed that the diffusion lengths are small with respect to the thickness of the respective doped regions. For the treatment of the problem where this condition is not fulfilled, see Example 5.1 in Chap. 5.

For finding the current in the short-circuited device, we put $V = 0$ into (3.1.21) and solve it, again with the assumption of unchanged depletion depth, finding

$$J_{\mathrm{sc}} = -qG_{\mathrm{L}}(L_n + L_p + d) \ .$$

3.1.4 Capacitance–Voltage Characteristics

In Sect. 3.1.2 we derived the current–voltage characteristics of a diode with the assumption of an abrupt junction and uniform doping in the two regions. These assumptions are arbitrary and do not in general correspond to reality. Capacitance–voltage measurements, in contrast, are often used to measure the doping profile of semiconductors. We therefore will derive the capacitance–voltage characteristics for an arbitrary (one-dimensional) doping profile. For simplicity (and because it corresponds to a common situation in detectors), we will restrict ourselves to an extremely asymmetric junction with very high p-doping on a low-doped n-substrate. In such a case we may assume that the space-charge region is fully contained in the low-doped substrate.

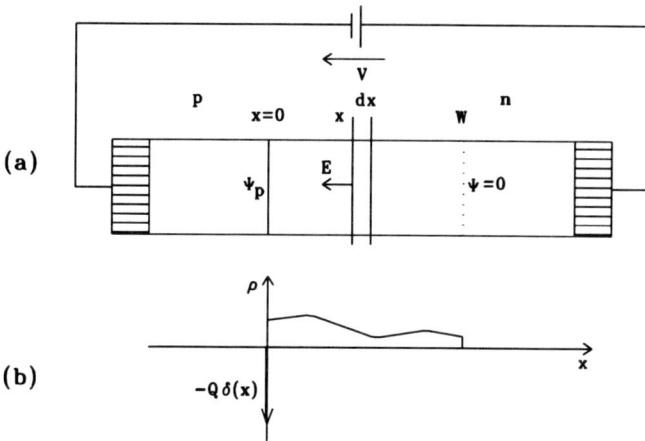

Fig. 3.5a,b. Finding the capacitance–voltage characteristics of a nonuniform doped p–n diode junction. The device structure with applied forward bias and space-charge region extending from $x = 0$ to $x = W$ (a) and the space-charge density (b) are shown

The situation is shown in Fig. 3.5 with the highly doped p region on the left of $x = 0$. Positive externally applied voltage V corresponds to forward-biasing of the diode. Variable W is the width of the space-charge region, and we define potential zero in the neutral n region (at the boundary towards the space-charge region).

In order to find the capacitance–voltage characteristics, we will consider the change in voltage and surface charge of the highly doped p region when the space-charge region is increased from width x to $x + \mathrm{d}x$, and perform an integration from $x = 0$ to W. Increasing the space-charge region by dx means that the surface charge will increase by $qN_{\mathrm{D}}dx$, thus causing an electric field change of $\frac{qN_{\mathrm{D}}dx}{\epsilon\epsilon_0}$ and a surface voltage change of $x\frac{qN_{\mathrm{D}}dx}{\epsilon\epsilon_0}$. We thus obtain

$$Q_p = -\int_0^W qN_{\mathrm{D}}(x)\,dx \tag{3.1.23}$$

$$\Psi_p = \int_0^W \frac{qN_{\mathrm{D}}(x)}{\epsilon\epsilon_0}x\,dx \ . \tag{3.1.24}$$

As was discussed previously, the potential difference across the junction Ψ_p is given by the difference between built-in voltage V_{bi} and applied bias voltage V ($V < 0$ for reverse-bias). V_{bi} was given in (3.1.1) as

$$V_{\mathrm{bi}} = \frac{kT}{q}\ln\frac{N_{\mathrm{A}}N_{\mathrm{D}}}{n_i^2} \ ,$$

where the doping concentrations have to be taken at the edges of the space-charge region.

In order to find the measured capacitance of the device, we have to take the derivative of charge Q_p with respect to the externally applied voltage V:

$$C = \frac{\partial Q_p}{\partial V} = \frac{\partial Q_p/\partial W}{\partial V/\partial W} \tag{3.1.25}$$

$$\frac{\partial Q_p}{\partial W} = -qN_{\mathrm{D}}(W) \tag{3.1.26}$$

$$\frac{\partial V}{\partial W} = \frac{-\partial \Psi_p}{\partial W} + \frac{\partial V_{\mathrm{bi}}}{\partial W} = -\frac{qN_{\mathrm{D}}(W)}{\epsilon\epsilon_0}W + \frac{kT}{q}\frac{1}{N_{\mathrm{D}}(W)}\frac{\partial N_{\mathrm{D}}(W)}{\partial W} \ . \tag{3.1.27}$$

We thus obtain, for the inverse capacitance,

$$\frac{1}{C} = \frac{W}{\epsilon\epsilon_0} - \frac{1}{qN_{\mathrm{D}}(W)}\frac{\partial V_{\mathrm{bi}}}{\partial W} = \frac{W}{\epsilon\epsilon_0} - \frac{kT}{q}\frac{1}{qN_{\mathrm{D}}^2}\frac{\partial N_{\mathrm{D}}}{\partial W} \approx \frac{W}{\epsilon\epsilon_0} \ . \tag{3.1.28}$$

The second term, being due to the variation of the built-in voltage with doping, usually will be negligible with respect to the applied (reverse-bias) voltage and thus can be neglected at least for reverse-bias. The width of the space-charge region can then be found from the capacitance measurement as

$$W = \frac{\epsilon\epsilon_0}{C} \ . \tag{3.1.29}$$

For the measurement of the doping concentration, we must look at the variation of the inverse square of capacitance with the applied voltage. Again, ignoring the variation of V_{bi} with W and making use of (3.1.29) and (3.1.24), we have

$$\frac{\partial(1/C^2)}{\partial V} = \frac{\partial(1/C^2)/\partial W}{\partial V/\partial W} = \frac{2W/(\epsilon\epsilon_0)^2}{qN_{\mathrm{D}}W/(\epsilon\epsilon_0)} = \frac{2}{qN_{\mathrm{D}}\epsilon\epsilon_0} \ . \tag{3.1.30}$$

The doping concentration at depth W is thus

$$N_D = \frac{2}{q\epsilon\epsilon_0 \frac{\partial(1/C^2)}{\partial V}} \quad .$$
(3.1.31)

3.1.5 Breakdown Under Strong Reverse Bias[12]

In Sect. 3.1.2 we showed through (3.1.7) and (3.1.14) that the reverse-bias current reaches saturation at high reverse-bias voltages. This result was obtained by considering the diffusion of minority carriers into the space-charge region of the reverse-biased junction. Further increase of the current is due to the growth of the space-charge region and the corresponding increase of the volume-generated current. That this behavior cannot continue *ad infinitum* is intuitively clear. At some point the electric field will become so high that an electrical breakdown will occur and the reverse-bias current increases drastically.

We will consider this breakdown, trying to understand the underlying mechanisms. It turns out that there are two mechanisms at work, one of them involving the interaction of the electric field with the covalently bound electrons. In terms of the bond representation, the electrons are liberated by the strong electric force; in the band model this *"Zener breakdown"* is described as the movement of electrons from the valence band to the conduction band by the electric field alone.

The other breakdown mechanism, *"avalanche breakdown"*, involves the charge carriers (electrons and holes) that are accelerated by the electric field between collisions with the lattice or crystal defects, some of them gaining enough kinetic energy to break up covalent bonds in the lattice when they collide with it. Two additional carriers, an electron and a hole, are created in an individual multiplication process as was described in Sect. 2.6.5.

We will now try to arrive at a semiquantitative description of these two mechanisms.

Zener Breakdown

The band structure of a highly doped, strongly reverse-biased *p–n* junction is shown in Fig. 3.6.

With the reverse-bias-voltage well above the band gap, the occupied energy levels in the valence band of the *p*-type semiconductor line up with unoccupied energy levels in the conduction band at the *n*-side. Movement of the valence electrons from the *p*-side towards the same energy state on the *n*-side is prevented by the approximately triangular-shaped potential barrier (Fig. 3.7) of height E_G and width

$$L = \frac{E_G}{q\mathcal{E}} \quad .$$
(3.1.32)

---

[12]This section may, if desired, be skipped in a first reading. We will refer to it in later Chaps. 9 and 11.

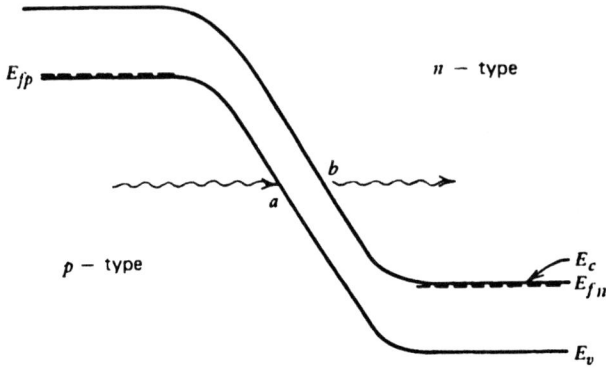

Fig. 3.6. Band structure of a strongly reverse-biased p–n diode junction. (After Muller-Kamins 1977, p. 200 Fig. 4.14)

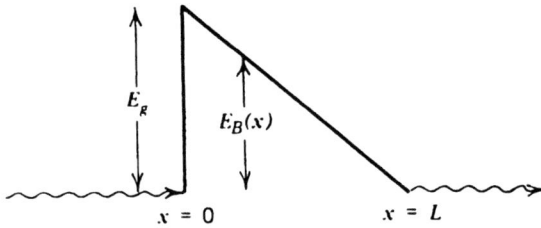

Fig. 3.7. Potential barrier preventing the flow of electrons from an occupied state of the valence band on the p side towards the empty state of same energy in the conduction band on the n side of the strongly reverse-biased p–n diode junction. Transition from n- to p-side is still possible by the quantum mechanical tunnel effect. (After Muller–Kamins 1977, p. 201 Fig. 4.15)

Such a barrier can be overcome by the quantum mechanical tunnel effect. The probability of tunneling Θ through a barrier of variable height is found using the WKB approximation:

$$\Theta \approx \exp\left[-2\int_0^L \sqrt{\frac{2mE_\mathrm{B}(x)}{\hbar^2}}\,\mathrm{d}x\right] \quad . \tag{3.1.33}$$

For the triangular-shaped barrier one obtains carrying out the integration

$$\Theta \approx \exp\left[\frac{-4\sqrt{2mE_\mathrm{G}}L}{3\hbar}\right] = \exp\left[\frac{-4\sqrt{2m}E_\mathrm{G}^{3/2}}{3q\hbar\mathcal{E}}\right] \quad . \tag{3.1.34}$$

Note the strong (exponential) dependence of the tunneling probability on the inverse electric field and the width of the barrier.

In order to estimate the Zener current, we have to know the number of valence electrons arriving at the barrier, that have an empty state of same energy at the other side, $\mathcal{N}vA$ (A being the cross-section of the device, $\mathcal{N}$ and v being the density and velocity of valence electrons respectively), and multiply it with the tunnel probability and the elementary charge:

$$I = qA\mathcal{N}v\Theta . \tag{3.1.35}$$

Example 3.5

Problem: *Estimate the magnitude of the electric field needed to obtain an appreciable Zener current in a reverse-biased silicon diode. Require a $10\,\mathrm{mA}$ current in a $10^{-5}\,\mathrm{cm}^2$ device. The density of valence electrons having a corresponding empty conduction band state of same energy is assumed to equal the atomic density of about $\mathcal{N} = 10^{22}\,\mathrm{cm}^{-3}$. As velocity, the thermal velocity of $v = 10^{-7}\,\mathrm{cm/s}$ is taken.[13] Find also the width of the tunnel barrier L and the relative change of the current with the electric field. Estimate the doping density required for a breakdown voltage of $10\,\mathrm{V}$.*

Solution: *Using (3.1.35) we find the required tunneling probability as*

$$\Theta = \frac{I}{qA\mathcal{N}v}$$

$$= \frac{10^{-2}\,\mathrm{A}}{1.6 \times 10^{-19}\,\mathrm{As}\cdot 10^{-5}\,\mathrm{cm}^2\cdot 10^{22}\,\mathrm{cm}^{-3}\cdot 10^{-7}\,\mathrm{cm/s}} = 0.6 \times 10^{-7} .$$

Using (3.1.34) we find the corresponding electric field as

$$\mathcal{E} = \frac{-4\sqrt{2m}E_G^{3/2}}{3q\hbar} \frac{1}{\ln\Theta} .$$

With the electron mass $m = 0.91095 \times 10^{-30}\,\mathrm{kg} = 0.91095 \times 10^{-30}\,\mathrm{Jm}^{-2}\mathrm{s}^2$, $E_G = 1.12\,\mathrm{eV} = 1.12\cdot 1.6 \times 10^{-19}\,\mathrm{J}$ and $\hbar = 1.0546\,\mathrm{Js}$, we can then establish that

$$\mathcal{E} = \frac{-4\sqrt{2\cdot 0.91095 \times 10^{-30}\,\mathrm{Jm}^{-2}\mathrm{s}^2}\,(1.12\cdot 1.6 \times 10^{-19}\,\mathrm{J})^{3/2}}{3\cdot 1.6 \times 10^{-19}\,\mathrm{As}\cdot 1.0546 \times 10^{-34}\,\mathrm{Js}}$$

$$\times \frac{1}{\ln(0.6 \times 10^{-7})}$$

$$= 4.9 \times 10^8\,\mathrm{V/m} .$$

The width of the barrier is found from (3.1.32) as

$$L = \frac{1.12\,\mathrm{eV}}{4.9 \times 10^6\,\mathrm{eV/cm}} = 2.3\,\mathrm{nm} .$$

The required doping density N can be estimated by setting the maximum electric field of a symmetrically doped abrupt p–n junction (see (3.1.2) with

---

[13]These assumptions are rather arbitrary and not well justified. They are made to get order-of-magnitude estimates. It is also unclear what "effective" mass should be taken for the electron.

$N_A = N_D = N$ and V_{bi} replaced by $V_{bi} - V$) equal to the electric field required for Zener breakdown:

$$\mathcal{E}_{max} = \sqrt{\frac{2q}{\epsilon\epsilon_0} N(V_{bi} - V)}$$

$$N \approx \frac{\epsilon\epsilon_0}{2q} \frac{\mathcal{E}_{max}^2}{V} = \frac{11.9 \cdot 8.854 \times 10^{-14} \, F/cm}{2 \cdot 1.6 \times 10^{-19} \, C} \cdot \frac{(4.9 \times 10^6 \, V/cm)^2}{10 \, V}$$

$$= 7.9 \times 10^{18} \, cm^{-3} \quad .$$

Avalanche Breakdown

Avalanche breakdown due to the process of charge multiplication described in Sect. 2.6.5 occurs when, for both types of charge carriers, the kinetic energy acquired by the electric field between collisions is high enough to trigger multiplication processes. Then the secondary generated charge carrier, moving in the opposite direction to the primary carrier, is again capable of initiating a multiplication process, and an avalanche thereby develops. The probability that a multiplication process is initiated is dependent not only on the electric field value but also on the spatial extent of this high field region as the charge carriers have to gain enough energy between collisions with mean free path length of the order of $10 \, \mu m$ (see Sect. 2.5) in order to initiate a multiplication process (Sect. 2.6.5).

We will try to estimate the onset of avalanche breakdown in an asymmetrically doped reverse-biased p–n^+ diode. Minority carriers will diffuse into the space-charge region. With the asymmetric doping, the rate of electrons diffusing from the undepleted p region into the space-charge region will be much larger than that of holes from the undepleted n region. Therefore, as a good approximation, the hole concentration at the n side boundary of the space-charge region can be taken as zero.[14] It will, however, be non-zero inside the space-charge region due to electron-induced multiplication processes.

In part of the space-charge region, the field may be high enough so that multiplication processes can occur. In such a case both electron (moving forward) and hole (moving backward) concentrations will increase by an equal amount. Introducing the field-dependent multiplication coefficients α_n and α_p for electrons and holes, we may write for the change of electron and hole concentrations in a thin region dx:

$$dn = dp = \alpha_n n(x)dx + \alpha_p p(x)dx \quad , \tag{3.1.36}$$

with α_n, α_p strong functions of x due to their dependence on the electric field, temperature, doping concentration, etc. Measured values for silicon and GaAs are shown in Fig. 3.8. Note the difference between electrons and holes.

As holes in our approximation are originating from multiplication processes only, we can set the hole concentration at position x equal to the difference of

---

[14]This non-essential assumption simplifies the analysis slightly.

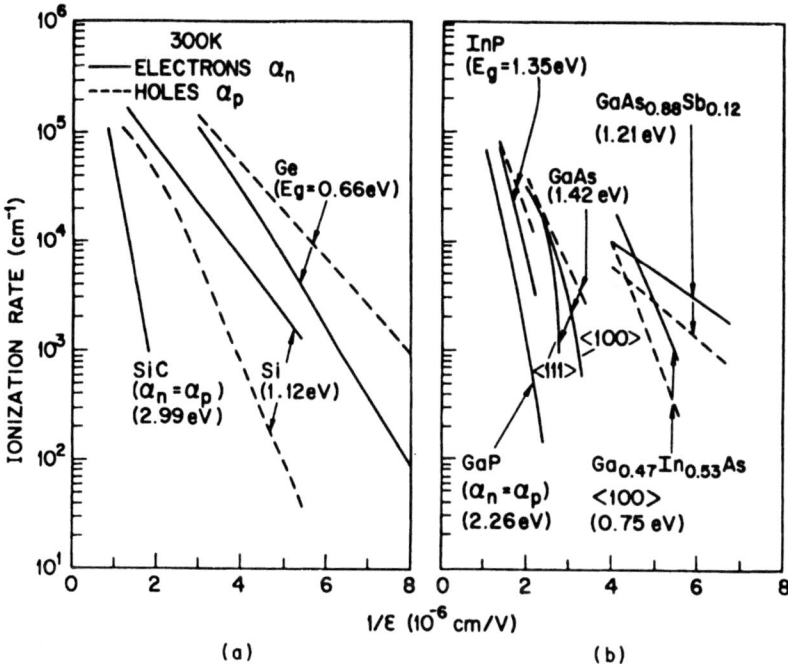

Fig. 3.8a,b. Measured multiplication coefficients for electrons and holes in silicon, germanium, GaAs and other compound semiconductors. (After Sze 1981, p. 47 Fig. 30)

electron concentrations taken at the n-side space-charge boundary ($x = L$) and at position x:

$$p(x) = n(L) - n(x) \ . \tag{3.1.37}$$

We then obtain, via (3.1.36),

$$\frac{dn}{dx} = (\alpha_n - \alpha_p)n(x) + \alpha_p n(L) \ . \tag{3.1.38}$$

An analytic solution to this equation is possible for the special case of $\alpha_n = \alpha_p = \alpha$. We then have

$$n(L) = n(0) + \int_0^L \frac{dn}{dx} = n(0) + n(L) \int_0^L \alpha(x)dx \ .$$

Taking as multiplication factor M the ratio between electrons leaving and electrons entering the space-charge region, we then derive

$$M = \frac{n(L)}{n(0)} = \frac{1}{1 - \int_0^L \alpha(x)dx} \ ,$$

and from that equation we notice that $M \to \infty$ for

$$\int_0^L \alpha(x)\mathrm{d}x = 1 \ .$$

This is the condition for avalanche breakdown.

3.2 Metal–Semiconductor Contact

The metal–semiconductor contact was one of the first practical semiconductor devices showing rectifying properties. In its early form it was a whisker pressed against a semiconductor surface. Now metal–semiconductor contacts (also known as Schottky barriers) are produced using a planar process, the technique now generally used in microelectronics.

Functioning of the metal–semiconductor contact (Fig. 3.9a) can be explained in the band model similarly to what has been done for the diode junction. Conductors differ from semiconductors by having a partially filled conduction band and therefore having the Fermi level – the level at which the occupation probability is one-half – inside the conduction band. Furthermore, the amount of charge carriers available in the metal is so huge that for static situations in good approximation the electric field inside the metal is zero and the interaction with the surrounding material can be described by a surface charge on the outside boundary of the metal.

Looking first at the situation of separated metal and semiconductor (Fig. 3.9b), we introduce the work function $q\Phi$, which is the energy necessary for moving an electron from the Fermi level to the vacuum. The value of Φ_{m} will depend on the type of metal and be different in value from the semiconductor work function Φ_{s}, which in addition is dependent on doping. The electron affinity $q\chi$, the difference between the conduction-band edge and the vacuum level, is an intrinsic property of the specific semiconductor independent of doping. We have assumed $\Phi_{\mathrm{m}} > \Phi_{\mathrm{s}}$, a necessary condition for obtaining rectifying properties in an n-type semiconductor–metal contact.

As metal and semiconductor are brought into contact, the Fermi levels have to line up and a voltage difference $V_{\mathrm{bi}} = \Phi_{\mathrm{m}} - \Phi_{\mathrm{s}}$ will build up across the junction and the band will be bent in the vicinity of the metal–semiconductor boundary. The charge carriers in the semiconductor will rearrange themselves in such a way that the thermal equilibrium condition is met at any point of the semiconductor, while at the metal side a surface charge compensating for the positive space-charge in the semiconductor boundary region will develop (Fig. 3.9d). From Fig. 3.9c we may read the barrier height, namely the threshold an electron with average energy in the metal has to overcome to reach the semiconductor region:

$$q\Phi_{\mathrm{B}n} = q(\Phi_{\mathrm{m}} - \chi) \ . \tag{3.2.1}$$

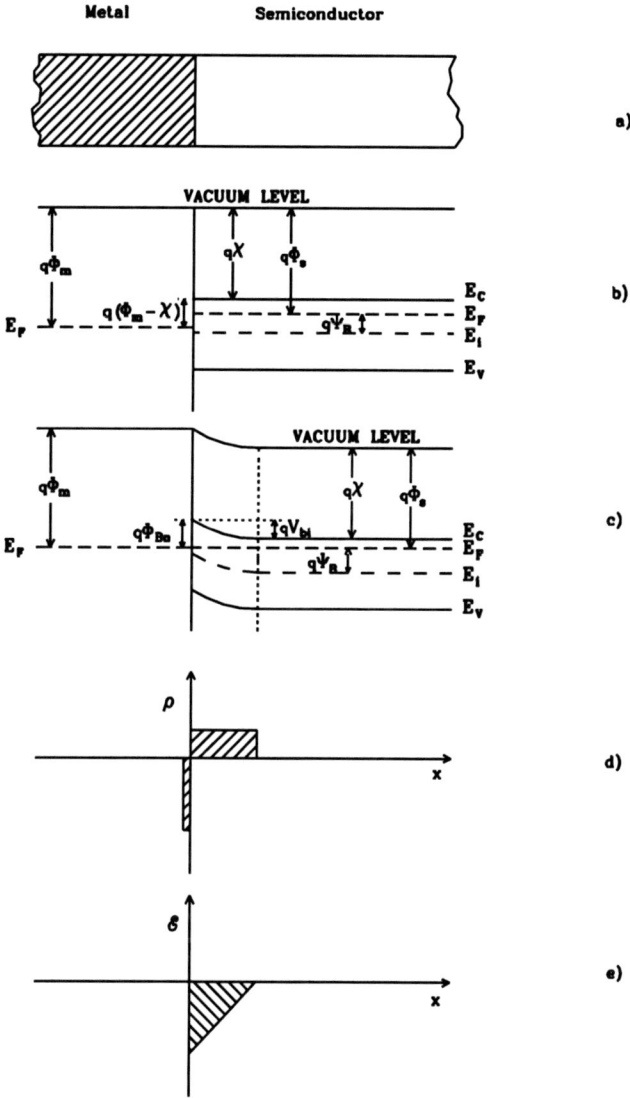

Fig. 3.9a–e. Metal–semiconductor contact (a). Description in the band model: metal and n–type semiconductor separately in thermal equilibrium (b); metal and semiconductor joined together (c); charge density (d); electric field (e)

The height of this barrier will remain unchanged if an external voltage is applied, while the threshold for electrons against movement from the semiconductor to the metal will change from the equilibrium value V_{bi} by the applied voltage. Thus we expect a rectifying behavior by the junction. The energy-band

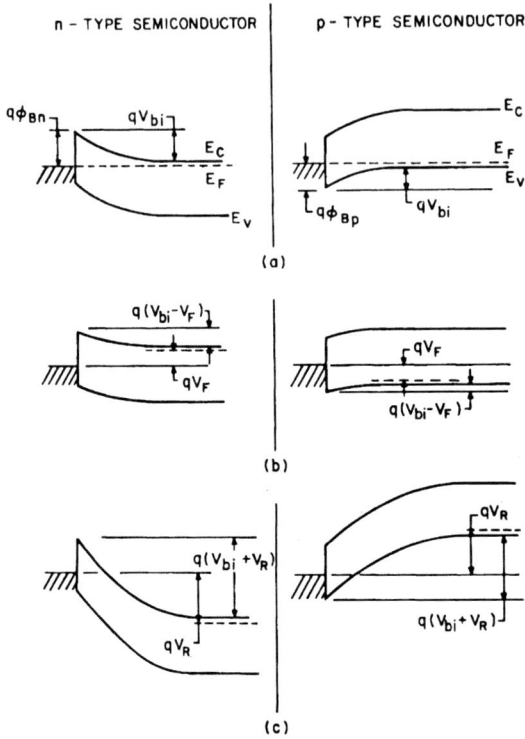

Fig. 3.10a–c. Energy band diagram for metal–n-type and p-type semiconductor junction: in thermal equilibrium (a); with forward bias (b); with reverse bias (c). (After Sze 1985, p. 164 Fig. 4)

diagram, with the application of an external voltage to metal n-type and p-type semiconductors, is shown in Fig. 3.10.

Measurement of the barrier height does not follow too closely the simple theoretical expectation. This fact is explained by the creation of a large number of interface states with energy levels in the band gap at the semiconductor–metal interface. Complicated chemical processes, such as the absorption of oxygen in surface barriers, also play a role and can even lead to an increase of barrier height.

3.2.1 Current–Voltage Characteristics

Contrary to the case of the p–n junction, where we were able to derive the current–voltage characteristics by considering minority carriers only in the two semiconductor regions, we have mainly to deal with majority carriers.

In order to arrive at the current–voltage characteristics, we first will consider in more detail the thermal equilibrium condition. The flow of electrons from the semiconductor to the metal will be exactly balanced by the opposite flow from the metal to the semiconductor. While the flow from the metal will be fixed (as it is expected to depend on the barrier height only), the flow from the semiconductor to the metal is expected to be proportional to the carrier concentration at the interface and thus depend on the applied voltage. Using – as was done already in the case of the p–n junction – (3.1.3) for an estimate of this density, we can expect again an exponential relationship between current and voltage:

$$J = J_s(e^{\frac{qV}{kT}} - 1) \ . \tag{3.2.2}$$

The saturation current J_s is found to be

$$J_s = A^* T^2 e^{-\frac{q\Phi_{Bn}}{kT}} \tag{3.2.3}$$

with A^* the effective Richardson constant. The values of A^* are $110\,\mathrm{AK^{-2}cm^{-2}}$ and $132\,\mathrm{AK^{-2}cm^{-2}}$ for n- and p-type silicon respectively (Sze 1981, p. 167). In addition to the majority carrier current, a usually much smaller minority carrier current is present. It can be estimated in exactly the same way as in the case of the p–n junction leading to (3.1.3).

3.2.2 Ohmic Contact

An ohmic contact is defined as a metal–semiconductor contact with a negligible contact resistance compared with the bulk or series resistance of the semiconductor. In the above treatment we have considered the thermic emission of electrons to be the main source of current. For very high doping concentrations (above $10^{19}\,\mathrm{cm^{-3}}$) the width of the potential barrier may become so small that the process of tunneling becomes important (Sze 1981, p. 171). Then the characteristic resistance of the junction becomes small and an ohmic contact is obtained.

3.3 Metal–Insulator–Semiconductor Structure

Besides the p–n diode junction the metal–insulator–semiconductor (MIS) structure is that where the properties have to be thoroughly understood in order to be able to work with semiconductor detectors. This structure is more frequently referred to as "MOS" or a Metal–Oxide–Semiconductor structure, since the common insulator in the case of silicon is SiO_2. The MOS structure forms the basis of most charge coupled devices (CCDs), which commonly are used for optical imaging. It also forms the essential part of MOS transistors, which are the building blocks of the most widely used microelectronics technology. An understanding of the MOS structure is also necessary for dealing with unwanted effects in the insulation structure of detectors and electronics, which in most cases is based on oxide. The MOS structure furthermore has proved to be an extremely useful structure for the investigation of semiconductor surface effects.

Metal Oxide   Semiconductor

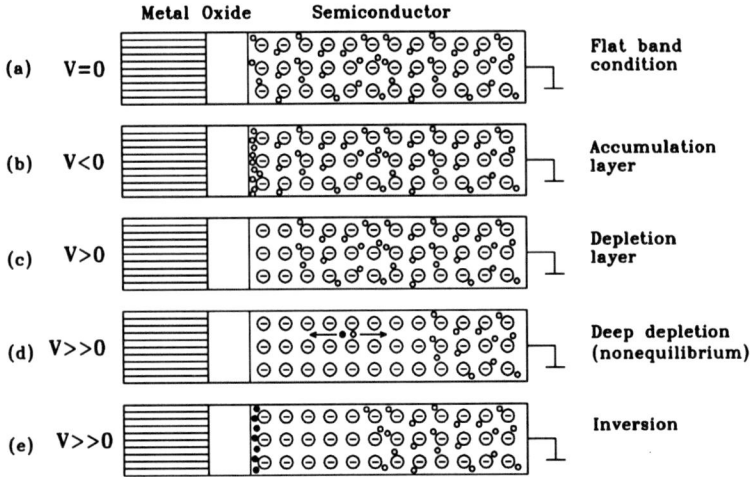

Fig. **3.11a–e.** States of an "ideal" *p*-type MOS structure: thermal equilibrium (a); accumulation (b); surface depletion (c); overdepletion (d); inversion (e)

In order to simplify the treatment, we will first consider the "ideal case" in which we assume identical work functions for metal and semiconductor and a complete absence of space-charges in the oxide. The description will then be generalized to include these effects.

Looking at the device first in a very naive fashion (Fig. 3.11), we have a homogeneously doped *p*-type semiconductor insulated by an oxide layer from the metal. Without the application of an external voltage, the holes will be uniformly distributed in the semiconductor and the electric field will be zero everywhere in the device. If a negative voltage with respect to the semiconductor bulk is applied to the metal, holes will be attracted to the semiconductor oxide interface and a thin layer of positive charge will form in the semiconductor at the boundary. The voltage will then appear across the oxide layer. This situation is called "accumulation".

The situation is more complicated if in our case a positive voltage is applied to the metal. Then the holes will be pushed away from the interface and a negative space-charge region will be formed. This is a stable situation if the voltage is small, and the process is called "depletion". If the voltage is further increased, the depletion width will increase also; however this "overdepleted" situation is not stable. Thermally generated electron–hole pairs will be separated by the electric field in the space-charge region, the holes moving towards the bulk and the electrons accumulating at the semiconductor–insulator interface. The thin negative charge layer is named an "inversion layer", and the status of the system is called inversion.

In the overdepleted state – although not in thermal equilibrium – the MOS structure can be used as detector. Electrons generated by ionizing radiation in the space-charge region will move towards the oxide–semiconductor bound-

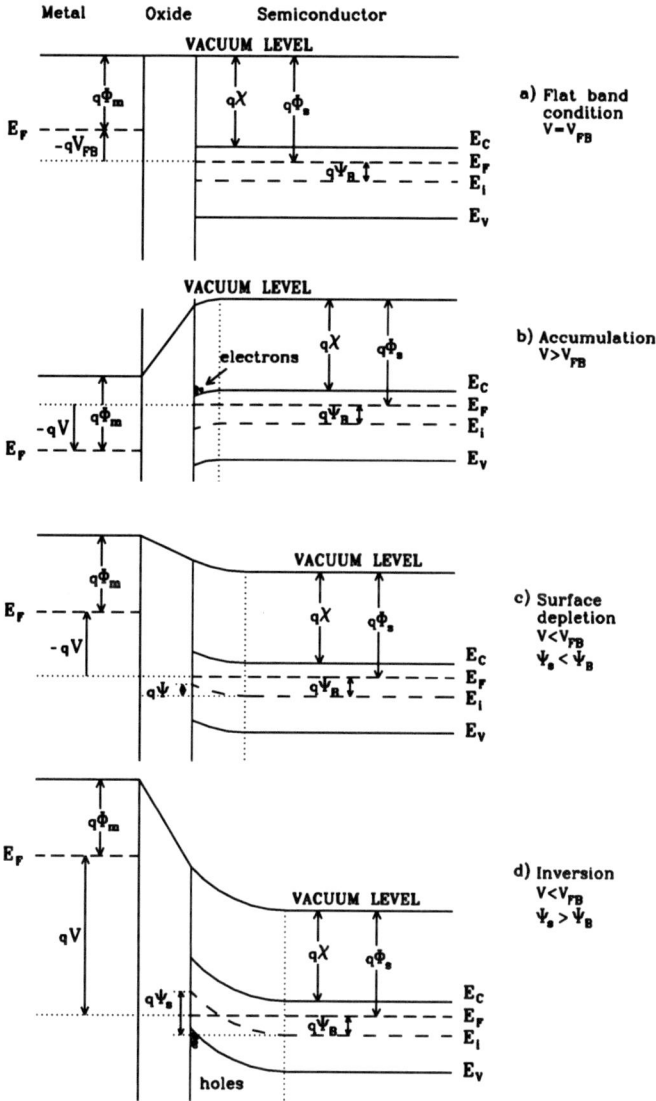

Fig. 3.12a–d. An n-type MOS structure in the band model: flat-band condition (a); accumulation (b); surface depletion (c); inversion (d). For simplification of the illustration it has been assumed that no oxide charges are present

ary and induce a charge of slightly smaller magnitude in the metal electrode. This induced charge can be measured. Electrons assembled at the insulator–semiconductor boundary can also be moved in a controlled fashion sideways towards a readout electrode – as is the case in most CCDs (see Sect. 6.6).

3.3.1 Thermal Equilibrium Condition

The same situations may be considered now in the energy-band model. Here, in contrast, we take as an example a homogeneously doped (doping concentration N_D) n-type semiconductor. As metal and semiconductor are separated by an insulator, charge carriers will have infinitely small probability for crossing between the semiconductor and the metal (and therefore the current is zero, even if an external voltage is applied to the device), so that we may consider thermal equilibrium separately in the two regions. We will right away consider the more general case, that the work function of metal and semiconductor are different (the work function of the semiconductor anyway depends on the doping level). Another very important aspect of the role of charges in the insulator is, however, postponed to the next section (Sect. 3.3.2).

We will use the band model (Fig. 3.12) and the abrupt change approximation (Fig. 3.13), which has already proved quite useful in the treatment of the diode, for a description of the operation modes of the MOS structure.

Flat-Band Condition

Figure 3.12a shows the so-called flat-band condition, the situation that the charge carrier concentration in the homogeneously doped n-type semiconductor is uniform up to the boundary with the oxide and the electric field strength throughout the semiconductor is zero. The Fermi level E_F (expressed as $q\Psi_B$, which is its distance from the intrinsic level) is obtained from (2.4.2) with the assumption of the complete ionization of the donor and acceptor states respectively. We have for n- and p-type semiconductors

$$q\Psi_{Bn} = E_F - E_i = kT \, \ln \frac{N_D}{n_i} \tag{3.3.1}$$

$$q\Psi_{Bp} = E_i - E_F = kT \, \ln \frac{N_A}{n_i} \;,$$

and the semiconductor work function can be obtained with the doping-independent electron affinity χ as

$$q\Phi_s = q\chi + E_C - E_i - kT \, \ln \frac{N_D}{n_i} \tag{3.3.2}$$

$$q\Phi_s = q\chi + E_C - E_i + kT \, \ln \frac{N_A}{n_i} \;.$$

As the oxide is free of charge, it is field-free also. Therefore the vacuum level will be constant throughout the device. As the work function $q\Phi$ (the energy necessary to move an electron from the Fermi level to the vacuum level) is different for the metal and the semiconductor, a flat-band voltage

$$V_{FB} = \Phi_m - \Phi_s \quad \text{(in the absence of oxide charges)} \tag{3.3.3}$$

has to be present to achieve this condition. It may be worthwhile pointing out that V_{FB} corresponds to the external voltage that has to be applied to

Metal Oxide   Semiconductor

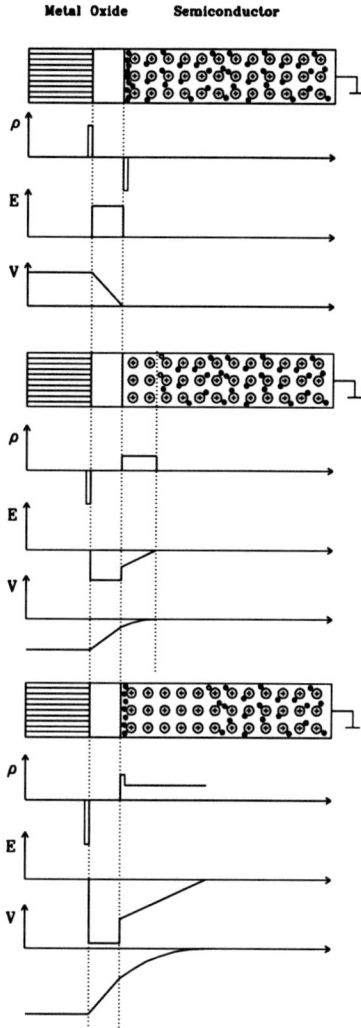

a) accumulation:

$$E_{ox} = -\frac{Q_{acc}}{\epsilon_{ox}\epsilon_0}$$

$$V - V_{FB} = E_{ox}d_{ox}$$

$$= -Q_{acc}\frac{d_{ox}}{\epsilon_{ox}\epsilon_0} = -\frac{Q_{acc}}{C_{ox}}$$

b) depletion:

$$E_s = -\frac{qN_D}{\epsilon_s\epsilon_0}d_s$$

$$\Psi_s = -\frac{qN_D}{\epsilon_s\epsilon_0}\frac{d_s^2}{2}$$

$$E_{ox} = -\frac{qN_D}{\epsilon_{ox}\epsilon_0}d_s$$

$$V - V_{FB} = -\frac{qN_Dd_s}{\epsilon_0}\left(\frac{d_s}{2\epsilon_s} + \frac{d_{ox}}{\epsilon_{ox}}\right)$$

c) inversion:

$$\Psi_s = -2\Psi_B = -\frac{qN_D}{\epsilon_s\epsilon_0}\frac{d_{max}^2}{2}$$

$$d_{max} = \sqrt{\frac{4\epsilon_s\epsilon_0\Psi_B}{qN_D}}$$

$$E_s = -\frac{qN_D}{\epsilon_s\epsilon_0}d_{max}$$

$$E_{ox} = -\frac{qN_D}{\epsilon_{ox}\epsilon_0}d_{max} - \frac{Q_{inv}}{\epsilon_{ox}\epsilon_0}$$

$$V - V_{FB} = -\frac{qN_Dd_{max}}{\epsilon_0}\left(\frac{d_{max}}{2\epsilon_s} + \frac{d_{ox}}{\epsilon_{ox}}\right) - \frac{Q_{inv}}{C_{ox}}$$

$$= -2\psi_B - \frac{1}{C_{ox}}\left(\sqrt{4qN_D\epsilon_s\epsilon_0\Psi_B} + Q_{inv}\right)$$

Fig. 3.13a–c. Charge density, electric field and potential of an n-type MOS structure in the abrupt change approximation: accumulation (a); surface depletion (b); inversion (c)

the device. This can be seen when making contact between the silicon bulk and a metal electrode of the same type as the gate. As shown in Sect. 3.2, the Fermi level is constant in the silicon–metal contact and a built-in voltage corresponding to the differences in work functions of silicon and metal develops. The difference between the Fermi levels between gate and silicon therefore equals the difference between the vacuum levels of metal gate and metal bulk contact.

Accumulation
If one applies a voltage more positive than V_{FB} across the structure (Fig. 3.12b and 3.13a) the potential at the semiconductor–oxide interface will also move in the positive direction, so that the energy bands will bend downwards in the boundary region. As the Fermi level gets closer to the conduction band edge, the electron concentration will increase in the vicinity of the interface. At each point inside the semiconductor the thermal equilibrium condition

$$\frac{n}{n_{\text{i}}} = e^{\frac{E_{\text{F}} - E_{\text{i}}}{kT}} \tag{3.3.4}$$

will be fulfilled.

Due to the exponential dependence of the electron concentration on the potential, the region of increased electron concentration is so thin that it can be approximated to by a surface-charge density at the interface (remaining at bulk potential). This simplification is similar to the assumption of an abrupt change between a space-charge and a neutral region in the case of the diode.

The surface charge density is then given by

$$Q_{\text{acc}} = -\epsilon_{\text{ox}}\epsilon_0 \frac{V - V_{\text{FB}}}{d_{\text{ox}}} = -C_{\text{ox}}(V - V_{\text{FB}}) \ , \tag{3.3.5}$$

where C_{ox} is the oxide capacitance per unit area.

Example 3.6
Problem: *Estimate the depth of the accumulation layer and express it as a function of the surface-charge density of the accumulation layer.*
Solution: *We will try to find the electron concentration $n(x)$ for an n-type semiconductor, x being the distance from the oxide–semiconductor interface. An estimate for the thickness of the electron layer will be obtained by taking the ratio of surface charge density $Q = q \int_0^\infty n(x)\mathrm{d}x$ and electron volume charge density at $x = 0$.*

Defining potential zero in the neutral region of the homogeneously doped n-type semiconductor with doping density N_{D}, the electron and hole densities are given as

$$n(x) = n_0 \exp\left(\frac{q\Psi(x)}{kT}\right) = N_{\text{D}} \exp\left(\frac{q\Psi(x)}{kT}\right)$$
$$p(x) = \frac{n_{\text{i}}^2}{n(x)} = \frac{n_{\text{i}}^2}{N_{\text{D}}} \exp\left(\frac{-q\Psi(x)}{kT}\right) \ .$$

As we are considering an equilibrium state, there are no currents to consider (drift and diffusion currents cancel each other out). The charge density is given by

$$\rho(x) = q\left[N_{\text{D}} - n(x) + p(x)\right] \ ,$$

and we use the (one-dimensional) Poisson equation

$$\frac{\partial^2 \Psi}{\partial x^2} = -\frac{\partial \mathcal{E}}{\partial x} = -\frac{\rho(x)}{\epsilon_s \epsilon_0}$$

$$= -\frac{q}{\epsilon_s \epsilon_0} \left[N_D - N_D e^{\frac{q\Psi(x)}{kT}} + \frac{n_i^2}{N_D} e^{\frac{-q\Psi(x)}{kT}} \right] \ .$$

We do not attempt to obtain an explicit solution for $\Psi(x)$ of this nonlinear equation but rather are interested in the electric field at the surface, from which we can obtain the total net charge within the semiconductor. We therefore multiply the Poisson equation by $\frac{\partial \Psi(x)}{\partial x}$ and integrate from x to infinity:

$$\int_x^\infty \frac{\partial^2 \Psi}{\partial x^2} \frac{\partial \Psi}{\partial x} dx = -\frac{q}{\epsilon_s \epsilon_0} \int_x^\infty \left[N_D - N_D e^{\frac{q\Psi(x)}{kT}} + \frac{n_i^2}{N_D} e^{\frac{-q\Psi(x)}{kT}} \right] \frac{\partial \Psi}{\partial x} dx \ .$$

Observing that $\frac{\partial^2 \Psi}{\partial x^2} \frac{\partial \Psi}{\partial x}$ can be written as $\frac{1}{2} \frac{\partial}{\partial x} \left(\frac{\partial \Psi}{\partial x} \right)^2$ and taking $\Psi = 0$ and $\frac{\partial \Psi}{\partial x} = 0$ at $x \to \infty$, we obtain through the integration of both terms:

$$-\frac{1}{2} \left(\frac{\partial \Psi}{\partial x} \right)^2 = \frac{q}{\epsilon_s \epsilon_0} \left[N_D \Psi(x) - N_D \frac{kT}{q} \left(e^{\frac{q\Psi(x)}{kT}} - 1 \right) - \frac{n_i^2}{N_D} \frac{kT}{q} \left(e^{\frac{-q\Psi(x)}{kT}} - 1 \right) \right] \ .$$

The surface charge density $Q(x)$, defined as the integral over the volume charge density $\rho(x)$ in the region x to ∞, can be calculated from the electric field $\mathcal{E}(x)$ as

$$Q(x) = \epsilon_s \epsilon_0 \frac{\partial \Psi}{\partial x} \tag{3.3.6}$$

$$= \pm \sqrt{2\epsilon_s \epsilon_0 q N_D} \sqrt{-\Psi(x) + \frac{kT}{q} \left(e^{\frac{q\Psi(x)}{kT}} - 1 \right) + \frac{n_i^2}{N_D^2} \frac{kT}{q} \left(e^{\frac{-q\Psi(x)}{kT}} - 1 \right)} \ .$$

Inspection of the equation shows that the first term within the square root corresponds to the donors atoms, the second to the electrons and the third to the holes. The square root has to be taken negative for $\Psi > 0$ (accumulation) and positive for $\Psi < 0$ (inversion).

Considering first the case of accumulation, we may neglect the holes and obtain for $\frac{q\Psi(0)}{kT} \gg 1$

$$Q(0) \approx -\sqrt{2\epsilon_s \epsilon_0 N_D kT e^{\frac{q\Psi(0)}{kT}}}$$

$$\rho(0) \approx -q N_D e^{\frac{q\Psi(0)}{kT}} \approx \frac{q}{2\epsilon_s \epsilon_0 kT} Q(0)^2$$

and the depth of the accumulation layer is given by

$$\hat{d} \approx \frac{Q(0)}{\rho(0)} \approx \frac{2\epsilon_s \epsilon_0 kT}{q Q(0)} \ . \tag{3.3.7}$$

The depth of the accumulation layer is thus shown to be inversely proportional to the accumulation surface charge density $Q(0)$ and proportional to the absolute temperature T. For room temperature $(300\,\mathrm{K})$ and electron density $N_{\mathrm{acc}} = 10^{12}\,\mathrm{e/cm^2}$ of the accumulation layer, we obtain $\hat{d} = 3.4\,\mathrm{nm}$.

Depletion

If a voltage $V \leq V_{\mathrm{FB}}$ appears across the n-type MOS structure, the bands will bend upwards and the electron concentration will decrease (Fig. 3.12c). Due to the exponential dependence on the distance between Fermi level and conduction-band edge, the charge-carrier concentration will fall over a very short distance to a value that is negligible compared with the doping concentration, so that the assumption of an abrupt change from space-charge to undepleted semiconductor region can be made. With this approximation (Fig. 3.13b), it is easy to find the electric field configuration. Using basic relations, depletion-layer surface-charge density Q_{B}, electric field $\mathcal{E}_{\mathrm{s}}$ and potential Ψ_{s} at the semiconductor surface can be expressed by the depletion-layer depth d_{s} as follows:

$$Q_{\mathrm{B}} = qN_{\mathrm{D}}d_{\mathrm{s}} \qquad \mathcal{E}_{\mathrm{s}} = -\frac{qN_{\mathrm{D}}}{\epsilon_{\mathrm{s}}\epsilon_0}d_{\mathrm{s}} \qquad \Psi_{\mathrm{s}} = -\frac{qN_{\mathrm{D}}}{2\epsilon_{\mathrm{s}}\epsilon_0}d_{\mathrm{s}}^2 \quad . \tag{3.3.8}$$

Here, zero potential in the neutral bulk region was assumed. The (constant) electric field in the oxide $\mathcal{E}_{\mathrm{ox}}$ is scaled from E_{s} by the ratios of dielectric constants:

$$\mathcal{E}_{\mathrm{ox}} = \frac{\epsilon_{\mathrm{s}}}{\epsilon_{\mathrm{ox}}}\mathcal{E}_{\mathrm{s}} = -\frac{qN_{\mathrm{D}}}{\epsilon_{\mathrm{ox}}\epsilon_0}d_{\mathrm{s}} \quad , \tag{3.3.9}$$

and the relation between applied voltage and depletion-layer depth is obtained as:

$$V - V_{\mathrm{FB}} = \Psi_{\mathrm{s}} + d_{\mathrm{ox}}\mathcal{E}_{\mathrm{ox}} = -\frac{qN_{\mathrm{D}}}{\epsilon_0}d_{\mathrm{s}}\left(\frac{d_{\mathrm{s}}}{2\epsilon_{\mathrm{s}}} + \frac{d_{\mathrm{ox}}}{\epsilon_{\mathrm{ox}}}\right) \quad . \tag{3.3.10}$$

The depth of the depletion layer (compare Fig. 3.12b) is thus given as

$$d_{\mathrm{s}} = \sqrt{\frac{\epsilon_{\mathrm{s}}\epsilon_0}{qN_{\mathrm{D}}}(V_{\mathrm{FB}} - V) + \left(\frac{\epsilon_{\mathrm{s}}}{\epsilon_{\mathrm{ox}}}d_{\mathrm{ox}}\right)^2} - \frac{\epsilon_{\mathrm{s}}}{\epsilon_{\mathrm{ox}}}d_{\mathrm{ox}} \quad , \tag{3.3.11}$$

and the bending of the bands at the interface is given by

$$\Psi_{\mathrm{s}} = -\frac{qN_{\mathrm{D}}d_{\mathrm{s}}^2}{2\epsilon_{\mathrm{s}}\epsilon_0} \quad . \tag{3.3.12}$$

Inversion

If the voltage is further decreased, the intrinsic level at the interface will reach – and eventually cross – the Fermi level ($\Psi_{\mathrm{s}} \leq \Psi_{\mathrm{B}}$). Then we shall have a majority of holes at the interface (Figs. 3.12d and 3.13c). This situation is called "weak inversion". If, with further decrease of the voltage, the hole concentration reaches or surpasses the electron bulk concentration, we arrive at the status of "strong inversion". This condition is reached when the surface potential has moved by twice the distance of the Fermi level from the intrinsic level in the undepleted bulk, i.e.

$$\Psi_s = -2\Psi_B \ . \tag{3.3.13}$$

The hole density at the surface equals the electron density in the bulk, and the depletion depth equals

$$d_{max} = \sqrt{\frac{4\epsilon_s\epsilon_0\Psi_B}{qN_D}} \ . \tag{3.3.14}$$

The electric field at the surface of the semiconductor is then

$$\mathcal{E}_s = -\frac{qN_D}{\epsilon_s\epsilon_0}d_{max} = -\sqrt{\frac{4qN_D\Psi_B}{\epsilon_s\epsilon_0}} \ , \tag{3.3.15}$$

and the threshold voltage is given by

$$V_T = V_{FB} - 2\Psi_B + \mathcal{E}_{ox}d_{ox} = V_{FB} - 2\Psi_B - \frac{d_{ox}}{\epsilon_{ox}\epsilon_0}\sqrt{4qN_D\epsilon_s\epsilon_0\Psi_B} \ . \tag{3.3.16}$$

Further increase above the corresponding threshold voltage in our approximation will only increase the strength of the inversion layer, the depletion depth remaining constant at d_{max}.

The surface-charge density of the inversion layer will be given as

$$Q_{inv} = (V_T - V)C_{ox} \ . \tag{3.3.17}$$

Example 3.7
Problem: *Estimate the depth of the inversion layer as a function of the surface charge density of the inversion layer.*
Solution: *We solve this example similarly to the accumulation case and start from (3.3.6), which was still of general validity. As before, we are interested in the surface-charge density (this time of the inversion layer) and the volume-charge density at the semiconductor–insulator surface. Assuming* $-\frac{q\Psi(0)}{kT} \gg 1$, *as is the case in inversion, we have*

$$Q(0) \approx \sqrt{2\epsilon_s\epsilon_0qN_D\left[-\Psi(0) + \frac{n_i^2}{N_D^2}\frac{kT}{q}\exp\left(\frac{-q\Psi(0)}{kT}\right)\right]}$$

$$\rho(0) = qp(0) = q\frac{n_i^2}{N_D}\exp\left(-\frac{q\Psi(0)}{kT}\right) \ .$$

From the first of the two equations we obtain

$$\frac{n_i^2}{N_D^2}e^{-\frac{q\Psi(0)}{kT}} \approx \frac{q}{kT}\left[\frac{Q^2(0)}{2\epsilon_s\epsilon_0qN_D} + \Psi(0)\right]$$

$$= \frac{1}{2\epsilon_s\epsilon_0kTN_D}\left[Q^2(0) + 2\epsilon_s\epsilon_0qN_D\Psi(0)\right] \ ,$$

which may be used by reexpressing the volume-charge density at the surface as

$$\rho(0) \approx \frac{q}{2\epsilon_s\epsilon_0 kT} \left[Q^2(0) + 2\epsilon_s\epsilon_0 q N_D \Psi(0) \right] \quad .$$

Estimating the inversion-layer thickness d by the ratio of surface-charge density of the inversion layer and volume-charge density at the surface, we obtain

$$d \approx \frac{Q(0)}{q(0)} \approx \frac{2\epsilon_s\epsilon_0 kT}{qQ(0)} \frac{1}{1 + \frac{2\epsilon_s\epsilon_0 N_D \Psi_0}{Q^2(0)}} \approx \frac{2\epsilon_s\epsilon_0 kT}{qQ(0)} \frac{1}{1 + \left(\frac{q N_D d_s}{Q(0)}\right)^2} \quad .$$

We have used (3.3.8) for expressing the surface potential $\psi(0)$ by the depletion depth d_s and notice that the last expression in the denominator is just the square of the ratios of charges in the space-charge region and in the inversion layer.

3.3.2 The Si–SiO$_2$ MOS Structure

The Si–SiO$_2$ MOS structure is the most extensively studied metal–insulator–semiconductor system. An important aspect of this system is the presence of oxide charges in the SiO$_2$ bulk and at the Si–SiO$_2$ interface. Before discussing the origin of these charges, their consequences for the static behavior of the structure will be described.

In that regard it is sufficient to obtain changes in the flat-band voltage due to these charges. As we have so far expressed the properties by the difference of the applied voltage from the flat-band condition, the previous relationships remain valid. The flat-band voltage change will depend on the amount and distribution of the charge inside the oxide.

A sheet of positive charge with surface charge density σ at distance x from the metal will cause a flat-band voltage change (Fig. 3.14) of $\Delta V_{FB} = -\frac{\sigma x}{\epsilon_{ox}\epsilon_0}$. Thus the combined effect of surface charge σ_{int} at the interface and a volume charge $\rho(x)$ will be

$$\Delta V_{FB} = -\frac{1}{\epsilon_{ox}\epsilon_0} [\sigma_{int} d_{ox} + \int_{x=0}^{d_{ox}} \rho(x)\, x\, dx] \quad . \tag{3.3.18}$$

This value has to be added to the workfunction difference between metal and semiconductor (see (3.3.3)).

The nature and properties of oxide and interface states will be discussed in more detail in Chaps. 4 (material properties) and 12 (radiation damage). Here and now, it is sufficient to make just a few remarks. Charges in the oxide may be due to holes trapped at defects in the oxide. These positive charges are fixed in space. The sites on which they may be trapped are especially numerous close to the Si–SiO$_2$ interface due to the strong lattice distortions in the transition region from oxide to silicon. Defects in this region are able to assume several charge states. If they are located close enough to the semiconductor, so that tunneling is possible, they can change their charge state by capturing

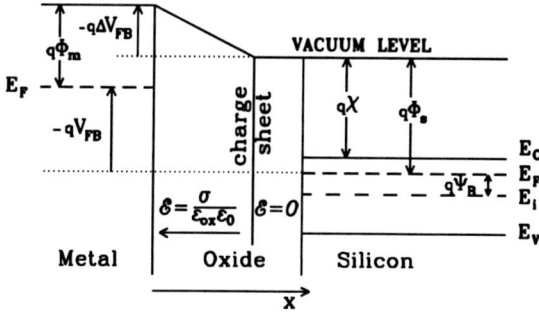

Fig. 3.14. Flat-band voltage change due to oxide charges

electrons or holes from the semiconductor. The charge state of these "interface states" therefore depends on the conditions on the semiconductor surface. Of a different nature are positively and negatively charged ions (e.g. Na^+), which can be distributed throughout the oxide and may move under the influence of an electric field. This movement is strongly temperature-dependent.

3.3.3 Capacitance–Voltage Characteristics

As was the case for the diode junction, the capacitance–voltage characteristics of the MOS structure provide diagnostic information, not only on the semiconductor but also on the insulator and the insulator–semiconductor interface.

As has been discussed in Sect. 3.3.1, there exist four characteristic situations for the MOS structure:

- *flat-band condition,* in which the semiconductor is in a uniform condition up to the insulator boundary;
- *accumulation,* in which the majority carriers are attracted towards the insulator–semiconductor interface, building a very thin conducting layer of the same type as the semiconductor;
- *surface depletion,* in which the majority charge carriers are repelled from the insulator–semiconductor interface. An insulating-depleted layer, whose thickness depends on the applied voltage, forms in the surface region of the semiconductor; and
- *inversion,* in which a thin conducting layer of minority carriers forms close to the insulator–semiconductor interface, followed by an insulating space-charge layer whose thickness is independent on the applied voltage.

Differential capacitance is measured by adding a small sinuisoidal voltage to the constant bias voltage of the device and measuring the sinuisoidal current.

The thermal equilibrium values for such important quantities as, for example, potentials, fields and surface charge densities have been calculated with the usual abrupt-change approximation used in Sect. 3.3.1. Since we now apply a bias varying with time, we have to consider also the time it takes until these equilibrium conditions are reached. In the accumulation case, varying the accumulation-layer charge is accomplished by attracting or repelling majority charge carriers from/towards the bulk. This is a rather fast process as it is governed only by drift and mobility. Similar considerations apply to the depletion case. The situation is very different in the inversion case. Here, the change in the inversion-layer charge density is accomplished by thermal generation and recombination in the space-charge and surface regions. For very good semiconductors, having large generation lifetime, the time constants for reaching equilibrium conditions can be extremely long (approaching milliseconds to seconds). Note, however, that this is true only in the absence of currents parallel to the interface. Such type of currents can be supplied by neighboring implanted electrodes, as is the case in MOS transistors (see Chap. 7). In such cases, much faster transients are possible.

Frequency-dependent measurements of capacitance–voltage characteristics therefore are able to give information not only on doping densities and flat-band voltage but also on carrier lifetimes. We will restrict ourselves to discussion of measurements at very low frequency – in which thermal equilibrium is reached at all times – and on the measurement at frequencies high enough, so as to prevent any significant change in the inversion-layer charge density.

At low frequencies we measure

- in the accumulation case, $C = C_{ox}$;
- in the depletion case, $C = \frac{C_{ox}C_s}{C_{ox}+C_s}$, the series capacitance of oxide and depletion layer capacitance; and
- in the inversion case $C = C_{ox}$, the depletion layer depth staying constant while the inversion-layer surface-charge density varies due to the applied sinuisoidal voltage.

In these cases, $C_{ox} = \frac{\epsilon_{ox}\epsilon_0}{d_{ox}}$ is the oxide capacitance, and $C_s = \frac{\epsilon_s\epsilon_0}{d_s}$ is the capacitance of the depletion layer.

At high frequencies the inversion-layer surface-charge density stays constant, while the depletion-layer depth varies with the applied voltage. The capacitance is then given by $C = \frac{C_{ox}C_s}{C_{ox}+C_s}$.

Between these idealized cases there exist continuous transistions whose description and quantitative investigations require significantly more elaborate treatment than is intended here.

3.3.4 Nonequilibrium and a Return to Equilibrium

Information on position-dependent semiconductor properties, such as doping densities and carrier lifetimes, can also be obtained by bringing a system into thermal nonequilibrium and observing the return to equilibrium. From the time behavior and its temperature dependence, quite sophisticated information can be extracted.

An example for this procedure is the measurement of the capacitance of a MOS structure as a function of time after it has been brought into deep depletion by the sudden application of a bias voltage. The capacitance will drop to a low value as the width of the depletion layer, given by (3.3.11), is above the thermal equilibrium value (see (3.3.14)) reached after the buildup of the inversion layer, whose asymptotic surface-charge density curve is given by (3.3.17).

The inversion-layer buildup is due to the thermal generation of electron–hole pairs in the depleted region (and also to the diffusion of minority carriers from the undepleted bulk into the space-charge region). As the depletion region shrinks during the process, it is possible to extract the carrier generation lifetime as a function of the position in the semiconductor.

The lifetime itself is dependent on the type and concentration of the crystal defects as well as on temperature. Its functional dependence on temperature together with complementary measurements as, for example, the volume generated current and many more sophisticated methods, are used for the investigation of defects in semiconductors.

Example 3.8
Problem: *A MOS stucture is built on top of detector-grade n-type silicon with doping concentration $N_D = 2 \times 10^{12} \, \text{cm}^{-3}$ and generation lifetime $\tau = 10 \, \text{ms}$. The oxide thickness is 2000 Å and the flat-band voltage $-2 \, \text{V}$. The gate voltage is suddenly changed from zero to $-20 \, \text{V}$. Find the thickness of the depletion layer immediately after application of the voltage step, and estimate the time constant for reaching thermal equilibrium.*

Solution: *The thickness of the depletion layer immediately after applying the voltage step is found from (3.3.11) using the dielectric constant of silicon $\epsilon_s = 11.9$, and is derived as $d_s = 76.4 \, \mu\text{m}$. The depletion thickness after reaching thermal equilibrium is found from (3.3.14) as $d_{max} = 13 \, \mu\text{m}$. The difference of total charge in the space-charge region immediately after application of the voltage step and after reaching thermal equilibrium is given by $\Delta Q = N_D \, (d_s - d_{max}) \, q = 2 \times 10^{12} \cdot 63.4 \times 10^{-4} \, q = 1.27 \times 10^{10} \, q \, \text{cm}^{-2}$. The volume-generated current immediately after application of the voltage step is $J = G d_s q = \frac{n_i}{\tau_g} d_s q = \frac{1.45 \times 10^{10}}{10^{-2}} \cdot 76.4 \times 10^{-4} \, q = 1.11 \times 10^{10} \, q \, \text{cm}^{-2} \text{s}^{-1}$.*

We take as estimate for the time constant the ratio $\tau = \frac{\Delta Q}{J} = \frac{1.27 \times 10^{10}}{1.11 \times 10^{10}} = 1.14 \, \text{s}$.

We have stressed here the observation of the return to equilibrium as a diagnostic tool. Nonequilibrium operation of a MOS structure, however, is also important in detector operation. This is the case in CCDs, for example, as will be described in Sect. 6.6.

3.4 The n^+-n or p^+-p Structures

In Sect. 3.1 considering the p–n junction, we observed that, even in thermal equilibrium, buildups of potential occur. These are caused by the diffusion of electrons into the p region and of holes into the n region, which creates an electrically charged region, the space-charge region at the boundary between n- and p-type doping. This internal potential buildup does not appear on external electrodes made of the same metals because of exact cancellation in the metal–semiconductor interfaces.

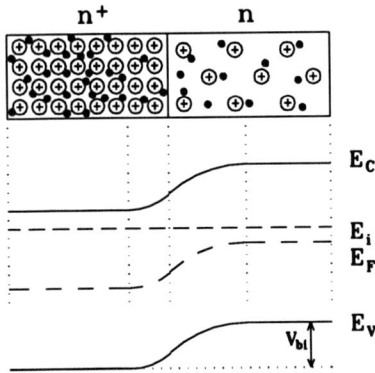

Fig. 3.15. Abrupt n^+-n junction in thermal equilibrium: spatial distribution of donors and electrons (*top*); representation in the band model (*bottom*)

A buildup of space-charge regions and potential differences also occurs for single-type semiconductors when the doping density changes, as is the case in n^+-n and p^+-p structures. The example of an abrupt n^+-n junction is shown in Fig. 3.15. Electrons from the n^+ region diffuse into the less doped n region, thus causing a surplus of negative charge on the n side and of positive charge on the n^+ side. The resulting electric field counteracts diffusion of further electrons. Considering the situation in the band model (bottom part of Fig. 3.15), the change in the Fermi level with doping concentration and the requirement that the Fermi level lines up in thermal equilibrium results in a built-in voltage V_{bi} equalling the difference of Ψ_B (see (3.3.1)) in the two regions.

As an example charge density, electric field and potential in an abrupt n^+-n junction with doping densities of 10^{14} and 10^{12} is shown in Fig. 3.16. The results have been obtained with a numerical simulation (see Chap. 12).

a)

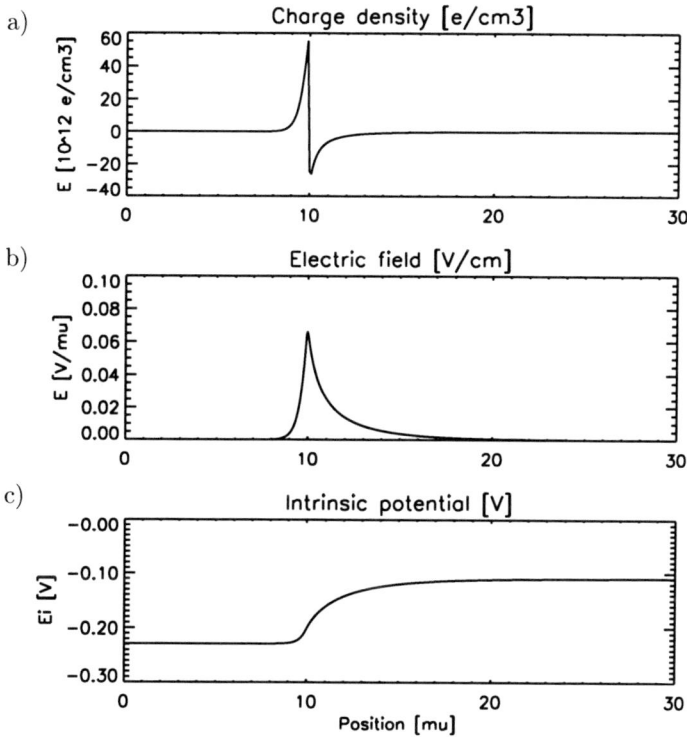

b)

c)

Fig. 3.16a–c. Numerical simulation of an abrupt n^+-n junction in thermal equilibrium: charge density (a); electric field (b); intrinsic potential (c)

3.5 Summary and Discussion

Based on semiconductor physics described in Chap. 2, the properties of basic semiconductor structures have been described in both a qualitative and a quantitative way. Approximations, such as the assumption of an abrupt change from an electrically neutral semiconductor to the complete absence of charge carriers in the space-charge region, have been made in the consideration.

Even in the absence of externally applied voltage, a p–n diode develops a space-charge region due to diffusion of the majority carriers into the other type semiconductor regions until the drift current exactly cancels the diffusion current. The built-in voltage thus generated is given by

$$V_{bi} = \frac{1}{q}(E_i^p - E_i^n) = \frac{kT}{q} \ln \frac{N_A N_D}{n_i^2} \tag{3.1.1}$$

and is exactly compensated for by the built-in voltages of the metal–semiconductor contacts if the contacts are of the same metal.

The current–voltage characteristics of the diode have been calculated in Sect. 3.1.2 under the assumption that the minority carrier density at the edge of the space-charge region is obtained from the majority carrier concentration in the opposite-type region with the same exponential dependence on the voltage that holds for thermal equilibrium.[15] The minority carrier currents at the boundary towards the space-charge region were then obtained by solving the diffusion equations inside the undepleted regions. For a reverse-biased diode, this procedure gives only the current caused by diffusion of minority carriers into the space-charge region; the additional contribution of charge generation in that region has to be added. At very high reverse bias, the reverse bias current increases dramatically. Two mechanisms can be responsible for this electrical breakdown: *"Zener breakdown"*, whose physical origin is due to the quantum-mechanical tunnel effect; and *"avalanche breakdown"*, which is due to charge multiplication in charge–carrier collisions within the lattice.

Irradiation of a diode with light or ionizing radiation will generate charge in the form of electron–hole pairs, which are separated in the electric field of the space-charge region. But also the neutral region will be sensitive, as the generated minority carriers have some chance of reaching the space-charge region by diffusion (Sect. 3.1.3).

The metal–semiconductor contact may – depending on its construction – also have rectifying properties. In many cases, however, interest centers on just providing an ohmic contact, which can be accomplished by using high doping concentrations so as to allow tunneling processes.

Depending on biasing, the metal–insulator–semiconductor structure (often a metal–oxide–silicon (MOS) structure) may be in one of four characteristic states:

- *Flat-band condition.* The potential of the metal $V_\mathrm{m} = V_\mathrm{FB}$ has been chosen so that the semiconductor all the way up to the insulator is in a homogeneous condition, as if it were extending to infinity.
- *Accumulation layer.* The metal potential has been changed from the flat-band condition in such a way as to attract majority carriers, which then assemble in a very thin sheet at the semiconductor–insulator boundary.
- *Depletion.* This can be a shallow stable space-charge region where a depletion of majority carriers occurs close to the insulator–semiconductor interface (surface depletion) or where a sudden large voltage change drives majority carriers deep into the semiconductor (deep depletion). The latter situation is not stable as the (thermally generated) electron–hole pairs separate in the electrical field of the space-charge region, leading to the fourth state (given below).
- *Inversion.* A layer of minority carriers is formed at the interface, separated from the majority carriers in the nondepleted bulk by a depletion region of width x_max that is independent of the applied voltage.

---

[15] This is equivalent to assuming separate Quasi-Fermi levels for electrons and holes (see Chap. 12), the Quasi-Fermi level of majority carriers remaining constant throughout the space-charge region.

Quantitative results have been derived using essentially simple electrostatics and some approximations, such as the abrupt appearance of an inversion layer, once the situation of strong inversion is reached.

The insulator in general contains charge. In SiO_2 this charge is usually positive and mainly due to holes trapped in defect sites. Defects are especially numerous in the vicinity of the Si–SiO_2 interface. Defects at the interface can assume different charge states depending on the presence of electrons or holes at the interface. Ionic charge in the oxide bulk is mobile, the mobility being strongly dependent on temperature.

The measurement of capacitance versus voltage and frequency can give very valuable information on oxide, interface and silicon properties. A thorough knowledge of the MOS structure is necessary for proper design of detectors even if the specific detector principle is not based on the MOS structure. As the oxide is the standard passivation material (used for surface protection) of silicon detectors, parasitic MOS or similar structures quite naturally appear and have to be designed so as not to generate breakdown problems or unwanted electrical connections between adjacent electrodes.

Part II

Semiconductor Detectors

4 Semiconductors as Detectors

Compared with other materials, semiconductors have unique properties that make them very suitable for the detection of ionizing radiation. Furthermore, semiconductors – especially silicon – are the most widely used basic materials for electronic amplifying elements (transistors) and more recently for complete microelectronics circuits. Thus part of the process technology that already existed in (micro) electronics could be taken or adapted for detector production. Integration of detector and electronics can be envisaged. In the following, the properties of important semiconductor materials will be discussed.

Also included in this chapter are data on other materials used in detector and electronics structures, such as insulators and metals.

4.1 The Properties of Intrinsic Semiconductor Materials

The uniqueness of semiconductor material properties can best be appreciated by comparing them with the most widely used radiation detectors that are based on ionization in gas. Values for silicon will be used in this comparison; properties of other important semiconductor materials are given in Tables 4.1 to 4.3.

- The small band gap (1.12 eV at room temperature) leads to a large number of charge carriers per unit energy loss of the ionizing particles to be detected. The average energy for creating an electron–hole pair (3.6 eV) is an order of magnitude smaller than the ionization energy of gases (~ 30 eV).
- The high density (2.33 g/cm^3) leads to a large energy loss per traversed length of the ionizing particle (3.8 MeV/cm for a minimum ionizing particle). Therefore it is possible to build thin detectors that still produce large enough signals to be measured. In addition, the very small range of δ-electrons prevents large shifts of the center of gravity of the primary ionization from the position of the track. Thus an extremely precise position measurement (of a few µm) is possible.
- Despite of the high material density, electrons and holes can move almost freely in the semiconductor. The mobility of electrons ($\mu_n = 1450$ cm^2/Vs) and holes ($\mu_p = 450$ cm^2/Vs) is at room temperature only moderately influenced by doping. Thus charge can be rapidly collected (~ 10 ns) and detectors can be used in high-rate environments.
- The excellent mechanical rigidity allows the construction of self-supporting structures.

Table 4.1. Intrinsic properties of selected semiconductors at $T = 300\,\mathrm{K}$ unless otherwise stated. Data are from the following references: a) Landolt–Börnstein vol. 22a (1987) and 17a (1982); b) Beadle, Tsai and Plummer (1985); c) Sze (1983); and d) Alig (1980). Theroretical values are marked with an asterisk

Substance	Si	Ge	GaAs	Diamond
lattice	diamond	diamond	zink-blende	diamond
lattice spacing [Å]	5.431^a	5.657^a	5.653^a	3.567^a
Optical transition	indirect	indirect	direct	indirect
Minimum energy gap [eV]	$1.12^{a,c,d}$ $1.17\,(0\mathrm{K})^a$	$0.664\,(291\mathrm{K})^a$ 0.67^b 0.735^d	$1.42^{a,d}$ $1.52\,(0\mathrm{K})^a$	5.48^a 5.47^d
$dE_g/dT \times 10^4$ [eV/K]	-2.3^a	-3.7^a	-3.9^a	-0.5^a
Effective density of state Conduction band N_C [cm^{-3}]	3.22×10^{19} ^{b} 2.8×10^{19} ^{c}	1.04×10^{19} ^{b}		
Valence band N_V [cm^{-3}]	1.83×10^{19} ^{b} 1.04×10^{19} ^{c}	6.0×10^{18} ^{b}		
Electron affinity [eV]	4.85 ^{b} 4.05 ^{c}			
Intrinsic carrier concentration [cm^{-3}]	1.02×10^{10} ^{a} 1.38×10^{10} ^{b} 1.45×10^{10} ^{c}	2.33×10^{13} ^{a} 2.4×10^{13} ^{b}	2.1×10^{6} ^{a}	
Mean energy for electron–hole pair creation ϵ_{pair} [eV]	3.63^d 3.63^{d*}	2.96^d 2.78^{d*}	4.35^d 3.90^{d*}	13.1^d 11.6^{d*}
Fano factor F	0.115^{d*}	0.13^{d*}	0.10^{d*}	0.08^{d*}
Drift mobility μ [cm^2/V·s] Electrons μ_n	1450 ^{a} 1500 ^{c}	3900 ^{a,c}	8800 ^{a} 8500 ^{c}	1800 ^{a,c}
Holes μ_p	505 ^{a} 450 ^{c}	1800 ^{a} 1900 ^{c}	320 ^{a} 400 ^{c}	$1600\,(290K)^a$ 1200 ^{c}
Saturation velocity [cm/s] Electrons $v_{s,n}$	$\sim1\times10^7$ ^{b}	6.2×10^6 ^{b}		
Holes $v_{s,p}$	$\sim8.4\times10^6$ ^{b}	5.7×10^6 ^{b}	10^7 ^{a}	
Intrinsic resistivity [Ωcm]	230×10^3 ^{b}	47^b		
Hall mobility [cm^2/V·s] Electrons			9200^a	
Holes	370^a	~2400 ^{a}	400^a	
Density of state effective mass Electrons m^*/m_0	1.18 ^{a} $1.0\,(4\mathrm{K})$ ^{b}	$0.55\,(4\mathrm{K})$ ^{b}	0.056 ^{a} 0.067 ^{c}	0.2 ^{c}
Holes m^*/m_0	0.81 ^{a} $0.591\,(4\mathrm{K})$ ^{b}	$0.29\,(4\mathrm{K})$ ^{b}	0.53 ^{a} 0.082 ^{c}	0.75 ^{a} 0.25 ^{c}

Table 4.2. Physical properties of selected semiconductors. Data are from the following references: a) Landolt–Börnstein vol. 22a (1987) and 17a (1982); b) Beadle, Tsai and Plummer (1985); and c) Sze (1983)

Substance	Si	Ge	GaAs	Diamond
lattice	diamond	diamond	zink-blende	diamond
lattice spacing [Å]	5.4307 [a]	5.657 [a]	5.653 [a]	3.5668 [a]
atomic number	14	32	31+33	6
average atomic mass	28.09	72.59	72.32	12.01
density $[g/cm^3]$	2.329 [a]	5.323 [a]	5.317 [a]	3.515 [a]
Coefficient of thermal linear expansion $[10^{-6}\,K^{-1}]$	2.56 [a] 2.6 [c]	5.90 [a] 5.8 [c]	6.86 [a,c]	1.0 [a]
Thermal conductivity [W/cm·K]	1.56 [a] 1.5 [b,c]	0.60 [b]	0.45 [a]	~10 [a]
Dielectric constant	11.9 [a,c]	16.2 [a]	12.9 [a]	5.7 [a]
Breakdown field [V/cm]	~3 10^5 [a]			
Index of refraction	3.42 [a,c]	3.99 [a]	3.25 [a]	2.42 [a]
Melting point [°C]	1392 [a] 1415 [b,c]	917 [a] 937 [b,c]	1220 [a] 1238 [c]	3907 [a]
Radiation length [cm] [g/cm²]	9.36 21.82	2.30 12.25		12.15 42.70

- An aspect completely absent in gas detectors is the possibility of creating fixed space charges by doping the crystals used. It is thus possible to create rather sophisticated field configurations without obstructing the movement of signal charges. This allows the creation of detector structures with new properties that have no analogy in gas detectors.
- As detectors and electronics can both be built out of silicon, their integration into a single device is possible.

Intrinsic properties of the important semiconductors silicon (Si), germanium (Ge) and gallium arsenide (GaAs) are collected in Table 4.1. Their physical properties are shown in Table 4.2. Diamond (C), which potentially has very interesting properties although at present sufficiently perfect material is not available, has also been included in these tables. Some caution has to be applied when using these tables, because parameters found in the literature are not always consistent, so that in some cases several numbers are quoted. Consulting the original literature when trying to judge the validity of the quoted numbers is recommended. Collection of data on semiconductors is taken from Landolt–Börnstein vols. 22a (1987) and 17a (1982), Beadle, Tsai and Plummer (1985), Sze (1983) and Alig (1980). Data on insulators and metals are from Landolt–Börnstein vol. 17c (1984). These collections usually give references to the original literature.

The most commonly used semiconductor materials are germanium and silicon but also other compound materials are used, such as GaAs and CdTl,

Table 4.3. Properties of compound semiconductors. Data are from Sze (1985) unless otherwise stated by d for Alig (1980). Theoretical values are marked with an asterisk

Semi-conductor	Lattice constant [Å]	Band-gap [eV]	Band	Mobility μ_n	μ_p	ϵ	Z	ϵ_{pair} [eV]	F
IV–IV SiC	3.08	2.99	I	400	50	10.0	14+6	6.9^d	0.09^{d*}
		2.86^d						6.88^{d*}	
III–V AlSb	6.13	1.58	I	200	420	14.4	13+51	6.88^{d*}	
GaAs	5.63	1.42	D	8500	400	13.1	31+33	4.35^d	0.10^{d*}
		1.42^d						3.90^{d*}	
GaP	5.45	2.26	I	110	75	11.1	31+15	6.54^d	0.09^{d*}
		2.22^d						5.37^{d*}	
GaSb	6.09	0.72	D	5000	850	15.7	31+51		
InAs	6.05	0.36	D	33000	460	14.6	49+33		
InP	5.86	1.35	D	4600	150	12.4	49+15		
InSb	6.47	0.17	D	80000	1250	17.7	49+51		
II–IV CdS	5.83	2.42	D	340	50	5.4	48+16	6.3^d	0.09^{d*}
								5.63^{d*}	
CdTe	6.48	1.56	D	10500	100	10.2	48+52	3.90^{d*}	0.10^{d*}
ZnO	4.58	3.35	D	200	180	9.0	30+8	7.50^{d*}	0.09^{d*}
ZnS	5.42	3.68	D	165	5	5.2	30+16	8.23^{d*}	0.08^{d*}
IV–VI PbS	5.93	0.41	I	600	700	17.0	82+16		
PbTe	6.46	0.31	I	6000	4000	30.0	82+52		

and in addition others may promise interesting possibilities in the future (e.g. diamond, SiC etc.). Germanium and silicon are indirect semiconductors, their most important difference being the factor of two in the band gap and the much shorter radiation length for germanium.[16]

Because of its high absorption probability, germanium is well suited for x-ray measurements, and also for near infrared detection due to its small band gap. However, because of the possibility of band-to-band thermal excitation of

---

[16]The radiation length is a measure for the energy loss of a high-energy charged particle due to bremsstrahlung. In a thin layer Δx, the energy loss is $\Delta E = E \frac{\Delta x}{X_0}$ with X_0 the radiation length. It is also a measure for the probability of high-energy photons (for which photon effect and Compton scattering is negligible compared with electron–positron pair production) interacting in the material. The mean free path of high-energy photons is approximately 9/7 times the radiation length. The radiation length is also related to the average scattering angle of a high-energy charged particle. For a thickness Δx, the average projected scattering angle is $\theta_{scatt} \approx \frac{15\,\text{MeV}/c}{P} \sqrt{\frac{\Delta x}{X_0}}$, with P representing the particle momentum.

electrons at room temperature and of correspondingly high reverse-bias currents, the detectors require cooling.

As silicon is the most commonly used material in the electronics industry, it has one big advantage with respect to other materials, namely a highly developed technology. For x-ray detection one problem is the limit in thickness that can be depleted with the application of reasonably low voltages, as intrinsic material cannot be manufactured.

In lithium-drifted silicon [Si(Li)] remaining acceptor concentrations are compensated for by interstitial lithium, which sticks in the vicinity of donor sites when drifted through the bulk of the material. However, technological processing of this material is very limited since high-temperature processes destroy the properties of the material.

GaAs so far has been used with limited success for nuclear radiation detection, where large sensitive volumes and a full charge collection are required. However, it is a common material in ultra-high-speed electronics, because of the high mobility of electrons ($\mu_n = 8800\,\mathrm{cm}^2/\mathrm{Vs}$), and in photonics, due to the direct semiconductor property. Interest in GaAs radiation detectors has risen in recent time in high-energy physics, as use in extremely high radiation environments is envisaged. High radiation tolerance with respect to reverse-bias currents is expected because of the large band gap of GaAs compared with silicon; thus the effect of crystal damage on the reverse-bias current is suppressed. There exists, however, the problem of incomplete charge collection due to trapping by (radiation-induced) defects. Use of GaAs, as well as other compound semiconductors, opens up the possibility of "band-gap engineering" as the width of the band gap ($E_\mathrm{G} = E_\mathrm{C} - E_\mathrm{V}$) can be controlled. This is the case, for example, in heterojunctions.

For special purposes, such as x-ray or infrared detection, semiconductors with very low radiation length and/or small band gap are used. For operation at elevated temperature and for radiation hardness at room temperature operation, materials with larger band gaps as (e.g. SiC and diamond) may become important in the future. Because of the extremely high mobility, diamond also looks like an excellent candidate for high-speed application. However, the technology for producing detector-grade diamond material is still in its infancy. We will concentrate here on the standard materials of silicon and germanium.

4.2 Properties of Extrinsic Semiconductor Materials

The properties of semiconductors can be changed dramatically (intentionally and unintentionally) by small deviations from a perfectly uniform crystal structure. This is intentionally done by adding tiny amounts of foreign atoms with more or less electrons in the outer shell than the proper atoms of the crystal (see Sect. 2.4). In this way it is possible to change the conduction type, particularly the ratio between electrons and holes, from equality in strong favour of either electrons or holes. Unintentionally one may also add further chemical elements and generate crystal defects, which can deteriorate the material properties.

The introduction of impurities and crystal defects may occur already during crystal growing, cutting or polishing, or be generated during processing or through radiation exposure (see Chap. 11). In the following text we will restrict ourselves to bulk properties. Emphasis will be on silicon, the most thoroughly studied and most widely used semiconductor material.

Surface defects are unavoidable, due to the break in symmetry near the boundary, where incomplete bonding is responsible for significant distortions of the lattice structure. This situation will be discussed in Sect. 4.3, in conjunction with insulators used for the passivation of the crystal surface.

4.2.1 Doping of Semiconductors

Doping of semiconductors, the addition of a tiny fraction of foreign atoms, can be performed during crystal growth or later on during device processing. The effect on electrical material properties will not only depend on type and concentration of these foreign atoms but also on the way they are built into the crystal structure. As interstitials they are squeezed between regular lattice sites, thus deforming locally the lattice and its chemical binding structure. Such structures, which usually are able to take on several charge states, are generated by ion implantation, for instance.

The ideal requirement for doping in a crystal is where the lattice is almost undisturbed apart from the substitution of a small fraction of the lattice atoms by the dopant atoms, causing only small distortions of the lattice and keeping the chemical binding almost unchanged. The additional or missing valence electrons are ejected into the conduction band or taken from the valence band.

When dopants are added during crystal growing, doping atoms will usually be built into regular lattice sites. They can also be brought into regular sites after, for example, ion implantation by heating the crystal. The dopants are then said to be activated, and the procedure is called "activation".

Localized energy levels will be created in the band gap that could not originally be entered. The energy levels of these fairly simple dopants can be estimated from a model in which a surplus electron moves in the field of a surplus nuclear charge in a donor atom, similar to the electron in the hydrogen atom. Scaling of the hydrogen results with the effective masses and the appropriate dielectric constant leads to order-of-magnitude agreement of the scaled hydrogen ground-state binding energy with the measured distances of the donor level from the conduction band, and similarly for a hole moving in the field of an acceptor atom, with the distance of the acceptor level from the valence band.

The energy levels for shallow dopants in Si, Ge and GaAs are compiled in Table 4.4. In group IV semiconductors (Si and Ge), substitution by group III elements creates shallow acceptors, and replacement by group V atoms creates shallow donors. In III–V compound semiconductors, doping with group IV elements will create shallow donors and acceptors, depending on the substitution of group III or group V elements. Standard dopants in silicon and germanium are boron (acceptor), phosphorous and arsenic (donors). In GaAs, the elements of Se, Si and S are used for n-type doping, and Zn, Cd, Be and Mg are used

Table 4.4. Shallow dopants in semiconductors. Data are from Landolt–Börnstein vol. 17a (1982)

Semi-conductor	Dopant	Z	Energy level [meV]	Charge states	Type
Silicon	Group III				
	B	5	E_v+45	0/-	substitutional acceptor
	Al	13	E_v+68	0/-	substitutional acceptor
	Ga	31	E_v+71	0/-	substitutional acceptor
	In	49	E_v+155	0/-	substitutional acceptor
	Tl	81	E_v+250	0/-	substitutional acceptor
Silicon	Group V				
	N	7	E_c-140	+/0	substitutional donor
	P	15	$E_c-45.3$	+/0	substitutional donor
	As	33	$E_c-53.7$	+/0	substitutional donor
	Sb	51	$E_c-42.7$	+/0	substitutional donor
	Bi	83	$E_c-70.6$	+/0	substitutional donor
Germanium	Group III				
	B	5	$E_v-10.8$ (4K,8K)	0/-	substitutional acceptor
	Al	13	$E_v+11.1$	0/-	substitutional acceptor
	Ga	31	$E_v+11.3$	0/-	substitutional acceptor
	In	49	$E_v+12.0$	0/-	substitutional acceptor
	Tl	81	$E_v+13.5$	0/-	substitutional acceptor
Germanium	Group V				
	P	15	$E_c-12.9$	+/0	substitutional donor
	As	33	$E_c-14.2$	+/0	substitutional donor
	Sb	51	$E_c-10.3$	+/0	substitutional donor
	Bi	83	$E_c-12.8$	+/0	substitutional donor
GaAs	Group II				
	Be	4	E_v+28		acceptor
	Mg	12	$E_v+28.8$		acceptor
	Zn	30	$E_v+30.7$		acceptor
	Cd	48	$E_v+34.7$		acceptor
GaAs	Group IV				
	C	6	E_v+27		acceptor
			$E_c-5.91$		donor
	Si	14	$E_v+34.8$		acceptor
			$E_c-5.84$		donor
	Ge	32	$E_v+40.4$		acceptor
	Sn	50	E_v+167		acceptor
GaAs	Group VI				
	S	16	$E_c-5.87$		acceptor
	Se	34	$E_c-5.79$		acceptor

for p-type doping. Under normal conditions the shallow dopants are almost completely ionized. The majority carrier concentration equals the difference between donor and acceptor concentrations.

4.2.2 Bulk Defects

Besides the intended doping of semiconductors, a large variety of mostly un-intended deviations from crystal symmetry are present in real semiconductors. They include a variety of other impurities and real defects, creating donor or acceptor states at deep positions in the band gap. An example of deep level impurity is the substitution of a group IV regular lattice atom by a group II atom (e.g. Zn in Si). In such a case a double donor with charge states neutral, negative and double negative and with deeper energy levels is created (Zn in Si $E_v + 0.316\,\mathrm{eV}$ and $E_v + 0.617\,\mathrm{eV}$). We consider the following as real defects: empty lattice sites (vacancies); additional atoms of the same or a foreign nature between regular lattice sites (interstitials); and complexes of interstitials next to vacancies (Frenkel defects). These "point defects" are symbolically presented in a two-dimensional lattice in Fig. 4.1.

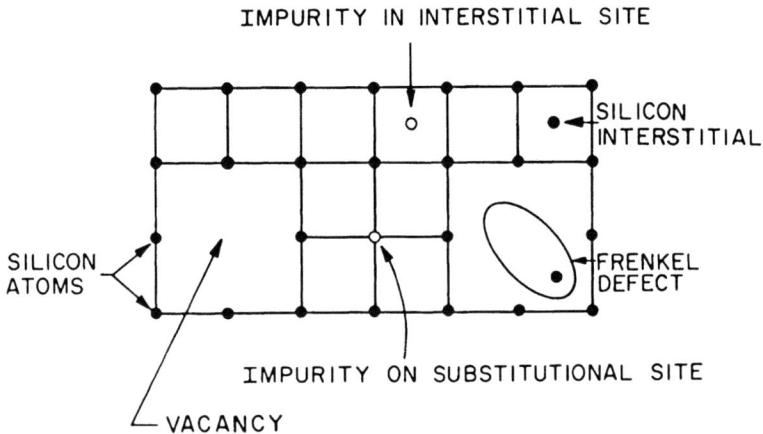

Fig. 4.1. Types of point defects in a simple lattice. (After Sze 1985, p. 317 Fig. 13)

Semiconductors composed of two elements, such as GaAs, in addition rather frequently have the wrong type of atom on a lattice site or two neighboring atoms of different type interchanged. Some of the defects are able to assume several charge states with correspondingly different energy levels. The density of defects in composed semiconductors therefore is usually larger than in single-element crystals such as Ge or Si.

The defects considered so far and schematically indicated in Fig. 4.1 are point defects. An example of a "line defect" is shown in Fig. 4.2.

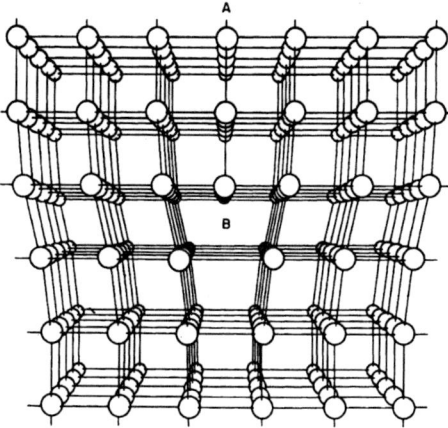

Fig. 4.2. Example of a line defect. (After Sze 1985, p. 317 Fig. 14)

Especially dangerous for detectors are energy states close to the middle of the band gap, as they are most effective in thermal creation of electron–hole pairs. Movement of an electron from the valence to the conduction band with their help may be done in two steps, through alternate hole and electron emission. Although the density of intermediate states is low with respect to those in the bands, this process is in general more likely than direct excitation from valence to conduction band due to the exponential energy dependence $\approx \exp(-\Delta E/kT)$ of the emission process.

In silicon, well known examples are heavy metals (Au, Ag and Cu) that possess several energy states, some of which are very close to the band gap center (Au donor at $E_c - 0.55\,\mathrm{eV}$).

"Defect clusters", a local high concentration of point defects, are expected to behave somewhat differently from uniformly distributed point defects. Such a defect cluster is capable of producing local charge concentrations and thus, depending on their type, of attracting or repelling charge carriers from the surrounding region. Defect clusters are expected to be formed by certain types of radiation that deposit large amounts of nonionizing energy in a small volume, for instance stopping protons or nuclei.

Defects are not necessarily immobile objects in the crystal. Depending on temperature, they may diffuse and on their way form defect complexes with other defects or impurities. A well known example is the formation of the phosphorous-vacancy complex (see Chap. 11), which has four charge states ranging from singly positive to doubly negative. Thus this so-called E-center simultaneously has donor and acceptor characteristics. Similarly, defect complexes may break up at elevated temperatures and the parts diffuse until they form different complexes. As high temperature steps are used during processing, the processes just described are important not only in crystal growth but also in detector fabrication. It is even possible to take advantage of defect mobility by artificially creating immobile defects on the surface to which the mobile impurities attach. This procedure is called "gettering" and the surface layer may be removed later on.

Connected with defect mobility is the formation of precipitates, the agglomeration of many impurity atoms of the same kind at a particular region of the crystal. The crystal has then an enclosure of tens to thousands of atoms of a particular kind. One may also have an agglomeration of vacancies, forming a hole inside of the crystal.

4.2.3 Effects on Material Properties

From the previous discussions it is clear that some important material properties will depend very strongly on imperfections. As well as carrier densities, other examples are mobility, generation lifetime, recombination lifetime, and trapping probability. These properties – which are important for detector performance – depend strongly on impurity and defect properties and concentrations.

All kinds of defects will decrease mobility. Generation and recombination lifetime will most strongly be reduced by defects with energy levels close to the band gap center. For trapping, the capture and delayed release of charge carriers by defects with medium-depth energy levels are dominant, the average capture time increasing exponentially with depth and being inversely proportional with the capture cross-section.

As there exists only incomplete knowledge on defect properties and as it is very difficult (and sometimes impossible) to determine the concentrations of impurities down to values that are low enough not to be of significance for the detector performance, it is not customary to specify in detail low-level impurities and microscopic defect concentrations. Usually the effective doping concentration (but not the degree of compensation between donors and acceptors) and the generation lifetime are specified.

4.3 Insulators and Metals

In semiconductor devices, insulators are used for the following purposes:

- as protection of the surface against chemical and mechanical influence;
- as defined termination of the outer surface of the crystal; and
- as part of electronic structures such as MOS structures and transistors.

Metals, and other conductors such as silicides and doped polysilicon, are used as part of devices and for interconnections. Polysilicon can also be used for the formation of resistors.

Table 4.5. Properties of selected insulators. Data are from a) Landolt–Börnstein vol. 17c (1984) and b) Beadle, Tsai and Plummer (1985)

Substance	SiO_2	Si_3N_4	$Si-O_x-N_y$
Structure	amorphous	amorphous	amorphous
Resistivity $[\Omega cm]$	$10^{14}-10^{16}$ [b]	$\sim 10^{14}$ [b]	
Density $[g/cm^3]$	2.27 [b]	3.1 [b]	
Dielectric constant	3.84 [a]	7.5 [b]	4.77–6.12 [b]
	3.8–3.9 [b]		
Dielectric strength $[V/cm]$	8×10^6 [a]	$\sim 1\times10^7$ [b]	$\sim 5\times10^6$ [b]
	$\sim 5\times10^6$ [b]		
Energy gap $[eV]$	8.8 [a]	~ 5.0 [b]	
	~ 8 [b]		
Coefficient of thermal linear expansion $[K^{-1}]$	5.0×10^{-7} [b]		
Thermal conductivity	0.014 [b]		
Index of refraction	1.46 [b]	2.05 [b]	1.6–1.88 [b]
Radiation length [cm]	10.2		
$[g/cm^2]$	27.05	26.28	

4.3.1 Insulator Properties

The properties of selected insulators used in detector fabrication are set out in Table 4.5. The material properties depend on details of growth techniques and in reality may deviate significantly from the values quoted in the table.

4.3.2 Semiconductor Surface Defects

Although the interior of a crystal may be close to symmetry, this symmetry naturally has to be strongly disturbed at the surface. Simply cutting the crystal at some crystal plane leaves many open (dangling) bonds which rearrange – in most cases involving chemical reactions with ambient gases – so as to distort the crystal lattice in the surface region. The occurrence of such kinds of defects is already unavoidable from a theoretical point of view.

Mechanical handling of the crystal, such as cutting, lapping and polishing, will introduce additional defects in the surface region, and in addition it brings about the danger of poisoning the surface with foreign atoms, which during processing can diffuse into the semiconductor.

The semiconductor surface is usually not left bare, but is intentionally covered with an insulating layer. This layer not only protects the semiconductor from chemical reactions with ambient gases and humidity, but can also be used as part of electronic and detector structures. Even if the crystal surface is left bare, it will be unintentionally covered with natural oxide once it gets into contact with air.

For silicon the natural insulation layer is SiO_2, which can be grown thermally by heating the wafer in ambient oxygen or an $O-H_2O$ mixture to 900–1200 °C. SiO_2 combines several good properties, which make it well suited as insulation or a passivation layer. It is mechanically strong and a very good electrical insulator with a high breakdown field. A kind of lucky coincidence is its matching with silicon: nearly all of the dangling bonds are saturated in the $Si-SiO_2$ transition region. For well grown oxides a positive fixed oxide surface charge density of $10^{10}-10^{12}$ e/cm^2 can be found, depending on crystal orientation and the growth technique. This has to be compared with the surface density of atoms at the crystal boundaries, which is of the order of 5×10^{15} cm^{-2}, as can be estimated from the lattice constant and the crystal structure shown in Fig. 2.1. A fixed oxide charge density one order of magnitude higher is obtained on $\langle 1,1,1 \rangle$ silicon than on the $\langle 1,0,0 \rangle$ orientation.

4.3.3 Metal Properties

Important electrical properties of selected metals are listed in Table 4.6. They include: resistivity as measured in thin sheets, which is important for noise and signal speed; the Schottky barrier height, which is essential for metal semiconductor contacts in relation to surface barrier detectors; and the metal work function Φ_m, which influences the flat-band voltage of the MOS structure.

Table 4.6. Properties of selected metals. Data are from Landolt–Börnstein vol. 17a (1982) and 17c (1984)

Substance	Al	Ag	Au	W	Cu	Ti
Atomic number	13	47	79	74	29	22
Density [gcm^{-3}]	2.702	10.5	19.29	19.3	8.29	4.52
Thermal conductivity [$Wm^{-1}K^{-1}$]	239	428	312	177	395	22
Thermal expansion [$10^{-6} K^{-1}$]	23.8	19.7	14.3	4.3	6.8	9
Resistivity [$\mu\Omega cm$]						
intrinsic	2.5	1.5	2.04	4.9.0	1.55	42
thin sheet	2.7	1.6	2.2	5.0	1.6	48
Schottky barrier height						
on n–Si	0.50	0.56	0.81	0.65	0.66	0.50
	0.75	0.78	0.73		0.77	
on p–Si	0.58	0.54	0.34		0.46	
on n–Ge	0.34	0.47	0.47	0.37		
	0.48	0.39	0.54	0.52		
Work function [eV]	4.13	4.97	5.06		4.87	
Melting point [°C]	660	961	1064	3422	1085	1668
Radiation length [cm]	8.9			0.35	1.43	3.56
[g/cm^2]	24.01			6.76	12.86	16.17

4.4 Choice of Detector Material

A variety of criteria may be applied in choosing the detector material. They will not only be connected with intrinsic semiconductor properties, such as the absorption length for the radiation under consideration, and the energy needed for electron–hole generation but also with technological feasibility and maturity, economic considerations, the possibility of integration of electronics, the density of crystal defects in the available material, and their effects on device properties and surface properties, which can have severe consequences on device functioning.

4.4.1 Interaction of Radiation with Semiconductors

The interaction of radiation with semiconductor materials causes the creation of electron–hole pairs that can be detected as electric signals. For charged particles, ionization may occur along the path of flight by many low-recoil collisions with the electrons. Photons have first to undergo an interaction with a target electron (photo or Compton effect) or with the semiconductor nucleus (e.g. pair conversion of photons). In any case, part of the energy absorbed in the semiconductor will be converted into ionization (the creation of electron–hole pairs), the rest into phonons (lattice vibrations), which means finally into thermal energy.

The fraction of energy converted into electron–hole pair creation is a property of the detector material. It is only weakly dependent on the type and energy of the radiation except at very low energies that are comparable with the band gap. For a given radiation energy, the signal will fluctuate around a mean value N given by

$$N = \frac{E}{\epsilon} \tag{4.4.1}$$

with E the energy absorbed in the detector and ϵ the mean energy spent for creating an electron–hole pair. The variance in the number of signal electrons (or holes) N is given by

$$\langle \Delta N^2 \rangle = F \cdot N = F \frac{E}{\epsilon} \tag{4.4.2}$$

with F the Fano factor (Fano 1947).[17]

Fano arrived at this expression by considering the probabilities of ionizing and nonionizing collisions of charged particles in gases, making some rather arbitrary assumptions in his model. His approach has been adapted to semiconductors by Shockley. Newer approaches with more realistic assumptions (Alig 1980, Fraser 1984) describe also the dependence of the mean energy needed for electron–hole pair creation and of the Fano factor on the energy of the radiation.

---

[17]Notice the formal similarity of this expression with Poisson's statistics when $F = 1$. This similarity, which has led to incorrect interpretations, is nevertheless coincidental.

Looking at the Fano factor in Table 4.1, one notices that the Fano factor is always significantly below unity. This is not surprising, since the reason for any fluctuation in the number electron–hole pairs for fixed energy is due to the variation in the fraction of energy that ends up in electron–hole separation and in phonon generation, which means eventual thermal energy. If all energy were converted into electron–hole generation, their numbers would be fixed and F would be zero. For very low radiation energy in the few-eV range, both the mean energy ϵ and the Fano factor F are expected to be energy-dependent. Measurements of this effect agree with this prediction (Lechner 1996).

Very important aspects of the detector material in spectroscopic applications are the penetration depth of charged particles and the absorption length of photons. A very small absorption length will result in a high probability of generating the signal close to the surface, where signal charge may only be partially collected because of surface treatment (e.g. doping), coverage with insensitive material (e.g. a naturally or artificially grown insulation layer) or deterioration in the semiconductor properties, which usually appears close to the surface due to distortion of the lattice. A very large absorption length leads to inefficiencies as radiation may traverse the detector without interaction.

The dependence of the absorption length on photon energy for silicon is given in Fig. 4.3.

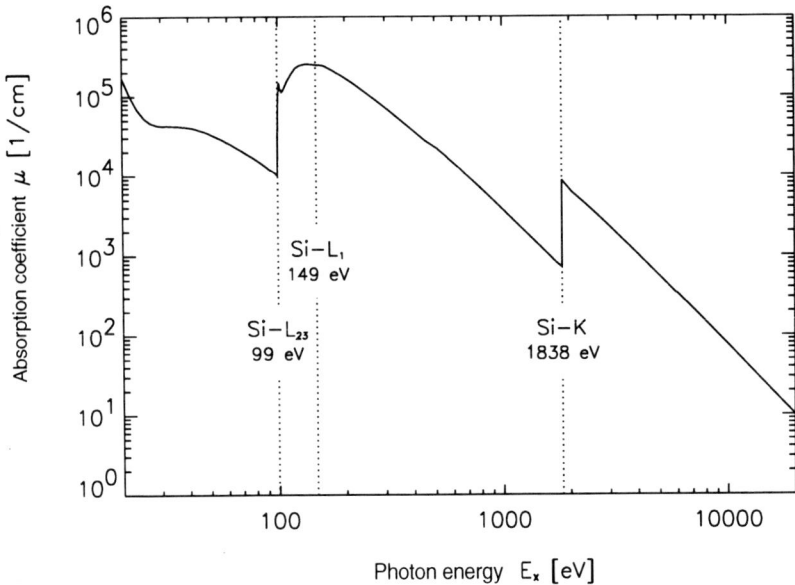

Fig. 4.3. Energy dependence of the photon absorption length in silicon due to the photo effect. Data are from Veigele (1973), Henke (1982) and Palik (1985)

4.4.2 Charge Collection and Measurement Precision

The proportionality of charge and absorbed energy in semiconductor detectors can be spoiled by incomplete charge collection. In addition, large leakage currents may add statistical fluctuations to the signal, such that the measurement precision deteriorates. Both of these effects are intimately connected with the imperfections of the detector material.

Charge collection is reduced by effects of trapping and recombination. Recombination requires the presence of both types of carriers simultaneously, which is not the case in the space-charge region. Therefore in most types of detectors in which the sensitive region is essentially restricted to the space-charge region, as for example in the case of a reverse-biased diode, recombination occurs between radiation-generated electrons and holes only as long as they are not separated by the electric field (and diffusion). In unbiased diodes, as used for instance in radiation-level measurements, the recombination lifetime will be a significant parameter.

Trapping of charge, the temporary capture and delayed release of charge carriers by local defects, is of extreme importance when signals of single interactions are measured. Consider, for example, a reverse-biased diode. There the signal observed by the creation of an electron–hole pair is proportional to the separation distance between electron and hole once they have both stopped, either when reaching the space-charge boundary or by being trapped at a defect position (see Sect. 5.2.1).

As electron and hole capture probabilities differ, trapping will not only lead to a reduction in signal and to fluctuations in signal height, but also to a positional (or ionization depth) dependence of the average signal.

While trapping occurs at defects with energy levels anywhere in the band gap – although with very different capture times – generation by alternative emission of electrons and holes is strongly dominated by defects with energy levels close to mid-gap. The resulting reverse-bias current adds shot noise (see Sect. 7.2) to the signal.

From the above discussion it seems clear that defects and imperfections are major selection criteria for semiconductor detector materials. It is worth noting that defect densities in compound semiconductors are orders of magnitude higher than in good group IV semiconductors such as Ge and Si. An extremely interesting material would be diamond; however, there one has the problem that large-sized single crystals cannot be grown at present. The technique of chemical vapour deposition (CVD) can be used to grow polycrystalline diamonds only. The boundaries between crystallites are probably responsible for incomplete charge collection in this material.

5 Detectors for Energy and Radiation-Level Measurement

In this section we concentrate on detectors that can only be used for energy, but not for position measurement. Of course, energy may also be measured – perhaps with even higher precision – with position-sensitive detectors. These detectors will be dealt with in Chap. 6.

The basic structure for the measurement of energy is a diode or a rectifying metal–semiconductor contact, and these have already been discussed in Chap. 3. The structures can either be used in an unbiased mode or with the application of a reverse bias.

Unbiased diodes are suitable for radiation-level measurements. The disadvantage of lower sensitivity due to the smaller sensitive volume (space-charge region) in many applications is offset by their property of producing no signal offset (dark current) in the absence of radiation.

We describe the more commonly used diodes before considering the historically older surface-barrier detectors.

5.1 Unbiased Diode

The operation of a diode under uniform irradiation has already been discussed in Sect. 3.1. Electron–hole pairs created in the natural space-charge region are separated by the electric field. If the voltage across the junction is kept at the original value, for instance by shortening the external leads of the device, a current proportional to the illumination will be obtained. If the current is prevented from flowing, for example by leaving the external connections open-ended, the voltage across the junction will drop until the minority carrier concentration at the edge of the space-charge region and the corresponding increase of the minority carrier diffusion current will compensate for the current generated by the illumination. In addition to the current from the space-charge region, the increase of minority carrier concentration in the illuminated neutral regions of the diode have also to be taken into account, and this gives rise to an additional diffusion current across the junction.

Although it is in principle also possible to use unbiased diodes for the measurement of single particles or radiation quanta, practical problems arise due to the thin space-charge region and the correspondingly small sensitive volume, as well as the high capacitive load at the amplifier input, which leads to noise problems.

In using an unbiased diode for radiation flux measurements, one encounters requirements that depend very strongly on the type of radiation. For radiation with small penetration depth (e.g. optical light or α particles) a large fraction of the signal can be lost in the entrance window and an extremely shallow diode implantation may be required. For more penetrating radiation (e.g. x-rays or other high-energy particles) the depth of the collecting region is important. That collecting region is composed of the space-charge region and part of the neutral bulk region. The width of the space-charge region is controlled by doping; the charge collection from the neutral region depends on the minority carrier lifetime (compare this with Example 3.3 in Sect. 3.1.3).

Example 5.1
Problem: *Estimate the current generated by a flux of 10^3 s^{-1} x-rays of energy $E = 10$ keV in a diode of 1 cm^2 built of an abrupt junction of a shallow ($T_p = 0.1$ μm) highly doped ($N_A = 10^{18}$ cm^{-3}) p region on a $T_w = 280$ μm thick low-doped ($N_D = 10^{12}$ cm^{-3}) n-bulk material. The minority carrier recombination lifetime is $\tau_p = 1$ ms in the n-doped bulk and $\tau_n = 100$ ps in the p-doped junction (see Fig. 5.1).*

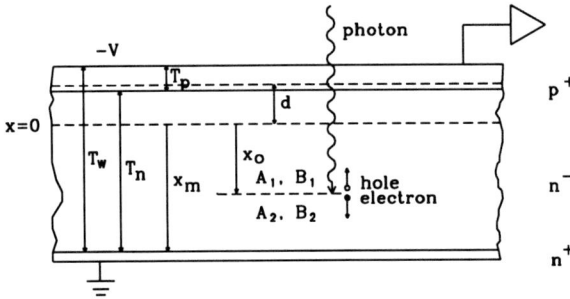

Fig. 5.1. A p–n diode junction detector as used in the example

Solution: *The width of the space-charge region of the unbiased diode can be obtained from (3.1.5) and (3.1.1) with $V = 0$ and $N_A \gg N_D$:*

$$d = \sqrt{\frac{2\epsilon\epsilon_0}{qN_D} V_{bi}} = \sqrt{\frac{2\epsilon\epsilon_0}{qN_D} \frac{kT}{q} \ln \frac{N_A N_D}{n_i^2}}$$

$$= \sqrt{\frac{2 \cdot 11.8 \cdot 8.854 \times 10^{-14}}{1.602 \times 10^{-19} \cdot 10^{12}} 0.0259 \ln \frac{10^{18} \cdot 10^{12}}{(1.45 \times 10^{10})^2}}$$

$$= 27 \times 10^{-4} \text{ cm} = 27 \text{ μm} \quad.$$

The width extends almost exclusively into the low-doped n region because of the very asymmetric doping. The diffusion lengths of minority carriers in the highly p-doped junction L_n and the lowly n-doped bulk L_p are given by

$$L_n = \sqrt{D_n \tau_n} = \sqrt{\frac{kT}{q} \mu_n \tau_n}$$

$$= \sqrt{0.0259\,\mathrm{V} \cdot 1500\,\mathrm{cm^2/Vs} \cdot 10^{-10}\,\mathrm{s}} = 6.2 \times 10^{-5}\,\mathrm{cm} = 0.62\,\mathrm{\mu m} \quad ;$$

$$L_p = \sqrt{D_p \tau_p} = \sqrt{\frac{kT}{q} \mu_n \tau_p}$$

$$= \sqrt{0.0259\,\mathrm{V} \cdot 500\,\mathrm{cm^2/Vs} \cdot 10^{-3}\,\mathrm{s}} = 0.114\,\mathrm{cm} \quad .$$

The efficiency for charge collection in the space-charge region is unity, while in the neutral regions it drops (for an infinitely thick wafer) according to (3.1.19) with

$$\epsilon = e^{-\frac{x_0}{L}} \quad ,$$

where x_0 is the distance in the neutral regions from the respective edges of the space-charge regions.

Here we encounter the problem that the diffusion length of the detector-grade silicon is much larger than the wafer thickness. We therefore need to reexamine the boundary conditions in Example 3.3 of Sect. 3.1.3 that have led to this equation. We will consider the n-side of the junction, x being the coordinate pointing from the edge of the space-charge region into the neutral n region. The general solution of the diffusion equation can be taken, with appropriate changes from electrons to holes, from (3.1.17), and we then have

$$p(x) = A e^{-\frac{x}{L}} + B e^{\frac{x}{L}} + p_{n_0} \tag{5.1.1}$$

$$F(x) = -D_p \frac{\partial p(x)}{\partial x} = \frac{D_p}{L}\left[A e^{-\frac{x}{L}} - B e^{\frac{x}{L}}\right] \quad . \tag{5.1.2}$$

The boundary condition at the edge of the space-charge region remains unchanged from the treatment in Sect. 3.1.3:

$$p(x = 0) = p_{n_0} e^{\frac{qV}{kT}} \tag{5.1.3}$$

with p_{n_0} the equilibrium hole concentration in the n-region and V the forward-bias voltage between the terminals of the diode.[18] The difference comes from the neutral region: here, the flux of minority carriers at $x \to \infty$ was assumed to vanish. This relationship has to be replaced by a condition at the wafer boundary $x = x_\mathrm{m}$. We will use physical reasoning for setting up this boundary condition. Close to the surface the density of defects of the crystal will very strongly increase. Furthermore, in many cases the surface will be strongly doped in order to produce an ohmic contact with the metallization. Thus one will have at the surface a thin region with the same or higher doping and a strongly reduced lifetime. We will approximate this situation by assuming a thin layer

---

[18]Although we are considering an unbiased diode, we introduce a bias voltage in the derivation. This will not only allow us to use the results in later sections of this text, when external biasing is explicitly introduced, but will also be useful in finding the voltage appearing at "open" and at resistively terminated terminals for the device under illumination.

with the same level of doping as the bulk but with a very short lifetime. Then the boundary condition at $x = x_m \approx T_n - d$ will be

$$p(x = x_m) = \tilde{p} \approx p_{n_0} \; . \tag{5.1.4}$$

In all other respects the consideration remains unchanged, so that at the position of irradiation $(x = x_0)$ we have continuity of hole density and a discontinuity of hole flux equalling the generation rate at $x = x_0$:

$$p(x_0 + \epsilon) = p(x_0 - \epsilon) \tag{5.1.5}$$

$$F(x = x_0 + \epsilon) - F(x = x_0 - \epsilon) = G_L \; . \tag{5.1.6}$$

These boundary conditions can be written in the following form:

$$\text{at} \quad x = 0 \quad : \quad A_1 + B_1 = p_{n_0}\left(e^{\frac{qV}{kT}} - 1\right) \tag{5.1.7a}$$

$$\text{at} \quad x = x_m : \quad A_2 e^{\frac{-x_m}{L}} + B_2 e^{\frac{x_m}{L}} = \tilde{p} - p_{n_0} \tag{5.1.7b}$$

$$\text{at} \quad x = x_0 \quad : \quad (A_2 - A_1)e^{\frac{-x_0}{L}} + (B_2 - B_1)e^{\frac{x_0}{L}} = 0 \tag{5.1.7c}$$

$$(A_2 - A_1)e^{\frac{-x_0}{L}} - (B_2 - B_1)e^{\frac{x_0}{L}} = \frac{L}{D_p}G_L \; . \tag{5.1.7d}$$

We are interested in the flux of minority carriers at the boundary of the space-charge region $(x = 0)$, which, according to (5.1.2) and (5.1.7a), is given as

$$F(x = 0) = \frac{D_p}{L}(A_1 - B_1) = \frac{D_p}{L}\left[2A_1 - p_{n_0}(e^{\frac{qV}{kT}} - 1)\right] \; .$$

We will first eliminate A_2 and B_2. Addition and subtraction of (5.1.7c) and (5.1.7d) leads to

$$A_2 = A_1 + \frac{L}{2D_p}G_L e^{\frac{x_0}{L}} \; , \qquad B_2 = B_1 - \frac{L}{2D_p}G_L e^{-\frac{x_0}{L}}$$

and inserting these values into (5.1.7b) gives us

$$A_1 e^{\frac{-x_m}{L}} + B_1 e^{\frac{x_m}{L}} = -\frac{L}{2D_p}G_L\left[e^{\frac{x_0 - x_m}{L}} - e^{-\frac{x_0 - x_m}{L}}\right] + \tilde{p} - p_{n_0} \; .$$

Combination with (5.1.7a) then yields

$$A_1\left(e^{-\frac{x_m}{L}} - e^{\frac{x_m}{L}}\right) + p_{n_0}\left(e^{\frac{qV}{kT}} - 1\right)e^{\frac{x_m}{L}}$$
$$= \frac{L}{2D_p}G_L\left(e^{\frac{x_m - x_0}{L}} - e^{-\frac{x_m - x_0}{L}}\right) + \tilde{p} - p_{n_0}$$

$$A_1 = \frac{1}{2\sinh\frac{x_m}{L}}\left[p_{n_0}(e^{\frac{qV}{kT}} - 1)e^{\frac{x_m}{L}} - \frac{L}{D_p}G_L \sinh\frac{x_m - x_0}{L} - (\tilde{p} - p_{n_0})\right] \; .$$

From the flux at $x = 0$, the hole current density becomes

$$J_h = qF(x = 0) = q\frac{D_p}{L}(A_1 - B_1) = q\frac{D_p}{L}\left[2A_1 - p_{n_0}\left(e^{\frac{qV}{kT}} - 1\right)\right]$$

$$= \frac{q}{\sinh\frac{x_m}{L}}\left[\frac{D_p}{L}p_{n_0}\left(e^{\frac{qV}{kT}} - 1\right)e^{\frac{x_m}{L}} - G_L\sinh\frac{x_m - x_0}{L} - \frac{D_p}{L}(\tilde{p} - p_{n_0})\right]$$

$$- q\frac{D_p}{L}p_{n_0}\left(e^{\frac{qV}{kT}} - 1\right) \ .$$

Setting the minority carrier concentration at the surface equal to the equilibrium concentration ($\tilde{p} = p_{n_0}$), we obtain with $x_m = T_n - d$

$$J_h = \coth\left(\frac{x_m}{L}\right)q\frac{D_p}{L}p_{n_0}\left(e^{\frac{qV}{kT}} - 1\right) - \frac{\sinh\frac{x_m - x_0}{L}}{\sinh\frac{x_m}{L}}qG_L \ . \tag{5.1.8}$$

The first term represents the (thermally generated diffusion) hole current from the undepleted n-region of the unilluminated diode, and the second term is the contribution from illumination. We can easily verify that (5.1.8) reduces to the hole-current equivalent of (3.1.18) in the limit $x_m \to \infty$. In comparison with an infinitely thick diode, the thermally generated diffusion current is increased by a factor $\coth(x_m/L)$. This is due to the enforcement of a higher minority carrier concentration at $x = x_m$ compared with the infinite-thickness case. The effect on the illumination-generated current goes in the opposite direction, decreasing by a factor $\frac{\sinh\frac{x_m - x_0}{L}}{\sinh\frac{x_m}{L}e^{-\frac{x_0}{L}}}$. The physical reason for this decrease is that the enhancement of the minority carrier concentration arising from illumination is forced down again at the surface, thus leading to a larger fraction of the illumination-generated minority carriers diffusing away from the diode's space-charge region.

The absorption length for 10 keV x-rays is $L_a \approx 120\,\mu m$. We can then integrate the product of absorption probability and charge collection probability over the full depth of the detector using this value, and multiply it with the signal charge generated by one photon ($q\frac{E}{E_{pair}} = 1.6 \times 10^{-19}$ As $\frac{E}{3.6\,eV}$ for silicon) and the flux of incident photons Φ:

$$I = \Phi q\frac{E}{E_{pair}}\left[\int_{\xi=0}^{T_p} e^{-\xi/L_a}\frac{\sinh\frac{\xi}{L_n}}{\sinh\frac{T_p}{L_n}}\frac{d\xi}{L_a}\right.$$

$$\left. + \int_{\xi=T_p}^{T_p+d} e^{-\xi/L_a}\frac{d\xi}{L_a} + \int_{\xi=T_p+d}^{T} e^{-\xi/L_a}\frac{\sinh\frac{T-\xi}{L_p}}{\sinh\frac{T-d-T_p}{L_p}}\frac{d\xi}{L_a}\right] \ .$$

Carrying out this integral is straightforward, with

$$\int e^{-ax}\sinh(bx)\,dx = \frac{1}{b^2 - a^2}e^{-ax}\left[b\cosh(bx) + a\sinh(bx)\right] + \text{const.}$$

and so for the current per unit area we obtain

$$I = \Phi q \frac{E}{E_{\text{pair}}} \left\{ \frac{L_a L_n}{L_a^2 - L_n^2} \left[\left(\coth \frac{T_p}{L_n} + \frac{L_n}{L_a} \right) e^{-\frac{T_p}{L_a}} - \frac{1}{\sinh \frac{T_p}{L_n}} \right] \right.$$

$$+ e^{-\frac{T_p}{L_a}} \left[1 - e^{-\frac{d}{L_a}} \right]$$

$$\left. + \frac{L_p L_a}{L_a^2 - L_p^2} \left[\left(\coth \frac{T - d - T_p}{L_p} - \frac{L_p}{L_a} \right) e^{-\frac{d+T_p}{L_a}} - \frac{e^{-\frac{T}{L_a}}}{\sinh \frac{T - d - T_p}{L_p}} \right] \right\} .$$

The first term in the bracket corresponds to the current generated in the undepleted p^+ junction, the second to the space-charge region, and the third to the undepleted bulk. With $T_p \ll L_a$ we may ignore any variation of the flux with depth and obtain for the first term in the bracket

$$\int_{\xi=0}^{T_p} e^{-\xi/L_a} \frac{\sinh \frac{\xi}{L_n}}{\sinh \frac{T_p}{L_n}} \frac{d\xi}{L_a} \approx \frac{L_n}{L_a \sinh \frac{T_p}{L_n}} \int_{\xi=0}^{T_p} \sinh \frac{\xi}{L_n} \frac{d\xi}{L_n}$$

$$= \frac{L_n}{L_a} \frac{\cosh \frac{T_p}{L_n} - 1}{\sinh \frac{T_p}{L_n}} = \frac{L_a}{L_n} \tanh \frac{T_p}{2L_n} .$$

Inserting the numerical values available, we can obtain the following:

$$\frac{I}{q\Phi} = \frac{10^4}{3.6} \left[4.155 \times 10^{-4} + 0.201 + 0.365 \right] = \frac{10^4}{3.6} 0.566 .$$

The ratio of absorbed flux in the three regions to the incident flux is thus 4.155×10^{-4}, 0.201 and 0.365 respectively, and so we can conclude that the charge-collection efficiency in the three regions is 0.04%, 20% and 36.5%.

5.2 Reverse-Biased Diode

The reverse-biasing of rectifying junctions increases the width of the depletion layer and therefore of the sensitive detector volume. Simultaneously the detector capacitance decreases. Detectors of this kind are used for the measurement of the energy of individual particles, for example in nuclear spectroscopy. The rectifying junctions can be either metal–semiconductor contacts, as is the case in the earlier surface barrier detectors, or p–n diode junctions produced by ion implantation or diffusion in the planar technology that now dominates.

The example of a p–n junction detector shown in Fig. 5.2 may be used as illustration of the working principle of reverse-biased detectors. The diode consists of a highly doped shallow p^+ region on a very lowly doped n^- substrate, the back portion of a highly doped n^+ layer. The purpose of this n^+ layer is twofold: it provides a good ohmic contact from the aluminum to the substrate and simultaneously it allows operation of the device in overdepleted mode.

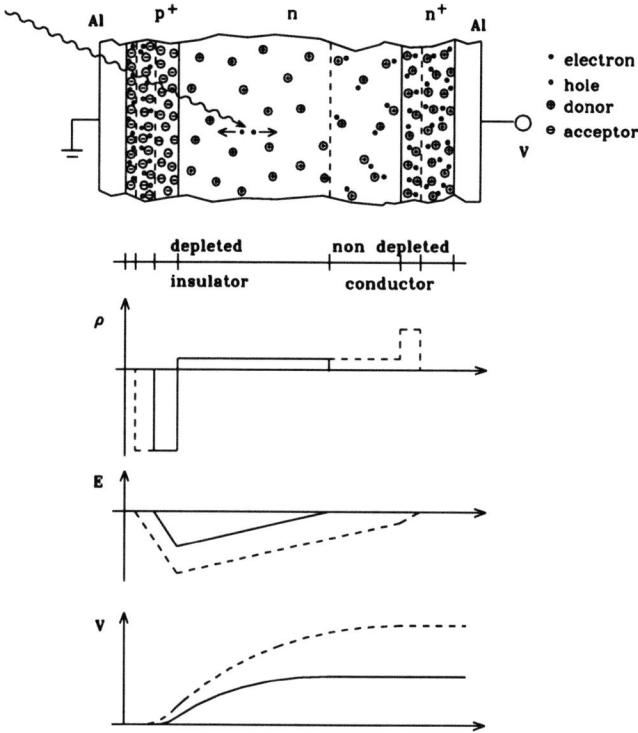

Fig. 5.2. A p–n diode junction detector: charge density, electric field and potential for partial (*continuous line*) and full (*dashed line*) depletion

Example 5.2

Problem: *Find the depletion-layer thickness and the detector capacitance for a diode detector consisting of an n-type bulk with resistivity of 5 kΩ cm together with a highly doped p^+ junction ($N_A = 10^{18}$ cm^{-3}) as function of the applied voltage. What is the minimum reverse bias voltage necessary for full depletion of a 300 μm-thick detector? What would change if p-type bulk of the same resistivity with a highly doped n-type junction were used?*

Solution: *The current flowing through a piece of semiconductor of length l and area A is given by*

$$I = qnA\mu_n \mathcal{E} = qnA\mu_n \frac{V}{l} \ .$$

Setting the electron density equal to the donor concentration, we find the relationship between resistivity r and doping concentration N_D to be given by

$$\frac{1}{r} = \frac{I}{V}\frac{l}{A} = qN_D\mu_n \ , \qquad N_D = \frac{1}{q\mu_n r} \ .$$

For the n-type bulk we find the doping concentration and the built-in voltage from (3.1.1):

$$N_D = \frac{1}{1.6 \times 10^{-19} \cdot 1500 \cdot 5000} = 8.3 \times 10^{11} \text{ cm}^{-3}$$

$$V_{bi} = \frac{kT}{q} \ln \frac{N_A N_D}{n_i^2} = 0.0259 \ln \frac{10^{18} \cdot 8.3 \times 10^{11}}{(1.45 \times 10^{10})^2} = 0.57 \text{ V} .$$

The width of the depletion region, with application of an external (negative for reverse bias) voltage (see (3.1.5)), reduces for very asymmetric doping to

$$d = \sqrt{\frac{2\epsilon\epsilon_0}{qN_D}(V_{bi} - V)}$$

$$= \sqrt{\frac{2 \cdot 11.8 \cdot 8.854 \times 10^{-14}}{1.6 \times 10^{-19} \cdot 8.3 \times 10^{11}}(0.572 + V_{rev}[V])}$$

$$= 3.96 \times 10^{-3} \sqrt{0.572 + V_{rev}[V]} \text{ cm} .$$

From this value of d we may also calculate the capacitance per unit area as

$$C = \frac{\epsilon\epsilon_0}{d} = \sqrt{\frac{\epsilon\epsilon_0 q N_D}{2(V_{bi} + V_{rev})}} = \frac{2.64 \times 10^{-10}}{\sqrt{0.572 + V_{rev}[V]}} \text{ F}$$

and the full depletion voltage of a detector with thickness d_{Det} is given by

$$V_{dep} = \frac{qN_D}{\epsilon\epsilon_0} \frac{d_{Det}^2}{2} - V_{bi} = 56.8 \text{ V} .$$

For a p-type bulk of same resistivity the doping concentration (and therefore also the full depletion voltage) is higher by approximately a factor of three than the above values, as shown by the ratio of mobilities:

$$N_A = \frac{1}{1.6 \times 10^{-19} \cdot 500 \cdot 5000} = 2.5 \times 10^{12} \text{ cm}^{-3}$$

$$V_{bi} = 0.60 \text{ V} \qquad V_{dep} = 172 \text{ V} .$$

5.2.1 Charge Collection and Measurement

Electron–hole pairs generated inside the space-charge region are separated by the electric field and move towards the electrodes. In the example shown in Fig. 5.2, holes will move towards the p^+ junction while electrons go to the backside n^+ electrode.

It may be worthwhile to point out that the signals at the detector will appear already before the arrival of the charges at the electrodes. During the process of separation, electrons and holes will induce unequal charges in the electrodes, due to their different distances, as indicated in Fig. 5.3. A hole will induce a total charge q in the top and bottom surface electrodes, split in the

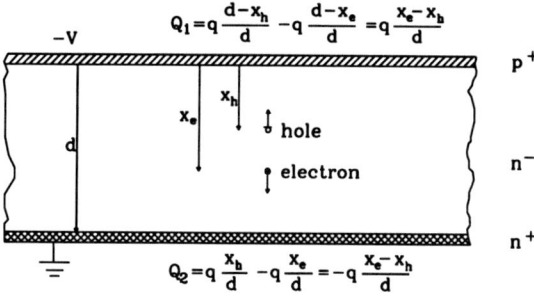

Fig. 5.3. Signal formation by the separation of electron–hole pairs due to the electric field in the space-charge region of the detector

ratio of the inverse distances. It is thus possible to obtain an induced charge $q\frac{d-x_{\rm h}}{d}$ on the top and $q\frac{x_{\rm h}}{d}$ on the bottom electrodes. Adding the charge induced by the electron, the total induced charge is determined as $q\frac{x_{\rm e}-x_{\rm h}}{d}$.

Example 5.3
Problem: *Find the signal form generated by a photon creating an electron–hole pair at a distance $x_0 = d/2$ from the highly p-doped entrance window of a 300 μm-thick n-type ($N_{\rm D} = 10^{12}$ cm^{-3}) silicon diode detector operated in 20% overdepletion mode.*
Solution: *The electric field as a function of the depth x is given as*

$$\mathcal{E}(x) = -\left[2\frac{d-x}{d^2}V_{\rm dep} + \frac{V-V_{\rm dep}}{d}\right] = -\left[\frac{V+V_{\rm dep}}{d} - 2\frac{xV_{\rm dep}}{d^2}\right] \quad ,$$

with V the applied voltage, $V_{\rm dep}$ the minimum voltage needed for full depletion, and d the detector thickness.

We will now proceed to derive the current induced in the bottom electrode separately for an electron moving to the bottom and a hole moving to the top surface. The electron drift velocity $v_n(x) = -\mu_n\mathcal{E}(x)$ gives the differential equation

$$\frac{dx_{\rm e}}{dt} = \mu_n\left[\frac{V+V_{\rm dep}}{d} - 2\frac{V_{\rm dep}}{d^2}x\right] \quad ,$$

which can easily be integrated. One obtains the following, with the boundary condition $x_{\rm e}(t=0) = x_0$, for the position (i.e. the depth as a function of time):

$$\begin{aligned}
x_{\rm e}(t) &= d\frac{V+V_{\rm dep}}{2V_{\rm dep}}\left[1-\left(1-\frac{x_0}{d}\frac{2V_{\rm dep}}{V+V_{\rm dep}}\right)e^{-2\mu_n\frac{V_{\rm dep}}{d^2}t}\right]\\
&= d\frac{V+V_{\rm dep}}{2V_{\rm dep}} + \left[x_0 - d\frac{V+V_{\rm dep}}{2V_{\rm dep}}\right]e^{-2\mu_n\frac{V_{\rm dep}}{d^2}t} \quad ;
\end{aligned}$$

and the related velocity is given by

$$\frac{\mathrm{d}x_e(t)}{\mathrm{d}t} = \mu_n \frac{V + V_{\text{dep}}}{d} \left(1 - \frac{x_0}{d} \frac{2V_{\text{dep}}}{V + V_{\text{dep}}}\right) e^{-2\mu_n \frac{V_{\text{dep}}}{d^2} t}$$

$$= \mu_n \left[\frac{2V_{\text{dep}}}{d^2} x_0 - \frac{V + V_{\text{dep}}}{d}\right] e^{-2\mu_n \frac{V_{\text{dep}}}{d^2} t} \ .$$

The same two equations hold for holes with the replacement of μ_n with $-\mu_p$.

The electron movement comes to a sudden stop when the surface electrode is reached ($x_e(t_e) = d$). The total electron drift time is then

$$t_e = \frac{d^2}{2\mu_n V_{\text{dep}}} \ln \left[\frac{V + V_{\text{dep}}}{V - V_{\text{dep}}} \left(1 - \frac{x_0}{d} \frac{2V_{\text{dep}}}{V + V_{\text{dep}}}\right)\right] \ .$$

Similarily, the hole will stop at the top surface ($x_h(t_h) = 0$) at

$$t_h = -\frac{d^2}{2\mu_p V_{\text{dep}}} \ln \left(1 - \frac{x_0}{d} \frac{2V_{\text{dep}}}{V + V_{\text{dep}}}\right) \ .$$

The current induced by a moving charge q is given by

$$i(t) = \frac{q}{d} \frac{\mathrm{d}x}{\mathrm{d}t} \ ,$$

as can be verified from the expression for induced charge. The current in our example then becomes

$$i(t) = i_e(t) + i_h(t) = \frac{q}{d} \left(-\frac{\mathrm{d}x_e}{\mathrm{d}t} + \frac{\mathrm{d}x_h}{\mathrm{d}t}\right)$$

$$= \frac{q(V + V_{\text{dep}})}{d^2} \left(1 - \frac{x_0}{d} \frac{2V_{\text{dep}}}{V + V_{\text{dep}}}\right)$$

$$\times \left[\mu_n e^{-2\mu_n \frac{V_{\text{dep}}}{d^2} t} \Theta(t_e - t) + \mu_p e^{2\mu_p \frac{V_{\text{dep}}}{d^2} t} \Theta(t_h - t)\right]$$

$$= \frac{q}{d^2} \left(2V_{\text{dep}} \frac{x_0}{d} - (V + V_{\text{dep}})\right)$$

$$\times \left[\mu_n e^{-2\mu_n \frac{V_{\text{dep}}}{d^2} t} \Theta(t_e - t) + \mu_p e^{2\mu_p \frac{V_{\text{dep}}}{d^2} t} \Theta(t_h - t)\right]$$

with $\Theta(x) = \begin{smallmatrix} 1 \text{ for } x \geq 0 \\ 0 \text{ for } x < 0 \end{smallmatrix}$.

The total induced current will thus be the superposition of an electron-induced current with falling exponential time behavior stopping at $t = t_e$ and a hole-induced current with rising exponential time behavior stopping at $t = t_h$, as indicated in Fig. 5.4.

In Example 5.3 we have taken into account the movement of charge due to the electric field (drift) only. Velocity saturation, in other words dependence of mobility on field strength, has been ignored. If diffusion is taken into account in addition, the times of arrival of electron and hole will spread about the calculated value. For a charge cloud consisting of many electrons and holes,

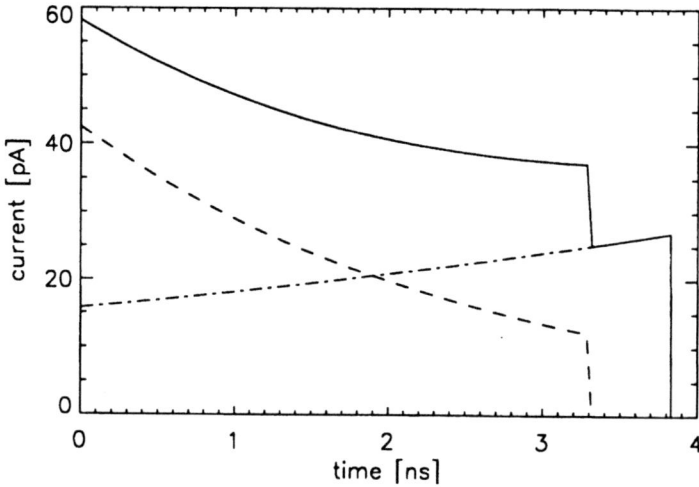

Fig. 5.4. Signal current formation induced by the separation of an electron–hole pair in the electric field of the space-charge region of the detector. The electron–hole pair is created in the center plane of a slightly (20%) overdepleted diode (see Example 5.2). Plotted are the electron-induced (*dashed line*), hole-induced (*dash-dot line*) and total (*continuous line*) currents

this will lead to a smoothing of the signal form shown in Fig. 5.4. Furthermore, for large signals – generated by alpha particles or heavy ions, for instance, – electrostatic repulsion has also to be considered.

For a partially depleted detector charge is not only collected from the space-charge region but also from the bulk. A smaller and much broader (slower) signal is expected in this case because part of the generated charge carriers will recombine, i.e. they will diffuse in the direction away from the space-charge boundary. The time for diffusing into the space-charge region has to be added to the drift time, and drifting of these carriers also starts in the low field region.

5.2.2 Surface Barrier Detectors

The cross-section of a silicon surface barrier detector is shown in Fig. 5.5. On the high-ohmic (low-doped) n-type silicon wafer, a rectifying Au–Si contact has been formed by evaporation of a thin ($\approx 200\,\text{Å}$) gold layer on the cleaned silicon wafer. The backside is covered by an evaporated layer of aluminum.

Surface barrier detectors in the past have shown stability and reproducibility problems, which probably were mostly due to the undefined conditions at the edges of the junction, making their behavior (e.g. breakdown voltage) dependent on the conditions of the general environment (e.g. humidity and temperature). Recently an improved technology, incorporating some features of the newer planar technology (Fretwurst 1990), has produced detectors with stable and reproducible performance.

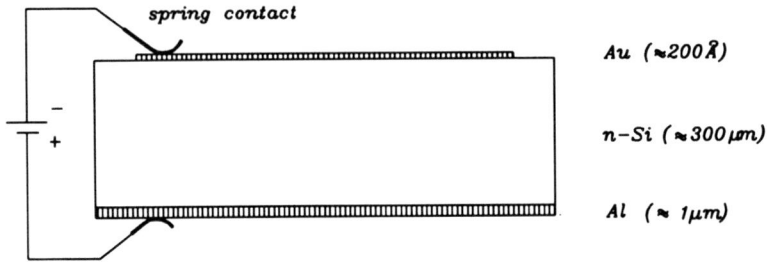

Fig. 5.5. Surface barrier detector

5.2.3 p–n Junction Detectors

Most of presently produced Si detectors are based on (1,1,1) n-type detector-grade material, which is easily available from industry as it is extensively used for the production of thyristors. Some detectors on p-type silicon have also been produced successfully (Beutenmüller et al. 1987).

Today's standard production method using planar technology was adapted from microelectronics to detector production by J. Kemmer (Kemmer 1980) and will be described in Chap. 10. Here, a simplified production sequence is illustrated in Fig. 5.6. The lowly doped n-type Si wafer is chemically cleaned and oxidized by heating it in an oxygen atmosphere at around 1000 °C. In a first photolithographic step, the junction layer is covered by a thin layer of photoresist, which is then illuminated through a mask in such a way that after development and chemical etching the SiO_2 is removed in the diode area. The procedure of transferring a structure from a mask onto the wafer by optical means is called photolithography. The remaining SiO_2 acts as a mask for the implantation of boron, so that the diode is only formed in the uncovered region. Implantation of arsenic over the entire backside is followed by the annealing step – heating at 600–800 °C – in order to repair damages to the crystal and to get the implanted atoms properly built into the lattice. Aluminization by evaporation or sputtering on both surfaces provides the ohmic back contact and the electrical connection to the diode. A photolithographic etching step for the top aluminum layer brings it into final shape, which in the example shown is a frame around the open diode, so that radiation has not to penetrate through aluminum before reaching the sensitive space-charge region. Heating to 420 °C, which is somewhat below the melting point of aluminum (660 °C), (sintering) in order to obtain a good connection between aluminum and silicon, completes the production sequence.

The depth of the inactive p^+ layer of the diode can be controlled by variation of the implantation energy and/or dopant dose, as well as by the subsequent high-temperature steps in the process, in which the dopants diffuse deeper into the crystal.

Somewhat delicate regions of the detector structure are the edges of the diode. The positive oxide charges, which are always present in SiO_2, are responsible for the generation of high electric fields once the detector is biased

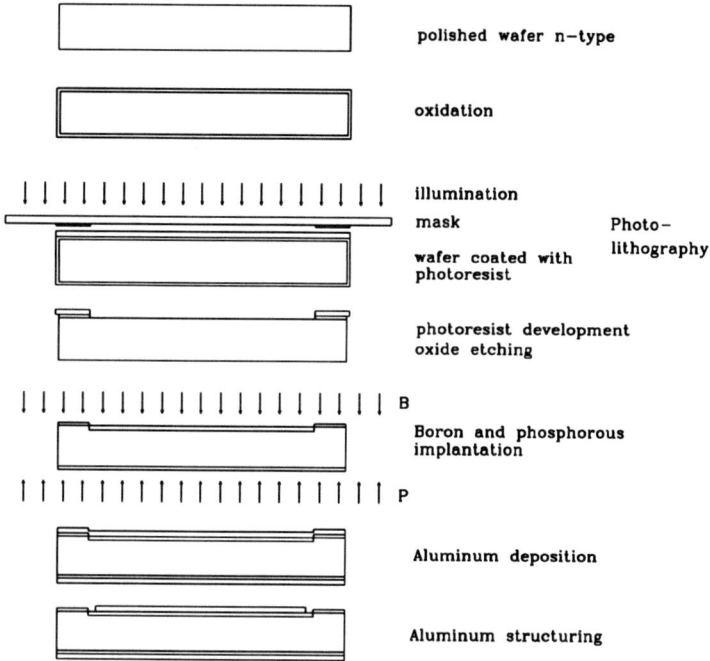

Fig. 5.6. Simplified production sequence of a p–n junction detector in planar technology

(compare Sects. 3.1.2 and 3.3.2), such that electrical breakdown becomes likely. Various structures and precautions have been invented and successfully applied to circumvent this problem. Examples are: guard rings; partial compensation of oxide charges by implantation; and covering of the oxide surface with high ohmic resistive layers in order to control and smoothly vary the potential at this site.

The problems connected with oxide charges and edge effects will be treated in greater depth in the chapters devoted to technology and device stability.

5.3 Summary

The working principles of large-area diode detectors, both in unbiased and reverse-biased condition, have been considered. The width of the space-charge region in homogeneously doped detectors rises with the square root of the applied reverse-bias voltage. Charge is collected not only from the space-charge region but also from the neutral region, as shown in Example 5.1, where the charge-collection efficiency has been determined as a function of position in the detector. The signal detected at the external electrodes is formed not at the time when electrons and holes reach the electrodes but already during their separation in the electric field. This is due to noncancelling charge induction of electrons and holes.

A short description of detector production has been given, although a more extended discussion can be found in Chap. 10.

Most of the physics described in this Chap. 5 applies also to, and is a prerequisite of, an understanding of position-sensitive detectors, which are described in the next chapter.

6 Detectors for Position and Energy Measurement

As the primary ionization in semiconductors is proportional to the energy loss of the incident radiation or particle, most semiconductor detectors will provide the possibility of energy measurement if the readout electronics measures the signal charge.[19] Position sensitivity may be obtained by creating a situation in which the signal charge is split amongst more than one readout electrode with the ratio of charges depending on the position. Alternatively, the detector may be segmented in many small subdetectors that are read out separately.

6.1 Resistive Charge Division

The principle of resistive charge division is shown in Fig. 6.1. The moderately doped p^+-layer of the diode has the function of a resistor also, dividing the signal charge between the two low input impedance amplifiers in the ratio of the inverse distances of the place of generation to the respective amplifiers.

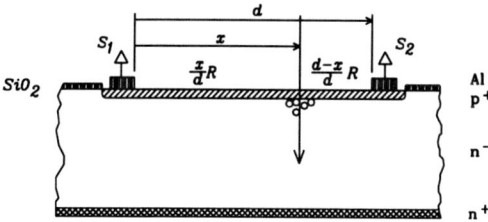

Fig. 6.1. Position measurement by resistive charge division

Thus the position may be inferred from the signals in the amplifiers in a linear approximation given by

$$x = \frac{S_2}{S_1 + S_2} d \ .$$

$(6.1.1)$

The position measurement error is then

---

[19]An exception is the avalanche diode when working in the avalanche region. There the behavior is similar to the Geiger–Müller gas detector.

$$\Delta x = (d - x)\frac{\Delta S_2}{S_1 + S_2} - x\frac{\Delta S_1}{(S_1 + S_2)} \tag{6.1.2}$$

and the expectation value for the mean-square deviation of the measurement from the true position is

$$\langle \Delta x^2 \rangle = \frac{1}{(S_1 + S_2)^2} \left[(d - x)^2 \langle \Delta S_2^2 \rangle \right.$$
$$\left. - 2x(d - x)\langle \Delta S_1 \Delta S_2 \rangle + x^2 \langle \Delta S_1^2 \rangle \right] . \tag{6.1.3}$$

ΔS_1 and ΔS_2 denote the errors in the two signals that are due to noise. Estimation of the noise signals requires knowledge of certain electronics theory, which will be dealt with in Chap. 7. Here it should be sufficient to point out that the dominant noise sources, which are the amplifier noise and the resistor noise that is generated in the resistive diode layer. While the noise due to the two amplifiers ΔS_a is almost uncorrelated $(\langle \Delta S_{a1} \Delta S_{a2} \rangle = 0)$, the noise generated by the resistor ΔS_R leads to complete correlation $(\Delta S_{R1} = -\Delta S_{R2} = \Delta S_R)$.

Assuming identical noise properties of the amplifiers

$$\langle \Delta S_1^2 \rangle = \langle \Delta S_2^2 \rangle = \langle \Delta S_a^2 \rangle$$

we obtain

$$\langle \Delta x^2 \rangle = \frac{d^2}{(S_1 + S_2)^2} \left[\langle \Delta S_a^2 \rangle \left(1 - 2\frac{x}{d} + 2\frac{x^2}{d^2} \right) + \langle \Delta S_R^2 \rangle \right] . \tag{6.1.4}$$

Thus the amplifier noise gives a square root of two larger contributions to the position measurement error for a signal generated at the edge of the device rather than at the center, while the resistor noise produces a position-independent measurement error. For a quantitative estimation of noise and optimization, which requires also consideration of signal risetime, the reader is referred to Chap. 7.

6.2 Diode Strip Detectors

Another, in principle simpler, way to measure position is obtained by dividing the large-area diode into many small (strip or even pixel-like) regions (Belau 1983a and 1983b) and to read them out separately (Fig. 6.2). The position of passage of the ionizing particle is then given by the location of the strip showing the signal.

The method of production of strip detectors may follow that of planar diodes (Kemmer 1980) described in Sect. 5.2.2 using the somewhat more complicated geometrical strip structures and the correspondingly higher required precision of masks and relative alignment of photolithographic steps in the procedure. It is of course also possible to produce strip detectors in surface barrier technology; however the early versions of semiconductor detector technology, in which the strip structure was obtained by evaporation through a mask (Heijne et al. 1980),

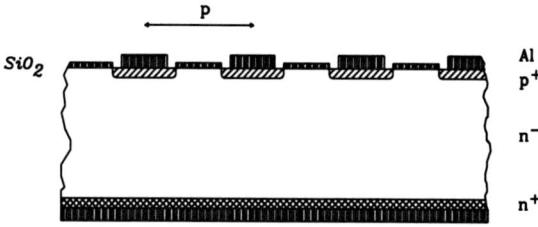

Fig. 6.2. Cross-section of a silicon diode strip detector

is not suitable for very fine structures. Thus surface barrier strip detectors do not play an important role in this context.

The measurement precision depends mainly on the strip spacing and the method of readout. As long as only digital information is used (taking the center position of the strip as the measured coordinate) and the effects arising from track inclination and charge diffusion during collection can be neglected, the measurement precision (root-mean-square deviation from the true coordinate) is given by the strip pitch p as

$$\langle \Delta x^2 \rangle = \frac{1}{p} \int_{-p/2}^{p/2} x^2 \mathrm{d}x = \frac{p^2}{12} \ . \tag{6.2.1}$$

Typical strip pitches are twenty to few hundred micrometers.

The measurement precision is substantially improved with analog readout if the strip pitch is chosen small enough so that the signal charge – due to diffusion – is collected on more than one strip and the coordinate is found by interpolation, e.g. by the center of gravity of the signal.

Example 6.1

Problem: *Estimate the width of the charge cloud arriving at the p-strip side surface of a $d = 300\,\mu m$-thick detector built on phosphor-doped silicon ($N_D = 10^{12}\,cm^{-3}$) operating at full depletion.*

Solution: *We will only attempt a very rough estimate, calculating the drift time from the average field and half-thickness of the detector:*

$$t_{\text{drift}} \approx \frac{d/2}{\mu_p V_{\text{dep}}/d} = \frac{d^2}{2\mu_p V_{\text{dep}}} = \frac{\epsilon\epsilon_0}{\mu_p q N_D} \ ,$$

with $V_{\text{dep}} \approx \frac{q N_D d^2}{2\,\epsilon\epsilon_0}$ according to (3.1.5). The holes will diffuse in a lateral direction, thereby assuming a Gaussian distribution with a root-mean-square value of the projected distribution given by

$$\sigma_x = \sqrt{2 D_p t_{\text{drift}}} = \sqrt{2 \frac{kT}{q} \mu_p t_{\text{drift}}} = \sqrt{\frac{kT}{q} \frac{2\,\epsilon\epsilon_0}{q N_D}} \ .$$

Putting into these expressions the numerical values $\mu_p = 505\,cm^2/Vs$, $\epsilon = 11.8$, $kT/q = 0.0259\,V$, $\epsilon_0 = 8.854 \times 10^{-14}\,F/cm$, we obtain:

$$V_{\text{dep}} = 69\,\text{V} \qquad\qquad t_{\text{drift}} \approx 12.1\,\text{ns}$$
$$\sigma_x \approx 5.81\,\mu\text{m}\ .$$

Matching the readout pitch to the diffusion width results in strip pitches of the order of ten micrometers. Reading out each individual strip in this high-density condition is very difficult. One approach to avoid this problem is the use of resistive or capacitive charge division which will be described below. In this case not every strip but only every k^{th} strip is read out, and the coordinate is found by interpolation.

6.2.1 Readout Methods

Digital (yes/no) readout may be used if no energy information is required and if the position accuracy given by the strip pitch is sufficient. One also does not lose position resolution compared with analog readout if the strip pitch is large with respect to the width of the diffusion cloud.

Analog (signal height) readout of every channel may lead to a substantial improvement of position measurement precision if the strip spacing matches the charge spread due to diffusion during collection.[20] In addition, the simultaneous measurement of energy loss becomes possible.

Charge division readout reduces the number of readout channels as only a fraction of the strips is connected to a readout amplifier. Charge collected at the other (interpolation) strips is divided between the two neighboring read-out channels according to the relative position. This can be accomplished by resistive or capacitive division.[21]

Resistive charge division works in the same way as described in Sect. 6.1 for a large-area diode, while the way the resistor is produced may be different. It is not necessary to have a large-area high-resistivity layer, but it is enough to produce individual resistors between the strips by e.g. sputtering a resistive strip perpendicular to the diode strips.

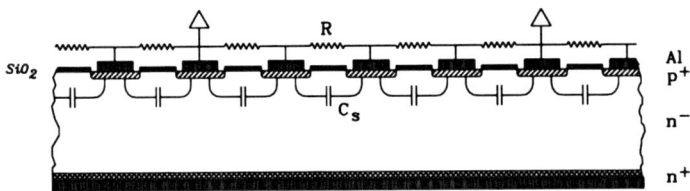

Fig. 6.3. Capacitive charge division readout. In the first detectors the high ohmic resistors were created by sputtering a thin strip like layer of silicon in an orthogonal direction to the diode strips across the whole wafer

---

[20]Charge spread can also be due to track inclination.
[21]The possibility of capacitive charge division was discovered when, in an early test of a surface barrier strip detector, one of the strips was not connected to the readout electronics and its signal charge was split between the two neighboring channels (Heijne 1980).

As seen in Sect. 6.1, these resistors will contribute to the noise and therefore degrade the position resolution. This is avoided with capacitive charge division (Fig. 6.3). It takes advantage of the built-in interstrip capacitances that are present automatically due to the geometrical structure of the device. Charge collected at the interpolation strips – in linear approximation – is divided according to the ratio of the series of interstrip capacitances to the neighboring readout strips into the two readout channels. The strip-to-ground capacitances (typically an order of magnitude smaller than the interstrip capacitances) and the capacitances between non-neighboring strips will distort the linearity of charge division, extend the signal cluster above two strips, and lead to a position-dependent loss of the collected signal.

One important aspect in capacitive charge division is the need of keeping the DC potential of the interpolation strips at the same value as the readout strips. This can be achieved, for example, by connecting them with very high resistors to the readout strips. If the intermediate strips were left floating, they would adjust themselves to a potential such that they would collect none of the signal charge, and thus charge division would cease to function.

Example 6.2
Problem: *Estimate the order of magnitude of interstrip resistance, required for capacitive charge division readout, of a strip detector of 5 cm strip length. Assume a strip-to-strip capacitance per unit strip length of 1 pF/cm, a signal-processing time of $\tau_{el} = 1\,\mu s$, and a dark current per unit strip length of 1 nA.*
Solution: *Two requirements have to be met that constrain the resistance. A lower limit is obtained from the requirement that the charge stays at the readout strip for a time considerably longer than the signal-processing time τ_{el}. This leads to $\tau = R \cdot C_s \gg \tau_{el}$ or*

$$R \gg \frac{\tau_{el}}{C_s} = \frac{10^{-6}\,s}{5 \times 10^{-12}\,F} = 200\,k\Omega \ .$$

In addition, an upper limit for R is obtained from the requirement that the intermediate strips stay at similar potential to the strips connected to the readout electronics. For infinite resistance, the voltage of the floating strips would change due to the dark current until the dark current would flow directly to the readout strips. As a consequence, the signal charge would be collected directly at the closest readout strip. If we allow a voltage drop between strips of $\Delta V = 0.1\,V$, the maximum allowed resistance value would be

$$R < \frac{\Delta V}{I} = \frac{0.1\,V}{5 \times 10^{-9}\,A} = 2 \times 10^{11}\,\Omega \ .$$

A further consideration relates to the electronic noise, which is influenced by the resistance. Noise considerations will push the preferred resistance value upwards. Electronic noise will be dealt with in Chap. 7.

6.2.2 Charge Collection and Measurement Accuracy

Treatment of measurement accuracy needs some knowledge of electronic behavior in connection with the detector. This will be dealt with in full in Chap. 7, and here we will restrict ourselves to effects intrinsic to the detector only.

The following effects influence the charge collection properties:

- *the spatial distribution* of electron–hole pairs created by the incident radiation;
- *the separation* of electrons and holes due to the electric field in the space-charge region of the detector;
- *diffusion* of the signal-charge carriers between the times of generation and arrival at the semiconductor surface; and
- *noncollinearity* between velocity and electric field in the presence of a magnetic field.

Most of these effects have been dealt with already in previous chapters. Here we will restrict ourselves to a semiquantitative investigation of two common cases.

Example 6.3
Problem: *An α-particle incident from the unstructured backside of a n-type strip detector is absorbed close to its entrance point. Determine the distribution of charge collected in strips of pitch p, assuming that the impact point is centered directly below one of the strips. Describe qualitatively the signal current in the center strip and in a strip roughly half the detector thickness away from the center.*
Solution: *Because the electron–hole pairs are generated very close to the backside surface, the short movement of the electrons can in good approximation be ignored and we are concerned only with the movement of holes. Holes will initially induce charge on the backside surface only, thereby compensating for the electron signal. While holes are moving towards the top surface, more and more positive charge will be induced in the strip electrodes located on the top surface, while the charge induction on the bottom decreases, as was shown in Sect. 5.1.3. The charge induced in the topside will be distributed over a region with radius comparable to the distance of the holes from the top surface. As the holes approach the top surface, this region will therefore shrink, while at the same time the total signal increases. Considering now the strip directly above the point of incidence, we will observe a continuously rising signal charge that reaches a maximum equaling the total number of holes created. A strip roughly half a wafer thickness away from the center strip will first see an increasing positive signal charge, and this will reach its maximum when the holes are somewhere in the upper half of the detector; the signal charge will then decrease to zero again, once the holes have reached the central strip.*

A precise calculation of the signal shape requires a numerical simulation that also should take into account other mechanisms such as diffusion, which may result in a spread of the asymptotic signal over more than one strip.

Example 6.4
Problem: *A high-energy charged particle traverses perpendicularily a strip detector embedded in a magnetic field B with orientation parallel to the strips. The detector is biased slightly above full-depletion voltage. Describe the spatial distribution of signal electrons and holes on the respective surfaces of the detector.*

Solution: *Assuming that we have an n-type strip detector with p-strips on the top, the high-field region will be on the topside. Thus all electrons will have to pass the low-field region of the space-charge region on the bottom, while this is the case only for those holes generated within this low-field region. The charge-collection time distribution for holes therefore will have its maximum at zero while the distribution for electrons will be peaked at the maximum electron drift time.*

As the charge carrier will diffuse during charge collection in the absence of a magnetic field, the spatial distribution will be given by a superposition of Gaussian distributions with widths rising with the square root of the drift time (see (2.7.3)). Therefore we could expect a fairly sharp peak and long tails in the position distribution for holes, while for electrons the distribution should more closely resemble a single Gaussian peak. The presence of a magnetic field makes charge carriers move at an angle (see Sect. 2.5.3) with respect to the electric field, and this angle is different for electrons and holes. The spatial distribution of charge carriers arriving at the surface is a superposition of laterally displaced Gaussian distributions whose widths are determined both by the charge-collecting time corresponding to the position along the track and also by the displacement given as the product of distance from the surface and the Hall angle.

6.2.3 Choice of Geometrical Parameters

As requirements on detector performance depend very much on application, no general recipe is available. Instead, some hints on criteria and relations will be given.

Performance optimization requires the simultaneous consideration of the detector and its electronics. Electronics will be considered in Chap. 7; here, we will only assume that a certain ratio of single-channel noise-to-signal (N/S) is present. This ratio will depend on detector capacitive load and detector leakage current. Deferring these relationships to a later point, an important criterion for detector optimization is position-measurement precision. This will depend on the type of readout and the geometrical parameters of the detector.

The measurement precision is defined as the root-mean-square distance of the measured coordinate from the true hit position. For an individual readout of each strip with essentially all charge collected on a single strip, one takes the hit-strip center as the measured coordinate and obtains as measurement precision the strip pitch divided by $\sqrt{12}$ (see (6.2.1)). For small strip distances, diffusion distributes charge between neighboring strips. If pulse height (ph) is measured, one can interpolate the position in between strips according to

$$x = x_1 + \frac{ph_2}{ph_1 + ph_2}(x_2 - x_1) = \frac{ph_1 x_1 + ph_2 x_2}{ph_1 + ph_2} \quad .$$

The position-measurement precision is then limited by the noise performance of the electronics and the (readout) strip pitch p, as follows:

$$\Delta x \approx (N/S)p \quad .$$

In order to provide charge division between strips, the width of the charge cloud has roughly to match the strip pitch.

Capacitive charge division allows interpolation over distances significantly larger than the diffusion width. There the (interpolation) strip pitch has to match the diffusion width so as not to limit measurement precision, and the measurement precision will be roughly given by the product of N/S and readout pitch.

Important aspects are the capacitance seen by the amplifier, which should be small in order to obtain small noise and – in the case of charge division – the fraction of charge lost to ground in the capacitive network of readout and floating interpolation strips and backplane. In order to keep this loss small, large interstrip capacitances are desirable, although this is in contradiction to the requirement of low capacitive load at the amplifier input.

The previous discussion is rather rough and incomplete. For a correct treatment, noise correlation has to be taken into account (Lutz 1991). In addition, other features such as efficiency and noise signals have to be considered. These may put limits on, for instance, the capacitive load that the detector represents to the electronics, and may therefore limit the length of the detector strips.

Further constraints may come from the requirement of reliable and stable operation of the detector, which requires the avoidance of high electric fields in the detector. This aspect will be discussed again in the chapter on device stability and radiation hardness (Chap. 11).

6.3 Strip Detectors with Double-Sided Readout

As electrons and holes in a strip detector are swept by the electric field to opposite sides of the wafer, it is possible to use both types of charge carriers for position measurement by providing charge-collection electrodes on both sides of the wafer (Fig. 6.4). This double-sided readout brings about the obvious advantage of providing twice the information for the same amount of scattering material.

With crossed strips on the two detector faces, projective two-dimensional measurement is obtained from one single detector. For a traversing particle, a spatial point can be reconstructed as both projections are taken from the same initial charge cloud. For absorbed radiation such as x-rays, two-dimensional

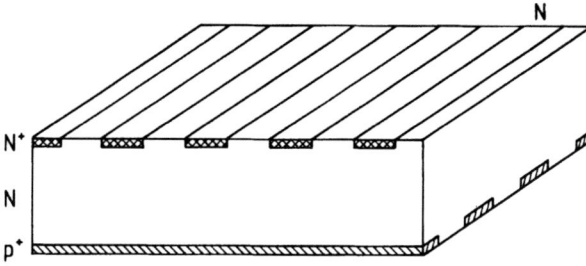

Fig. 6.4. Double-sided strip detectors: naive solution. (After Kemmer and Lutz 1988, p. 592 Fig. 9)

measurement becomes possible. With analog readout it is furthermore possible (to some degree) to correlate signals from the two sides, making use of Landau fluctuations and the exact equality of positive and negative charge created by each ionizing particle. This can be of interest for resolving ambiguities when several particles simultaneously hit the detector.

A problem in producing the double-sided kind of detector is the insulation of each of the strips from the others simultaneously on both sides of the detector. This problem and several solutions have been described in original papers (Sedlmair 1985, Lutz 1986). The naive solution of only providing highly doped n and p strips on the two sides of the detector (Fig. 6.4) fails because of the buildup of an electron-accumulation layer (an inversion layer on p-type material) between the N^+ strips below the insulating oxide (Fig. 6.5a). This layer of electrons produces an electrical shortening of neighboring strips. It is caused by the positive charges that are always present at the silicon–oxide interface.

There are three possibilities for curing the problem:

- *large-area p-type surface doping.* In this case the oxide charges are compensated for by the negative acceptor ions and the buildup of the electron layer is prevented (Fig. 6.5b). This method requires a delicate choice of p-type doping concentration and profile. Too-large doping results in high electric fields and possible electrical breakdown at the strip edges, where moderately high p-doping joins directly the highly doped n^+ strips. This potential problem is alleviated by the other two solutions presented below:
- *disruption of the electron layer by implantation of p strips* between the n^+ charge-collection strips (Fig. 6.5c); and
- *disruption of the electron layer by a suitably biased (negatively with respect to the n^+ strips) MOS structure* (Fig. 6.5d). For moderate biassing neither electrons nor holes will be present underneath the MOS structure, while for high negative bias a hole layer (inversion on n-type, accumulation on p-type silicon) will form.

Results from the first detectors produced using the implantation method (see Fig. 6.6) and the MOS-structure method are shown in Figs. 6.7 and 6.8. Only one n^+ strip was biased when the capacitance between this strip and the

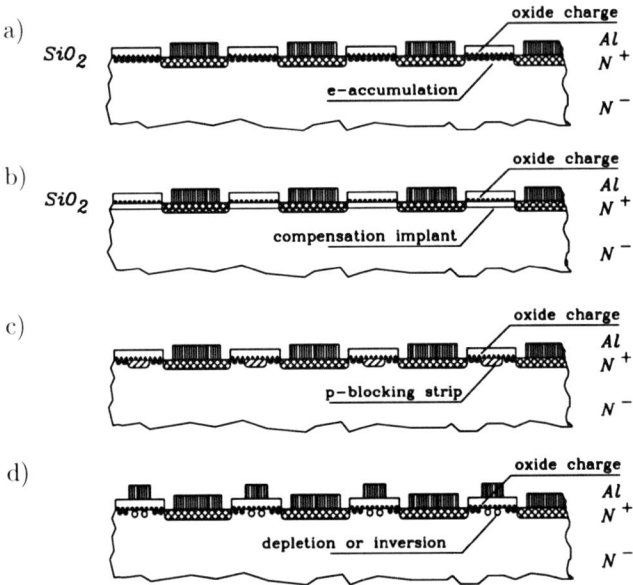

Fig. 6.5a–d. Insulation problem for n^+ strips in silicon, due to electrical shortening by the electron accumulation layer (a) and three methods of solution: Large area p-implantation (b); interleaved p strips (c) and negatively biased MOS structures (d)

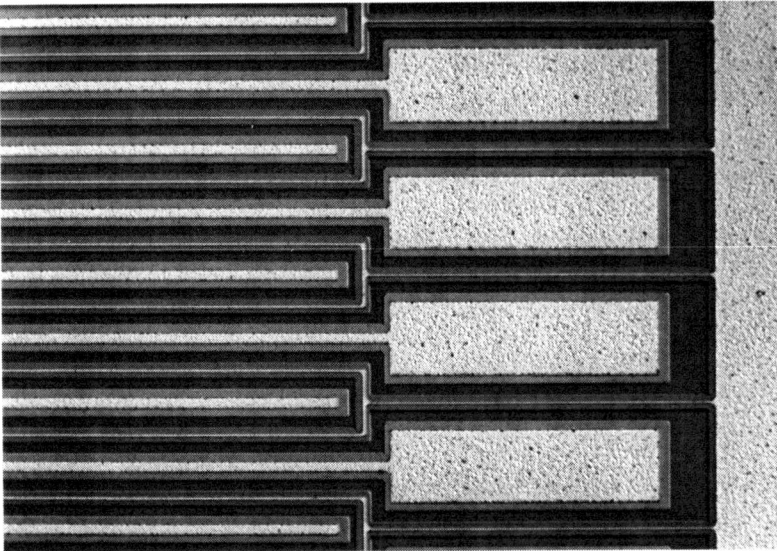

Fig. 6.6. Microphotograph of n-strip side of a detector with p^+ insulation structures. (After Kemmer 1988, p. 593 Fig. 12)

a)

b)

c)

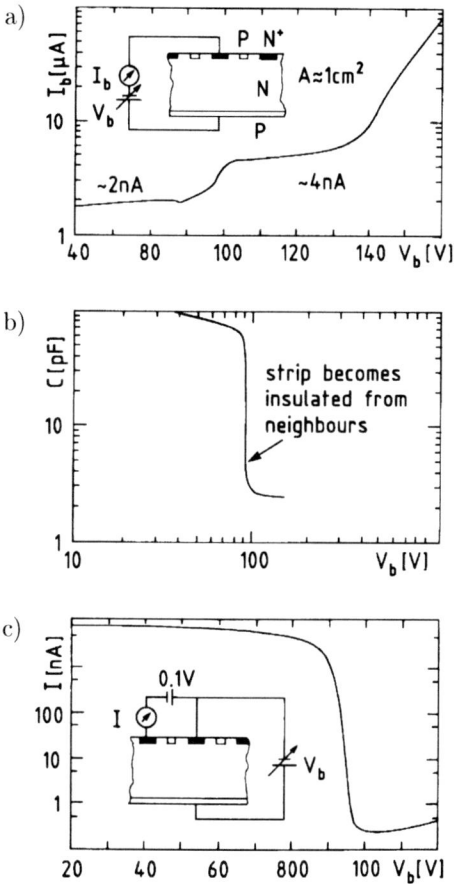

Fig. 6.7a–c. Test results of detectors with insulation by structured p-implantation: reverse-bias current of strip detector with electron readout (a); N^+ strip to ground capacitance as a function of bias voltage (b); Current between neighboring strips as function of bias voltage (c). (After Kemmer 1988, p. 593 Fig. 13)

backside was measured, either as a function of the detector bias (Fig. 6.7) or as function of the MOS gate voltage with an already fully depleted bulk (Fig. 6.8).

The sudden drop of the capacitance marks the condition under which isolation from the neighboring n^+ strips occurs. This is confirmed by a direct measurement of the resistance between neighboring strips (also shown in the same figures).

Comparing the three methods of insulation, one should realize that the voltage on the p regions on the side of the n-strips (compensation implant, p-type insulation strips, or a hole layer underneath the MOS gate) cannot be chosen freely because, for too large positive voltages, a punch-through hole current will flow from the insulation structure across the bulk to the p side. One normally uses this fact to automatically bias the p regions on the n side, connecting the bias only to the p side and the n strips and leaving the p regions on the n side floating. A true two-dimensional consideration or simulation is needed to find the voltage to which the floating insulation structures will adjust themselves. The result will not only depend on the doping conditions but also on the geometry of the detector, in particular on strip widths and pitch.

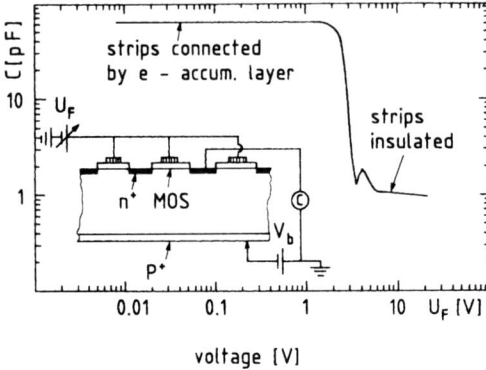

voltage [V]

Fig. 6.8. Capacitance between n-strip and diode as a function of field-plate voltage. The bias voltage was constant and above depletion during the measurement. (After Kemmer 1988, p. 593 Fig. 14)

6.4 Strip Detectors with Integrated Capacitive Readout Coupling

Capacitively coupled readout (Fig. 6.9, right) has the obvious advantage of shielding the electronics from dark current, which with direct coupling (Fig. 6.9, left) can lead to pedestal shifts, a reduction of the dynamic range, and may even drive the electronics into saturation.

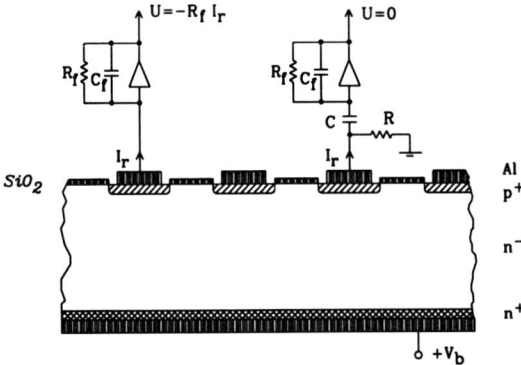

Fig. 6.9. Direct and capacitive coupling of electronics to the detector. With direct coupling (*left*) the detector reverse bias current I_r has to be absorbed by the electronics. With capacitive coupling (*right*), only the AC part of the detector current reaches the electronics, while the DC part goes into a bias circuit, here shown as a simple resistor

As it is difficult to fabricate high ohmic resistors and almost impossible to produce sufficiently large capacitors on LSI electronics, it seemed natural to integrate these elements into the detector. This has been done in a collaborative effort by a CERN group with the Center of Industrial Research in Oslo (Caccia 1987), where the detectors were produced. Capacitances have been built by separating implantation and metallization of the strips by a thin oxide layer. Biasing resistors were made in polysilicon. The detectors gave very satisfactory results. Detectors of this design have been used in the vertex detector of the DELPHI experiment at the electron–positron collider (LEP) at the European Center of Nuclear Research (CERN) in Geneva.

A new method of supplying the bias voltage to the detector has been invented and used for double-sided readout by a Munich group (Kemmer 1988). It leads to a considerable simplification of the technology as it does not require resistors but only uses technological steps that were already required for DC coupled detectors. The polysilicon technology is avoided altogether; instead, the voltage is supplied through the silicon bulk.

A detector configuration with crossed strips on the two wafer sides is shown in Fig. 6.10. It demonstrates two slightly different methods of integration of biassing and capacitive readout structures on p and n side of the wafer. These structures may of course also be used for single-sided readout.

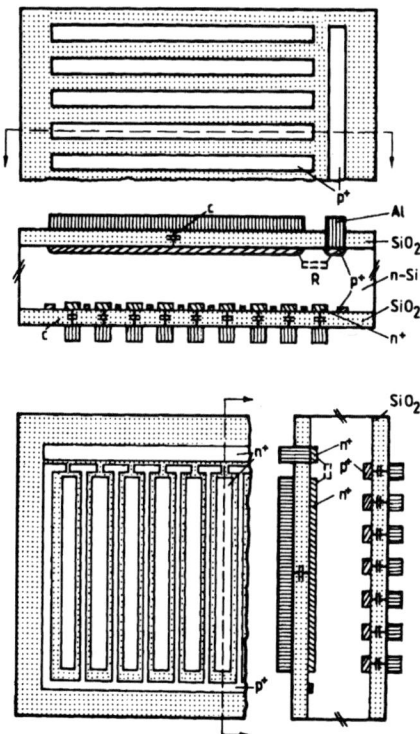

Fig. 6.10. Double-sided strip detector with integrated biassing structures and coupling capacitors. The strips on the two detector sides have orthogonal directions. Shown are a top view of the p strip side (*top*), a cross-section along the middle of a p strip (*middle*), a bottom view of the n-strip side (*bottom left*) and a cross-section along the middle of an n strip (*bottom right*). The symbol C indicates the coupling capacitance built by strip implant, insulator and strip metal, while R stands for the biassing structure, which on the p side is a punch-through structure and on the n side is an electron-accumulation layer resistor whose dimensions are defined by the enclosing p-type insulation structure. (After Kemmer 1988, Fig. 15)

As before, the coupling capacitor for each strip is constructed by interleaving a thin oxide layer between implantation and metallization. The n^+ strips at the bottom are insulated from each other by the surrounding p implant. This implant does not completely enclose the strip in each case, however. At the strip ends, a narrow passage towards a directly connected n^+ biassing electrode is left free. The electron-accumulation layer in this path builds the biassing resistor

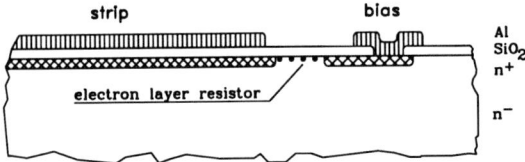

Fig. 6.11. n-strip biassing by electron accumulation-layer resistors. The diagram shows a cut along the strip direction. The electron layer is caused by the oxide charges. It is sidewise enclosed by p implants. Bias and strip implants are at nearly the same potential

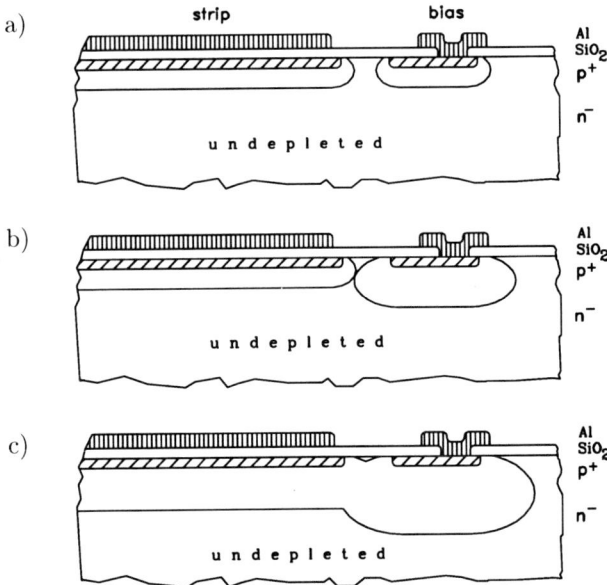

Fig. 6.12a–c. p strip punch-through biassing. The diagrams show a cut along the strip direction: (a) before application of a bias voltage, where the space-charge regions around the strip and the bias implant are separated from each other and no current is flowing; (b) at onset of punch-through, where the space-charge region around the bias implant has grown so as to just touch the other region. The potential barrier between strip and bias implants has diminished, but is just large enough to prevent the thermal emission of holes towards the bias strip; (c) at larger bias voltage, where the space-charge region has grown deeper into the bulk. Holes generated in the space-charge region and collected at the strip implant are thermally emitted towards the bias strip. The voltage difference between strip implant and bias depends on geometry, doping and bias voltage. A weaker dependence on oxide charge is also present

for the strip. The resistance may be chosen by varying the length and width of the gap. Clearly the insulating p structure may also be replaced by a suitably biased MOS structure.

The biassing structure on the diode side (p-strip side) is somewhat simpler. Here a directly connected p^+-strip close to the ends of the readout strips is supplying the bias. One relies on the fact that two nearby diodes will only withstand a difference of a few volts before a current starts flowing between them through the bulk. The implanted part of the readout strips will therefore stay at a potential slightly more positive than the bias electrode.

Cuts along the strip direction of these biassing structures, as well as some other punch-through biassing possibilities, are shown in Figs. 6.11 to 6.13. Their properties are to some degree influenced by the potential that develops on the oxide surface. This will depend on the potential of the nearby metal electrodes and the surface's conduction properties. It is also possible to define the oxide surface potential by providing a separate MOS gate, as shown in Fig. 6.13b. Punch-through biassing is also referred to as reach-through and FOXFET biassing. The latter expression is used when the oxide surface is covered with a gate. A comparison of punch-through current–voltage characteristics with their simulation results is shown in Fig. 6.14.

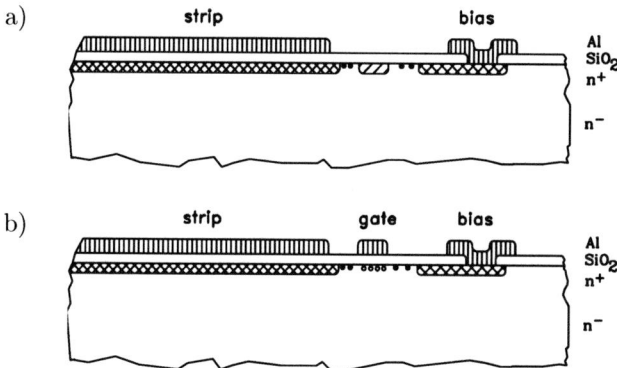

a)

b)

Fig. 6.13a,b. Punch-through biassing of n strips, with interruption of the electron surface layer by: (a) a p-type implant barrier; (b) a MOS-type barrier, the gate voltage being sufficiently negative with respect to the bias, so as to cause a hole layer below the gate. A cut along the strip direction is shown. For p-type substrate, the working mechanism can be understood in analogy to the the p-type device shown in Fig. 6.12 because in both cases the depletion region starts from the diode side. For n-type doping, a natural depletion region will surround the p-doped insulation structure or the hole inversion layer. With applied reverse-bias voltage, a space-charge region will grow from the backside diode (not shown in the figure) such that, when it reaches the natural space-charge region of the insulation structure, the potential of the insulation layer (p-doped region or inversion layer) will follow the change in reverse-bias voltage. Simultaneously, strip implants will be insulated by a potential barrier from the bias contact. The implanted strips therefore will adjust to such a voltage that the average electron current collected on the strips equals the electron current that is thermally ejected over the potential barrier towards the bias contact

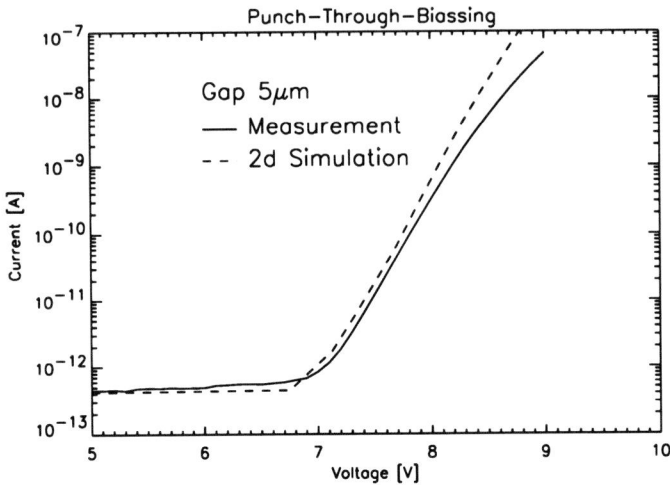

Fig. 6.14. Measurement and simulation of the p-strip punch-through characteristics

Besides the advantages of capacitive coupling, which have already been mentioned, an arrangement like this has several additional attractive properties;

- *easy testing of total dark current*, with only one connection on each surface;
- *automatic biassing of floating strips* in the charge division readout; avoidance of separate high-resistance layers for this purpose; and
- *electronics on both sides of the detector* that may stay at ground level if the oxide layer is good and thick enough to hold the bias voltage.[22]

Detectors following this concept have been produced and were implemented in the ALEPH experiment at CERN (Decamp 1990 et al.; Busculic 1995 et al.; Mours 1996 et al.). The electronics on one side of the detector has been operated with offset ground potential in order to keep the voltage across the oxide low during operation.

Simple and economic production of strip detectors has become an important issue in recent times as applications in high-energy physics and other fields require many square meters of (double-sided) strip detectors. Simplification of the detector concept has been brought to an extreme in a proposal requiring only three photolithographic steps for double-sided capacitively coupled detectors with integrated biassing structures (Kemmer and Lutz 1993). The structure of such a simplified detector is shown in Fig. 6.15, where strips on both surfaces are formed and insulated from each other by alternating accumulation and inversion layers obtained by suitable biassing of the aluminum strip structures. Punch-through biassing from a biassing ring is used on both sides. The biassing ring on the diode side, again formed as an inversion layer, is directly connected

---

[22]There is, however, in this operational mode a danger of avalanche breakdown due to high electric field regions (see also Chap. 11).

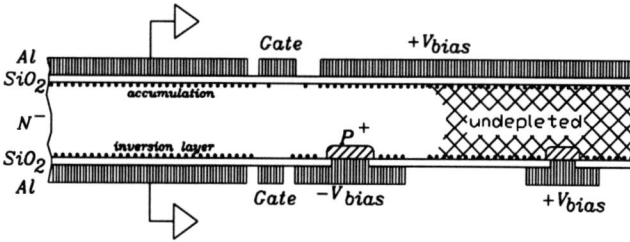

Fig. 6.15. Simplification of strip detectors: cross-section along the strip direction. This double-sided detector with integrated biassing structures and coupling capacitors can be built with only three photolithographic steps

through an implanted diode. The accumulation layer bias ring on the opposite side is connected through the undepleted bulk by a forward-biased diode at the edge of the diode side. Photolithography therefore is needed only for contact holes on the diode side and metallization on both sides.

6.5 Drift Detectors

The ingenious semiconductor drift detector was invented by E. Gatti and P. Rehak (1984). First satisfactorily working devices were built in a collaborative effort by J. Kemmer at the Technical University Munich, the Max Planck Institute in Munich and the inventors (Rehak et al. 1985).

The working principle may be explained by starting from the diode (Fig. 3.1 and Fig. 6.16a) if one realizes that the ohmic N^+ contact does not have to extend over the full area of one wafer side but can instead be placed anywhere on the undepleted conducting bulk (Fig. 6.16b). Then we have space to put diodes on both sides of the wafer (Fig. 6.16c). At small voltages applied to the N^+ electrode, we have two space-charge regions separated by the conducting undepleted bulk region (hatched in Fig. 6.16). At high enough voltages (Fig. 6.16d) the two space-charge regions will touch each other and the conductive bulk region will retract towards the vicinity of the N^+ electrode. Thus it is possible to obtain a potential valley for electrons in which thermally or otherwise generated electrons assemble and move by diffusion only, until they eventually reach the N^+ electrode (anode), while holes are drifting rapidly in the electric field towards the P^+ electrodes.

Based on this double-diode structure with sidewards depletion, it is easy to arrive at a construction of the drift detector if one adds an additional electric field component parallel to the surface of the wafer in order to provide for a drift of electrons in the valley towards the anode. This can be accomplished by dividing the diodes into strips and applying a graded potential to these strips on both sides of the wafer (Fig. 6.17).

Other drift field configurations (e.g. radial drift) can be obtained by suitable shapes of the electrodes. Drift chambers may be used for position and/or energy

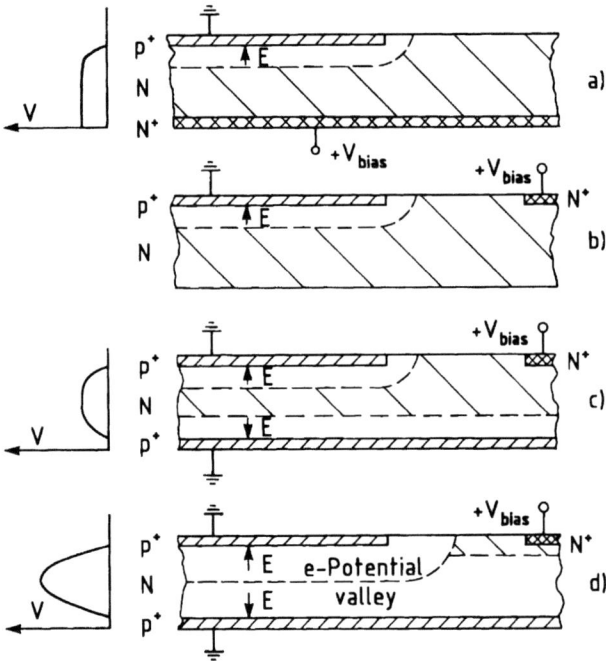

Fig. 6.16a–d. Basic structures leading towards the drift detector: diode partially depleted (a); diode with depletion from the side (b); double diode partially depleted (c); double diode completely depleted (d)

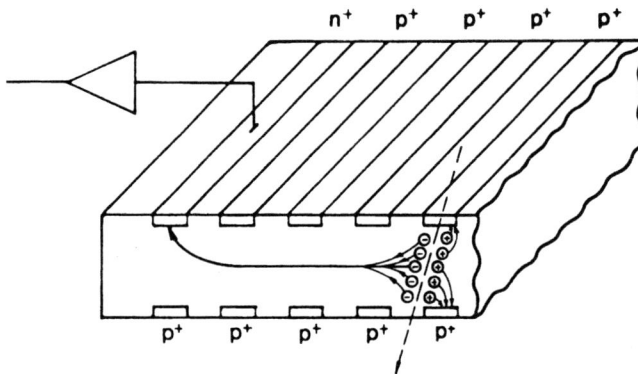

Fig. 6.17. Principle of the semiconductor drift detector. An ionizing particle traverses the detector and creates electron–hole pairs along its path. Holes are swept out to the p^+-doped strips. Electrons move towards the potential valley and drift towards the collecting anode. (After Rehak et al. 1986, Fig. 1)

measurement of ionizing radiation. In the first case the position is determined from the drift time. Furthermore, dividing the N^+ strip anode in Fig. 6.17 into pads, one achieves two-dimensional position measurement (see Sect. 6.5.2).

6.5.1 Linear Drift Devices

Although linear devices seem to be the most straightforward application of the drift detector principle, one encounters some nontrivial problems. They are due to the finite length of the biassing strips and the continually rising potential to be applied to these strips, which in total leads to a very large voltage of several hundred or (for very large drift length) a few thousand volts to be applied to the device. Therefore from the beginning nontrivial guard structures had to be implemented in order to provide a controlled transition from the high voltage to the undepleted wafer region at the edge of the device.

A schematic drawing and a photo of the first operational drift detector (Rehak et al. 1985) are shown respectively in Figs. 6.18 and 6.19. Anodes placed on the left and right side of the drift device collect the signal electrons generated by ionizing radiation. The most negative potential is applied to the field-shaping electrode in the center. Electrons created to the left (right) of this

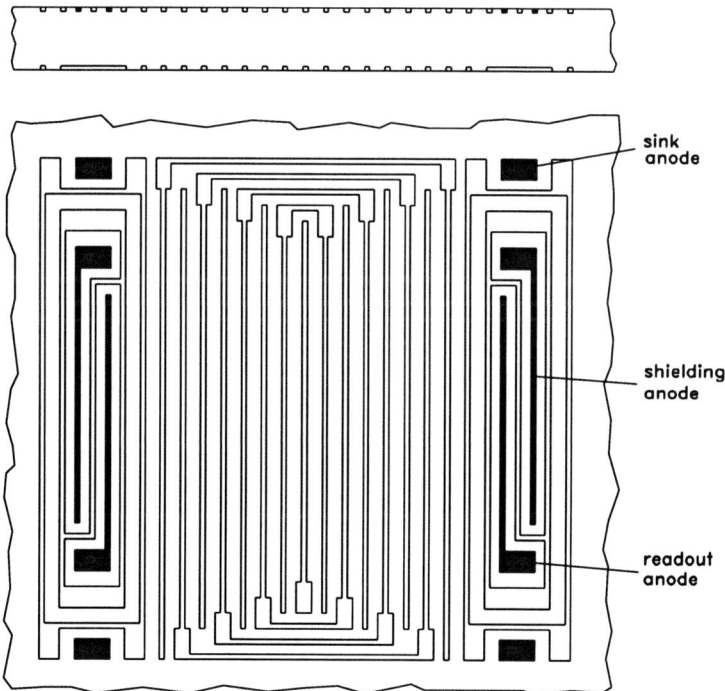

Fig. 6.18. Schematic cross-section and top view of a linear drift detector with p-doped field-shaping electrodes (*light*) and two n-doped (*double*) anodes (*dark*)

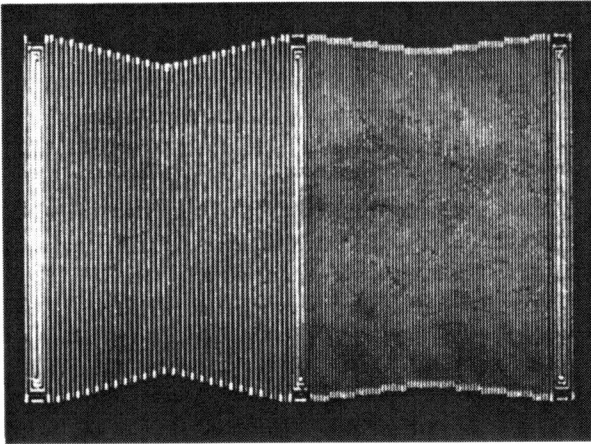

Fig. 6.19. Top view of the first working (linear) drift detector. Two different drift detectors are positioned side by side and share one anode region in the center of the figure. In the left-hand part of the device the potential of each field strip was provided externally. In the right-hand part, with the higher field strip density, a punch-through mechanism (via the "chain effect") was used for biassing. (After Rehak et al. 1985, p. 224 Fig. 2)

electrode will drift to the left (right) anode. The p^+-doped field electrodes do not simply end on the side, but some of them are connected to the symmetrical strip on the other half of the detector. In this way one insures that the high negative potential of the field strips drops in a controlled manner towards the potential of the undepleted bulk on the rim of the detector.

Looking closely at the anodes (Fig. 6.18), it can be seen that there are pairs of n-doped strips present. Each pair is surrounded by a p-doped ring, which also functions as the field-shaping electrode closest to the anode. The two n strips

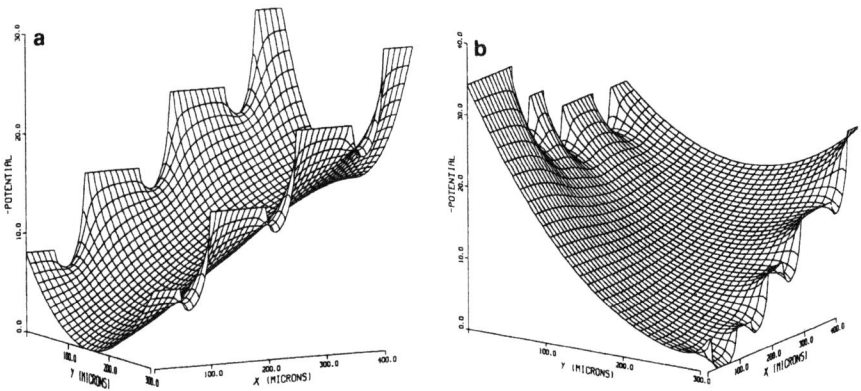

Fig. 6.20a,b. Electron-potential distribution in the linear region (a) and close to the anode region where the potential valley is directed towards the surface (b). (After Strüder et al. 1989, Fig. 3)

are separated from each other by a p-doped strip that also connects to the ring surrounding the anode region. Surrounding the n strips completely by p-doped regions ensures that the electron-accumulation layer adjacent to the n-doped anodes does not electrically connect them to each other (see also Sect. 6.3) or to some other region of the detector (such as the undepleted bulk). Having a pair of anodes will result in the collection of electrons in a single anode incident from one side only. Considering the pair located at the rim of the device, the outermost n strip may be used to drain away electrons from the high voltage protection region, while the other strip is used for the measurement of signals created in the active detector region.

The opposite side of the silicon wafer is for the large part identically structured; differences are only seen in the anode region. Here the n implantation has been omitted and replaced by p-doped strips. In the main part of the detector, the strips on opposite sides of the wafers have been kept at the same potential, thus assuring a symmetrical parabolic potential distribution across the wafer (Fig. 6.20a). Near the anode, however, the increasing potential difference between the two wafer surfaces moves the potential valley for electrons to the front side until it ends at the anode (Fig. 6.20b).

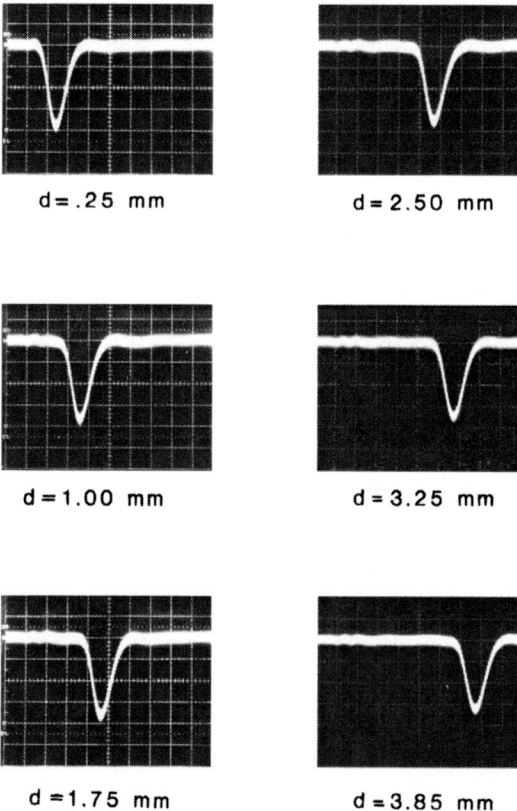

d = .25 mm

d = 2.50 mm

d = 1.00 mm

d = 3.25 mm

d = 1.75 mm

d = 3.85 mm

Fig. 6.21. Output pulse signal (after amplification and shaping) for constant drift field as a function of the position of the light pulser, providing a signal charge of roughly 22 000 electrons. The time scale is 200 ns/div, the drift field 265 V/cm. (After Gatti et al. 1985, Fig. 4)

The Si drift detector has been tested with a light pulser in the laboratory and in a high-energy pion beam (Rehak et al. 1985) using a telescope consisting of silicon strip detectors to determine the position of incidence for each pion traversing the drift detector. In the beam test, the time of incidence was provided by a scintillation counter. Thus the drift detector parameters (drift velocity and the time offset) could be fitted and the measuring precision could be determined.

Figure 6.21 shows the output pulse signal, after amplification and Gaussian shaping, induced by a light pulser positioned at six different locations. In Fig. 6.22 the position of the pulser has been held constant at 3.5 mm from the collecting anode while the drift field is changed in a range from 425 V/cm (a) to 67 V/cm (i). Not only the shifting of the position of the pulse can be observed but also its widening due to diffusion of the charge cloud during the drift time.

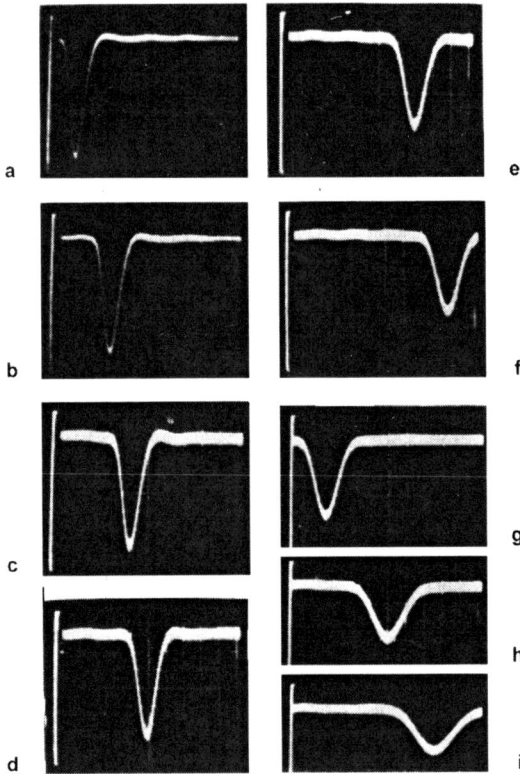

Fig. 6.22a–i. Output pulse signal (after amplification and shaping) as a function of the drift field for constant position of the incident particle. The time scale is 200 ns/div, and the origin has been shifted by 0.4 μs for plots (g) to (i). The values of the drift field are 425 V/cm (a), 265 V/cm (b), 210 V/cm (c), 185 V/cm (d), 130 V/cm (e), 105 V/cm (f and g) 80 V/cm (h) and 67 V/cm (i). (After Gatti et al. 1985, Fig. 5)

The position extracted from the drift time in the beam test measurement is compared in Fig. 6.23 with the prediction of the beam telescope. An almost linear relationship is observed between predicted and measured positions, the small deviations probably being due to position-dependent variations of the drift field caused by inhomogeneous doping of the silicon wafer. The position measurement precision, including its dependence on the drift field, has been investigated with a light pulser in the laboratory and also in a high energy pion beam (Fig. 6.24). The measurement precision depends rather strongly on

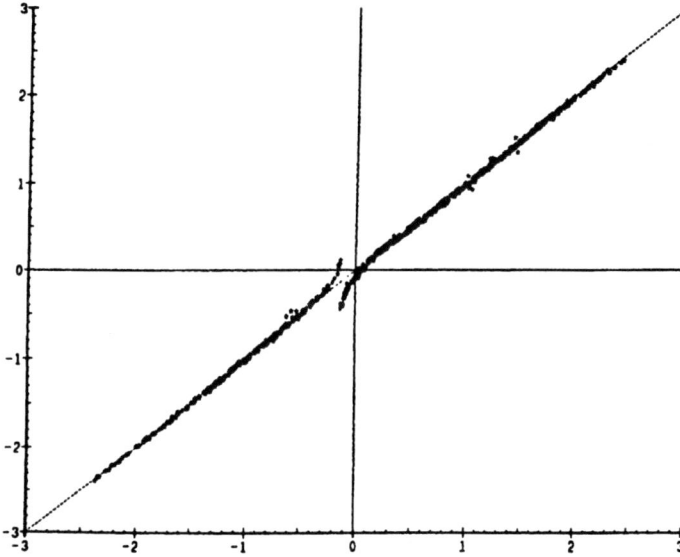

Fig. 6.23. Scatter plot of position as extracted from drift-time measurement, versus position as predicted from a silicon-strip detector beam telescope

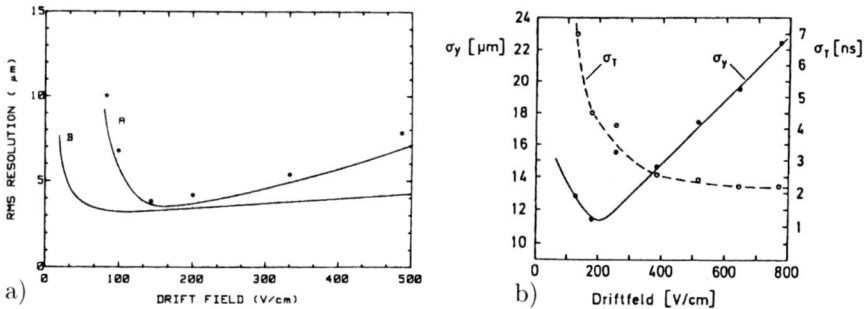

Fig. 6.24a,b. Position-measurement precision as a function of the drift field, from data taken in the laboratory with a light pulser of intensity corresponding to one minimum ionizing particle (a) and from data taken in a high energy pion beam (b). The variable σ_y is the position-measurement precision in the drift direction, and σ_t is the same measurement converted to time resolution

signal processing (see Sect. 6.5.5). Solid line (A) in Fig. 6.24a is the result of a prediction with fixed electronic shaping time, as used in the test. Line (B) in Fig. 6.24a is the result expected for a shaping that is optimized for the respective field strength. Results in the beam for the position measurement precision in the drift direction σ_y were found to be significantly worse because of some system problems (particularly temperature instability, stray light and electronic pickup noise); however, the results were still comparable with those reached with strip detectors.

6.5.2 Matrix Drift Devices

Dividing the anode of a linear drift detector into pads (Fig. 6.25) naturally leads to two-dimensional position measurement. One coordinate is obtained from the drift time, the other by the pads on which the signals appear. If not only the time and pad number are determined but also the analogue information on the signal height, the second coordinate may be further improved by interpolation using the signal levels in neighboring pads. The signal will be distributed over more than one pad if the diffusion during the drift time leads to a charge cloud that is comparable to the spacing of the anode pads.

Example 6.5
Problem: *Calculate the drift time for a 1 cm-long drift device with an applied voltage of 500 V. Estimate the width of the electron charge cloud arriving at the anode region when drifting over the full length of the drift detector (1 cm) and for 1 mm only.*

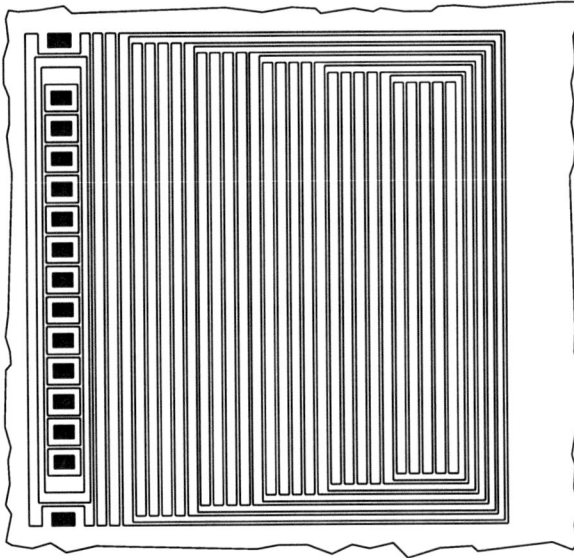

Fig. 6.25. Matrix drift detector with anode strip divided into pads. The dark pad anodes are imbedded in a p-doped grid that provides insulation between neighbors

Solution: *The drift velocity is given by*

$$\nu_n = \mu_n \mathcal{E} = \mu_n \frac{V}{l} = 1500 \frac{\mathrm{cm}^2}{\mathrm{Vs}} \frac{500\,\mathrm{V}}{1\,\mathrm{cm}} = 7.5 \times 10^5\,\mathrm{cm/s}\ .$$

The drift times for $1\,\mathrm{cm}$ *and* $1\,\mathrm{mm}$ *respectively are then* $t_1 = 1.33\,\mu\mathrm{s}$ *and* $t_2 = 0.133\,\mu\mathrm{s}$. *The root-mean-square value of the Gaussian distribution is* $\sigma = \sqrt{D_n t} = \sqrt{\frac{kT}{q}\mu_n t}$, *resulting in* $\sigma_1 = \sqrt{0.0259\,\mathrm{V} \cdot 1500 \frac{\mathrm{cm}^2}{\mathrm{Vs}} \cdot 1.33 \times 10^{-6}\,\mathrm{s}} = 72\,\mu\mathrm{m}$. *Similarly,* σ_2 *is calculated as* $22.7\,\mu\mathrm{m}$.

For very long drift distances and/or low drift fields, the signal charge will be spread out over more than two readout pads. This is an undesirable feature when reading out closely spaced signals. Lateral diffusion can be suppressed by creating deep strip-like p-implanted regions parallel to the nominal drift direction (Castoldi 1996). In this way deviations from the designed drift direction due to nonuniform doping of the silicon are also avoided.

6.5.3 Radial Drift Devices

Radial drift devices are in some sense simpler to design than linear devices because the problem of proper termination of the field-shaping strips does not occur.

Radial devices are especially interesting for energy measurement. A small point-like anode with extremely small capacitance may be placed into the center of the device (Fig. 6.26). Small capacitance results in low electronic noise, as will be shown in Chap. 7.

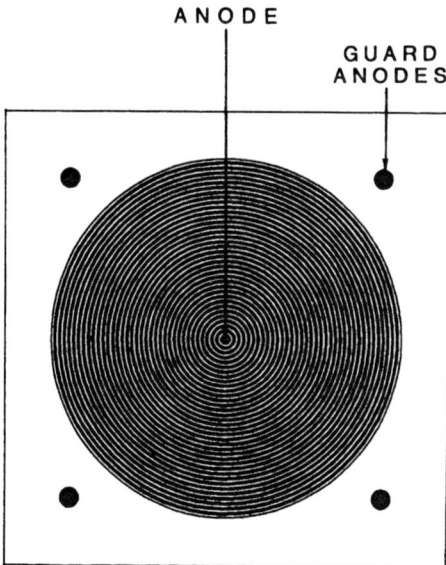

ANODE

GUARD ANODES

Fig. 6.26. Radial drift detector with point anode of $200\,\mu\mathrm{m}$ in the center. (After Gatti et al. 1985, Fig. 14)

Radial drift to the outside is also possible. If the circular anode is divided into pads, again two-dimensional position measurement is possible. An interesting feature of such an arrangement is the high measurement precision at small radius in the direction perpendicular to the radius. The position in this second coordinate is obtained from the charge distribution measured in the anode pads by projecting it back in a radial direction.

A large-area device of this type (Chen et al. 1992), with a hole in the center for the passage of the particle beam, has been produced for the CERES particle physics experiment at CERN. The device also uses a method to drain the current generated at the oxide–silicon interface between the field-shaping rings to an n-doped drain contact, separated from the signal-collecting anode (Rehak et al. 1989). In this manner the anode leakage current is reduced and the measurement precision is increased.

6.5.4 Single-Sided Structured Devices

Standard drift detectors require finely structured processing on both wafer sides. Besides requiring a sophisticated technology including front to backside alignment, this leads to nonuniform absorption of incident radiation caused by the varying dead-layer thickness over the entrance window. This, and the necessity of providing electrical connections on both sides of the wafer, leads to problems in some applications, such as the measurement of lowly penetrating radiation (e.g. low-energy x-rays) or in optical applications.

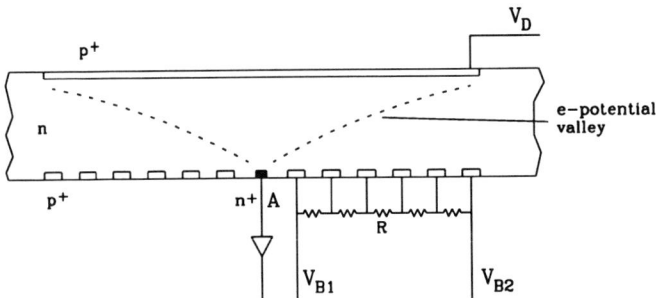

Fig. 6.27. Cross-section of a single-sided structured drift detector. The voltage of the drift rings and the large-area radiation entrance window at the top are provided externally

Considerable simplification, which at the same time brings about an improvement in performance, is obtained by the introduction of single-sided structured drift devices (Kemmer and Lutz 1987). The principle is shown in Fig. 6.27. The top (entrance) side of the wafer is formed by a single large-area diode, while the bottom side is structured into strip- or ring-like diodes and a collecting electrode of the same doping type as the bulk. Putting suitable voltages on the electrodes, one can shape the potential valley for majority carriers (electrons) in such a way that it continuously slopes down from a starting point close to the top large-area diode towards the collecting electrode. In the operation of such a device, the attainable drift field is restricted by the condition that the voltage between diodes placed opposite each other on the two wafer surfaces has to be lower than the wafer's full-depletion voltage.

The maximum drift field is obtained when the voltage between outermost and innermost field-shaping electrodes on the bottom side is slightly below twice the depletion voltage, while the potential of the large-area top diode is roughly halfway in between. Such a condition can be reached by externally supplying voltages to each electrode. However, one may also use an integrated resistive voltage divider for supplying these voltages. In such a case only four connections (and their voltages) have to be supplied to the device: the readout connection to the collecting anode, the large-area bias, and the innermost and outermost field-shaping electrodes.

Further simplification in operation can be achieved by using punch-through biasing both between adjacent field-shaping electrodes and also between back and front sides of the wafer (Fig. 6.28). Punch-through between closely spaced p^+ regions on n-type bulk has already been explained in connection with strip

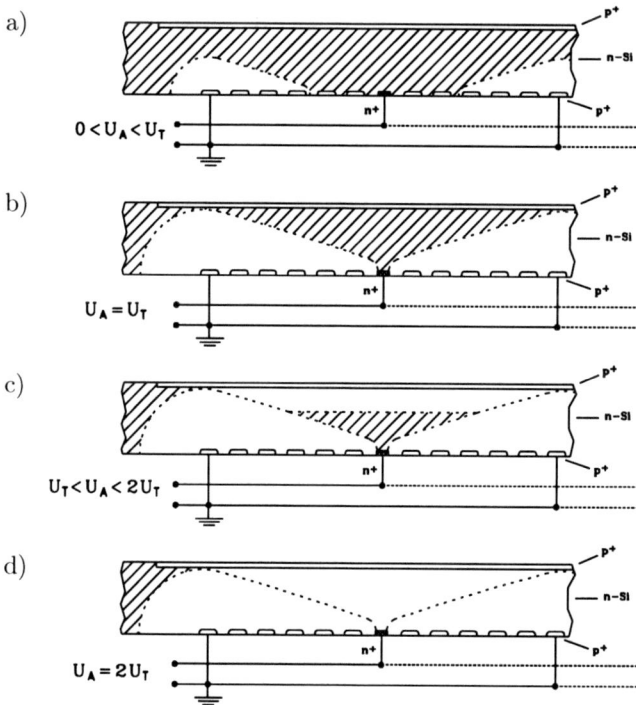

Fig. 6.28a–d. Single-sided structured drift detector with punch-through biassing of the field-shaping electrodes and large-area diode. This setup requires only two electric connections on the backside. The sequence of drawings (a–d) shows the growth of the space-charge region with the application of a reverse bias voltage between the center n^+ collecting anode and the outermost field-shaping electrode. (After Kemmer et al. 1987, Fig. 1)

detector biassing in Sect. 6.4 (Fig. 6.12). As seen in Fig. 6.14, a pair of closely spaced diodes can stand only a limited relative voltage before a rapidly rising current starts to flow. The onset of this current appears when the space-charge regions of the two diodes touch each other. Through keeping the outermost field-shaping ring at ground potential and applying a rising negative voltage to the anode of the device shown in Fig. 6.28, inner rings will follow this voltage only up to the onset of punch-through towards the neighboring ring (Fig. 6.28a). Once the space-charge region reaches the top-side (Fig. 6.28b) the potential of the large-area diode is fixed by the punch-through mechanism across the wafer, and with further increase of the bias voltage the space-charge region grows from the top downwards (Fig. 6.28c) until the device is fully depleted (Fig. 6.28d).

Such a device can therefore be operated without electrical connection on the radiation entrance side on the top. However, care has to be taken to capacitively ground the floating electrodes, so as to get a signal response independent of the position of incidence.

6.5.5 Readout of Drift Devices and Measurement Precision

As has been seen already, drift devices are very well suited for both position and energy measurement. Compared with other position-sensitive detectors, such as strip detectors, they offer the advantage of a smaller number of readout channels while providing approximately the same position resolution (although at limited particle rates).[23] Compared with other energy-sensitive detectors, such as large-area diodes, they have an extremely small detector capacitance, and this leads to low noise and hence to excellent energy resolution.

Position is calculated from the time between primary ionization (the passage of the particle) and collection of the signal charge at the anode, while the energy is found from the total collected charge. As a consequence, optimization of electronics will depend strongly on the type of information required.

As the charge-transfer loss in drift devices can be made very small (Rehak et al. 1986), energy-measurement resolution can be derived in the same way as for planar diodes but taking into account the very small detector capacitance. We therefore will concentrate here on position-measurement precision.

In order to measure the position, the readout electronics will in general convert the charge signal sensed at the readout electrode into a pulse or a bipolar signal.[24] The center of this pulse, or the zero crossing of the bipolar signal, will be taken as a measurement of the arrival time.

The position measurement precision will be limited by effects within the detector and by the electronic noise generated in the readout electronics. Effects of the first kind are

---

[23]One gets problems with ambiguities if more than one signal is generated during the drift time.

[24]Using bipolar signals and zero crossing makes the timing measurement independent of signal height.

- variation of the drift time with the depth of generation of the signal charge;
- widening of the signal peak due to diffusion and electrostatic repulsion, and the corresponding statistical fluctuations; and
- dark current flowing in the detector.

The first of these effects is intrinsic to the detector and its influence on the measurement precision is independent of the properties of the readout system. Reasons for this effect are inhomogeneities in the field distribution due to (for example) the finite width of the field-shaping electrodes and nonuniformities in crystal doping.

The second effect depends on the drift time and therefore on the position of incidence and electric field strength, while the dark current in the detector is to a large extent dependent on the quality of the technological process. In addition, it is a strong function of temperature. It can be reduced markedly when draining the surface-generated current to an electrode separated from the signal electrode (Rehak et al. 1989).

The question of matching a detector with its readout electronics, and in particular of choosing optimal shaping of the signal in order to obtain maximum measurement precision, has been treated in detail in the literature (Gatti et al. 1984) We will restrict ourselves here, therefore, to a few remarks. The width of the peak will be a convolution of the signal width due to diffusion and the shaping of the electronics. Superimposed on the signal is the electronic noise created by the dark current of the detector and by its electronics. Longer shaping times will result in a decrease of the electronic noise and at the same time to a widening of the signal, whose center position will be shifted by the noise from the detector dark current and the electronics. As the diffusion width is dependent on the drift field and the position of incidence of the radiation, each operating condition of the device would in principle require its own signal shaping condition in order to obtain optimum measurement precision. Furthermore, this optimum shaping is dependent on the point of incidence, the amount of ionization charge, and on the angle of incidence of ionizing particle tracks.

6.6 Charge Coupled Devices as Detectors

Charge coupled devices (CCDs) have for a long time been used as electronic elements for the storing and transfer of charge and – more important – as optical sensors, most noticeably as image devices in video cameras. Some years ago they also found their application as particle detectors in elementary particle physics (Bailey et al. 1983; Damerell et al. 1987 and 1990), where specially selected optical CCDs were used. Meanwhile a large detector system has been constructed. It measures tracks in electron–positron collisions (Abe et al. 1997).

More recently, $p-n$ CCDs for the special purpose of particle and x-ray detection have been developed (Strüder et al. 1990, 1993 and 1997). They are based on the principle of sidewards depletion of a double-diode structure, which is also used in the semiconductor drift chamber. Their first use is in two space-based x-ray telescopes: XMM (Lumb et al. 1997) and ABRIXAS (Richter et al. 1996).

CCDs are nonequilibrium detectors. Signal charge is stored in potential pockets within a space-charge region, the content of which are then transferred to a collecting readout electrode. In order to retain the thermal nonequilibrium condition, thermally generated charge that also assembles in the potential pockets has to be removed from time to time. Usually this is done automatically during the readout cycle of the device.

While in conventional MOS CCDs minority carriers are collected, the p–n CCDs are majority carrier devices. The conventional MOS CCDs described in Sect. 6.6.1 for didactic purposes store and transfer the charge directly at the semiconductor–insulator interface, but these devices are in practice not used anymore. They have been replaced by buried-channel CCDs, in which the store-and-transfer region is moved a small distance away from the surface. In p–n CCDs, this region is moved a considerable distance into the bulk.

6.6.1 Three-Phase "Conventional" MOS CCDs

This type of CCD serves best to demonstrate the working principles involved in CCDs. A cross-section along the direction of movement of charge is shown in Fig. 6.29.

Similarly to the strip detector, this structure is derived from the simple MOS structure (Fig. 3.5) by dividing the metal electrode into strips. The bulk is connected by a backside ohmic contact. A diode strip (N^+ on P^- substrate) is added so as to be used for the collection of the minority carriers (Fig. 6.29a).

Every third metal electrode is kept at the same potential and the bulk is put at a negative voltage ($-V_{bias}$) so that the device is operated in overdepleted mode (see Sect. 3.3). Putting the electrodes on different potentials creates local energy minima for electrons at the $Si-SiO_2$ interface, just below the electrodes with the highest applied voltage (ϕ_1 in Fig. 6.29). Electrons produced in the space-charge region, for instance by light or ionizing radiation, will move towards the oxide and assemble below these electrodes. The partial filling up of electron potential minima will lead to local shrinking of the depletion-layer thickness and to a reduction of the depth of the corresponding electron potential minima (Fig. 6.29b). The charge can now be moved towards the readout electrode by a periodic change of the voltages ϕ_1, ϕ_2 and ϕ_3, as shown in the figure. First ϕ_2 is increased to the same level as ϕ_1 and the signal charge will spread between ϕ_1 and ϕ_2. If now ϕ_1 is lowered, the signal charge will transfer below the electrodes ϕ_2. If this procedure is followed for ϕ_2 and ϕ_3 and then again for ϕ_3 and ϕ_1, the signal charge is transferred by a complete cell size. After several cycles the charge will finally arrive at the diode, where it can be measured.

The device shown in Fig. 6.29 serves only to demonstrate the charge collection and transfer mechanism. However, as a practical device it will not work. One of the reasons is the presence of gaps between the metal electrodes with bare oxide, where the potential at the outer surface is not defined. This may

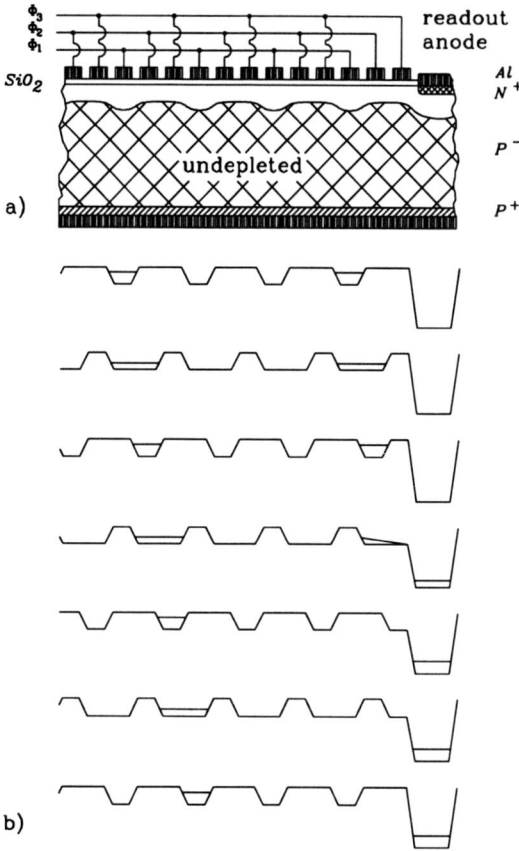

Fig. 6.29a,b. Working principles of a three-phase MOS CCD: schematics (a); charge-transfer (b). Every third gate electrode is connected to the same potential (ϕ_1, ϕ_2, ϕ_3) so that a periodic (electron) potential appears below the gates at the Si–SiO$_2$ boundary. Electrons are collected in the minima of these potential distribution. They can be shifted towards the readout anode by suitable change of the potentials, as shown in part (b) of the figure

lead to potential barriers that hinder the transfer of charge. Therefore practical devices have overlapping gates with an insulator in between. The gates are usually made of polysilicon rather than metal.

Secondly, the signal charge has to be prevented from spreading in the direction perpendicular to the plane of the drawing, along the gate electrodes. This can be accomplished by a variation of the doping at the surface. Most frequently this is done by implanting p strips, so-called "channel stops", perpendicular to the gate electrodes.

In this description we have considered the CCD as a detector, storing signal charges that are created in the device itself. However, it may also be used as a purely electronic device for the storage and transfer of charge. Then the signal charge has to be injected into the device with the use of special structures.

6.6.2 Linear and Matrix CCDs

So far, a simple linear structure has been described. Putting many of these linear structures next to each other and using common gates, one arrives at a matrix CCD. Usually it is combined with a linear CCD perpendicular to it. The signal charge is then transferred line by line into the linear CCD, from where it can be shifted cell by cell into a single output electrode (Fig. 6.30).

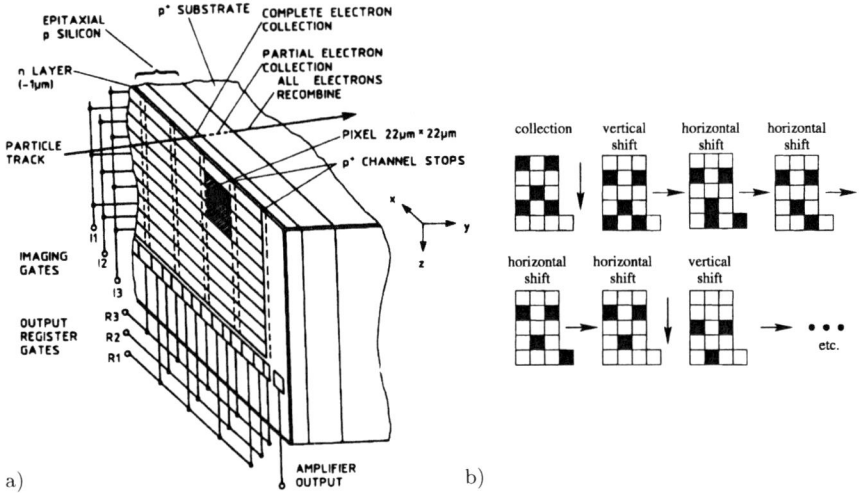

Fig. 6.30a,b. Matrix CCD on the basis of a three-phase MOS CCD: schematics (a); charge-transfer sequence (b). Charge is shifted in the vertical direction with all pixels of the matrix in parallel, the lowest row being transfered into a horizontal linear CCD. This horizontal CCD is then read out through a single output node

6.6.3 Charge Collection and Charge Transport

As mentioned earlier, signal charge is collected not only from the space-charge region but (due to diffusion) also from the undepleted bulk of the detector. For optical CCDs the space-charge region typically has a depth of 1–$3\,\mu m$. Even with an increase of this value to around $10\,\mu m$ still yields a rather small value for penetrating radiation such as x-rays or high-momentum charged particles. The depth of charge collection in the undepleted bulk is given by the diffusion length $L = \sqrt{D\tau_r}$ (see (3.1.13)). As the collection time for these charge carriers is much longer than for those produced in the space-charge region, they will also have a much wider spatial distribution and may spread over several pixels.

Example 6.6
Problem: *Estimate the amount of charge collected for a minimum ionizing particle in an n-channel CCD with $10\,\mu m$ depletion depth if the recombination*

lifetime is $\tau_n = 0.1\,\text{ms}$. Estimate the width of the charge distribution for charge carriers produced in the undepleted bulk. Assume uniform bulk properties over the wafer depth of $d = 300\,\mu\text{m}$.

Solution: All charge carriers produced in the $10\,\mu\text{m}$-thick depletion layer will be collected. With an average energy loss of $3.8\,\text{MeV}/\text{cm}$ and an average energy of $3.63\,\text{eV}$ needed for creation of an electron–hole pair, 1046 electrons will be created in the space-charge region. Their spatial distribution will be narrow as they are collected rapidly by the drift field.

Added to this total will be charge carriers diffusing out of the undepleted region. Charge collection out of the undepleted region has already been considered in Example 3.3 in Sect. 3.13 when considering a p–n diode under irradiation with light, and in Example 5.1 in Sect. 5.1 when considering the effect of a continuous photon flux incident perpendicular to the surface of a unbiased diode.

In the first example the neutral semiconductor region was assumed to be large compared with the diffusion length $L = \sqrt{D\tau}$, while in the second example this assumption was not made. With $D_n = \frac{kT}{q}\mu_n = 0.0259\,\text{V} \cdot 1450\,\frac{\text{cm}^2}{\text{Vs}} = 37.6\,\frac{\text{cm}^2}{\text{s}}$ and $\tau_n = 0.1\,\text{ms}$, we obtain $L_n = \sqrt{37.6\,\frac{\text{cm}^2}{\text{s}} \cdot 10^{-4}\,\text{s}} = 613\,\mu\text{m}$, larger than the wafer thickness. We therefore have to resort to the approach used in Example 5.1 and will need to reuse and rewrite some of the results obtained there.

Equation (5.1.8) defines the efficiency of charge collection as a function of distance in the undepleted bulk (maximum distance x_m) by taking the ratio of current J_n and charge production rate G_L at depth x_0:

$$\eta(x_0) = -\frac{J_n}{qG_\text{L}} = \frac{\sinh\frac{x_\text{m}-x_0}{L_n}}{\sinh\frac{x_\text{m}}{L_n}} \ .$$

Averaging over the depth, we obtain the fraction of charge collected for a m.i.p. as

$$\bar\eta = \frac{\int_0^{x_\text{m}} \sinh\frac{x_\text{m}-x}{L_n}\,dx}{\sinh\frac{x_\text{m}}{L_n}} = L_n\frac{\cosh\frac{x_\text{m}}{L_n} - 1}{\sinh\frac{x_\text{m}}{L_n}} \ .$$

With $x_\text{m} = 290\,\mu\text{m}$, $L_n = 613\,\mu\text{m}$, we obtain $\bar\eta = 0.232$. And we may define as effective charge collection length from the undepleted region $d_\text{eff} \equiv \bar\eta x_\text{m} = 67.4\,\mu\text{m}$.

Charge produced in the depth of the neutral region will not only be partially lost, it will also take some time until it diffuses into the space-charge region so as to be collected in the pixels. During this time it will also diffuse in a lateral direction, thereby having a chance to reach the neighboring pixels. Calculating the time distribution of the electron cloud reaching the space-charge region, and thus also the lateral width of the charge cloud, is possible but tedious. For getting only a rough estimation, we will make the very rough assumption that all charges originate from the effective charge collection depth $d_\text{eff} = \bar\eta x_\text{m}$ and that they all reach the space-charge region when the diffusion width $\sigma \equiv \sqrt{2D_n t}$

equals the effective charge collection depth d_{eff}. As diffusions in longitudinal and lateral directions are independent, this assumption gives a root-mean-square width for the lateral diffusion equalling the effective charge collection depth $d_{\mathrm{eff}} = 67\,\mu m$. The average time for diffusion into the space-charge region is then estimated as $\bar{t} = \frac{d_{\mathrm{eff}}^2}{2D_n} = 0.6\,\mu s$.

Transfer efficiency is an extremely important parameter of a CCD, since charge has to transit through many cells before arriving at the readout node. Three mechanisms govern the transfer of charge: electrostatic repulsion, drift, and diffusion. Depending on the geometry, operating conditions and size of signal charge, the relative importance of these three effects will be different. Electrostatic repulsion plays a role only for large signal charges and small pixel size in the transfer from an energetically higher region to a lower one. Diffusion will be a relatively slow process, dominating electrostatic repulsion when most of the charge has already been transferred. The most desirable mechanism is drift; however, with the thin oxide (that is, thin compared with the width of the gate electrodes) the electric field component parallel to the detector surface at the Si–SiO$_2$ interface remains very small.

Trapping is another very important effect for transfer efficiency. It is due to crystal defects leading to local energy states (e.g. within the band gap) that act as trapping centers. If an electron (hole) passes close to this center, it has a chance of being trapped for a while before it is restored to the conduction (valence) band by thermal emission. Thus in our device it may be prevented from being moved towards the next cell as quickly as it might.

As crystal defects are especially abundant in the surface region, it is desirable to move the charges a certain distance away from the Si–SiO$_2$ interface into the bulk. At the same time, an increase of speed of charge transfer also

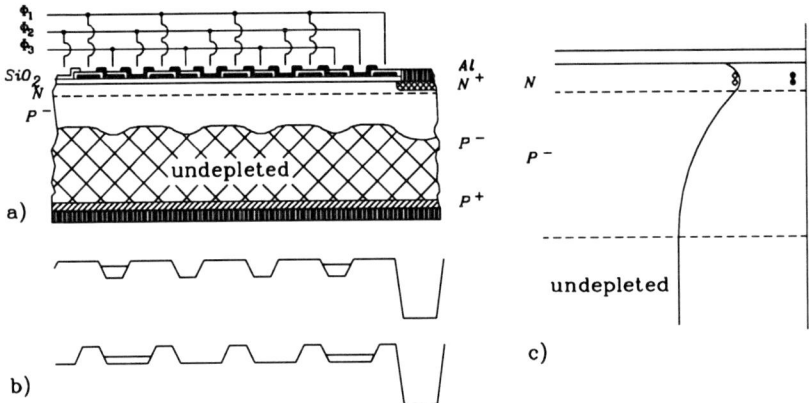

Fig. 6.31a–c. Principle of a buried-channel MOS CCD: cross section (a); potential along the shift direction (b); and potential perpendicular to the surface (c)

becomes possible. This is due to an increase in the electric field components parallel to the surface in the regions below the centers of the gates. One thus does not have to rely on the slow diffusion process for moving the charge from below the gates towards the drift region but instead by means of a drift field extending over the full transfer region.

This movement at a small distance from the surface is accomplished in a buried-channel CCD as shown in Fig. 6.31. A n doping in the surface region shifts the energetic minimum for electrons away from the surface into the bulk. The n doping has to be chosen in such a way that the space-charge region still reaches all the way to the interface, so that the effect of the gate electrodes is not shielded by a conducting n layer. Electrons generated in this n-doped surface layer are collected at the same position as those originating in the bulk. Holes generated in this layer flow away perpendiculary to the drawing plane towards p-doped channel stops.

6.6.4 Signal Readout

CCDs have the remarkable property that – in principle – completely isolated charge buckets are transferred to the output electrode. Whatever the noise on the transfer gates, it will not change the amount of charge that finally arrives at the readout electrode. The size of this readout electrode can be made very small, so that the detector capacity as seen by the readout electronics is very low (typically 10–100 fF). In order to profit from this small capacitance in the noise performance of the device (compare Chap. 7), it is necessary to integrate at least part of the electronics onto the detector so that the stray capacitances of the leads (typically several pF) can be avoided.

In the simplest – and usual – case, the electronics consists only of a single input transistor used as a source follower, together with an external current source and a reset switch. The latter is also in the form of a single transistor and has very low drain-gate capacitance (Fig. 6.32).

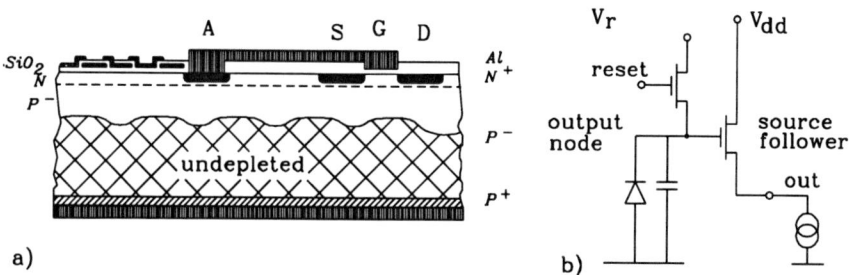

Fig. 6.32a,b. CCD readout circuit consisting of a source follower and a reset switch: cross-section with collecting anode A and source follower transistor (a); circuit diagram with the detector readout electrode symbolically represented by a reversely biased diode in parallel to a capacitor (b)

6.6.5 Other Types of MOS CCDs

There exist a large variety of MOS CCDs that have been developed either for increased performance (e.g. speed of operation, improved charge-transfer efficiency, increased sensitive depth) or simplified operation and production technology.

One example, the buried-channel CCD – designed for improved transfer efficiency and speed – has already been mentioned in Sect. 6.6.3. An increase in the detector's sensitive volume (for x-ray detection, for example) may be obtained by using weakly doped material. A reduction in the number of phases necessary for the charge-transfer can be achieved with a modified gate structure.

In the buried-channel device shown in Fig. 6.33, this reduction has been accomplished by a variation of the surface doping below the gate electrodes in such a way that the electron energy level below the left side of the gate is always higher than at the right. Thus electrons will always move to the right of the gate and movement of the two gate potentials with respect to each other will result in a transfer of the signal charge towards the readout electrode. Contrary to the three-phase CCD, where the direction of the charge movement could be controlled by the phases of the gate potential, this device has a built-in transfer direction.

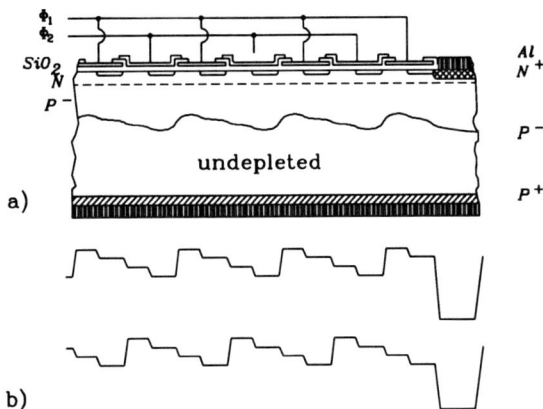

Fig. 6.33a,b. Two-phase CCD with built-in transfer direction: cross-section (a); potential along the transfer direction (b)

6.6.6 Fully Depleted p–n CCDs

A fully depleted p–n CCD works on a different principle from the previously described MOS-based devices, resulting in some rather interesting properties. While in MOS-based devices minority carriers are collected, these devices use the majority carriers, which are produced in the completely depleted and therefore also radiation-sensitive bulk material. The working principles can be explained from the drift chamber. The possibility of using the drift-detector principle for the CCD has been mentioned already in the first publication by its inventors (Gatti et al. 1984).

Working Principle

Looking at the sidewards depletion structure (see Fig. 6.16) we notice that the electron potential valley can be moved towards the top surface by biassing the lower diode negatively with respect to the top diode. Through the application of a periodic potential to the strips of the drift detector rather than a continually rising one, the valley will be structured in depth. The lower diode may be formed as a single large-area structure, and from that one obtains the device shown diagramatically in Fig. 6.34.

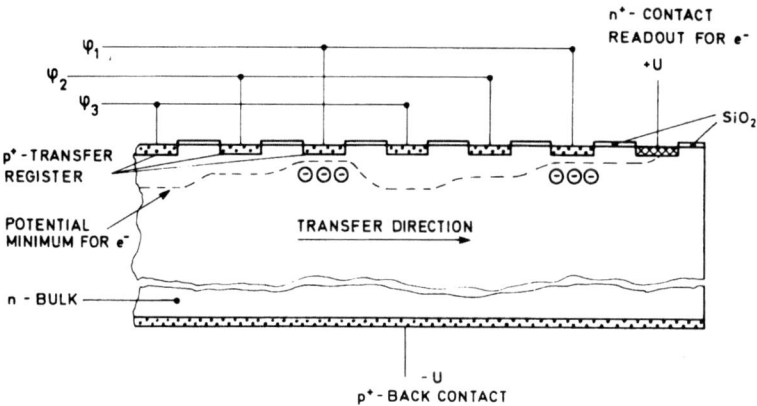

Fig. 6.34. Fully depleted p–n CCD: cross-section along channel. (After Strüder 1989, Fig. 6)

This device however still has some problems. In order to obtain sufficient modulation of the valley depth, the valley has to be at a distance of the order of the gate width away from the gates. This reduces (with reasonable bulk doping) the threshold for holes between front and backside of the device to such a low level that a large hole-current will flow from the top gates towards the backside diode. Furthermore, spread of the signal charges along the gates is not prevented.

Both problems can be solved simultaneously with an increase of doping in the surface region if this is done in a strip-like fashion, with strips perpendicular to the gate direction. These strips may be called "channel guides", the narrow gaps between the strips have the function of channel stops (Fig. 6.35).

Device Properties

Compared with standard CCDs, the fully depleted p–n CCDs have several important advantages when used as detectors:

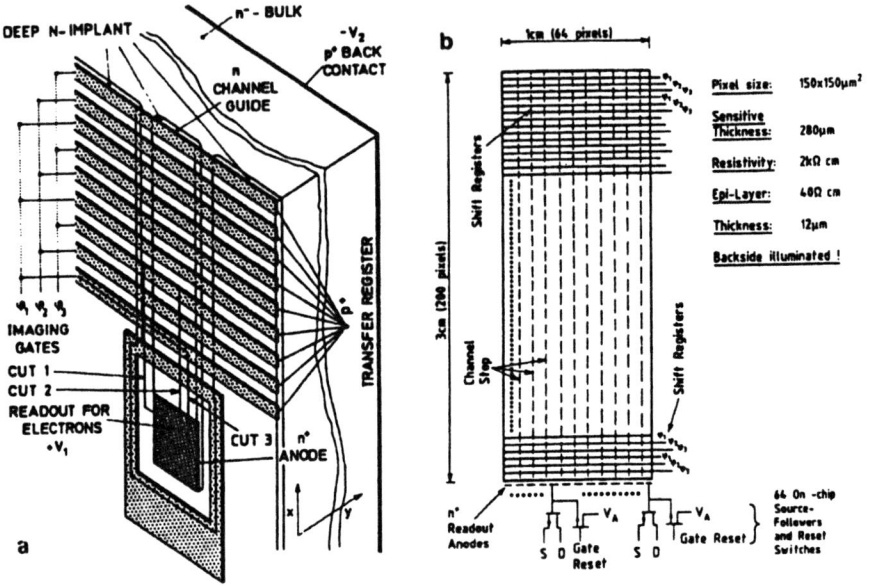

Fig. 6.35a,b. Perspective view of one column in a fully depleted p–n matrix CCD (a) and the parallel readout structure with one readout node for every column (b)

- *enhanced sensitivity* as the full volume is depleted. This is especially important for the measurement of x-rays at higher energies and for high energy charged particles. The efficiency for x-rays can be estimated from the energy dependence of the absorption coefficient, which was given in Sect. 4.4.1. High-energy charged particles give typically an order of magnitude higher signal than that registered for them in an optical MOS CCD;
- *uniform response* with backside illumination as the backside consists of a large-area diode. The window may be made very thin in order to get a good response for short-range radiation;
- *high speed of operation* due to charge transfer at a moderate distance from the surface;
- *the possibility of building large cell structures*. For MOS CCDs this is not possible since, with the transfer very close to the surface, the drift field would become too small to allow fast charge transfer; and
- *Improved radiation hardness*, because radiation-sensitive MOS structures do not play an essential role in the function of the device (compare Chap. 11 on radiation hardness).

Example 6.7
Problem: *Estimate the detection efficiency for 10 keV x-rays in a 300 μm thick fully depleted p–n CCD and compare it with an optical CCD of a factor 10 shorter in its charge-collection depth.*

Solution: *From Fig. 4.3 the inverse of the absorption length in silicon is read as* $a = 75\,\text{cm}^{-1}$. *For the fully depleted p–n CCD, the charge-collection depth equals the wafer thickness. The fraction of x-rays absorbed in the detector is*

$$f = 1 - e^{-a\,d_{\text{coll}}} = 1 - \exp(-75\,\text{cm}^{-1} \cdot 300 \times 10^{-4}\,\text{cm}) = 1 - e^{-2.25} = 0.9\ .$$

For a factor 10 shorter charge-collection depth, the probability of interaction in the sensitive volume drops to $f = 0.2$.

Experimental Applications

A $6 \times 6\,\text{cm}^2$ device has been developed for the use in two space-based x-ray astronomy experiments (Strüder et al. 1993 and 1997; Soltau et al. 1996). They are put in the focal plane of Wolter telescopes, making use of total x-ray reflection at grazing inclination angles. The pixel size of $150 \times 150\,\mu\text{m}^2$ is matched to the resolving power of the telescope. Besides other x-ray applications (such as medical imaging), and synchrotron radiation experiments, the device may also be useful for optical applications as well as for the measurement of charged particles.

Readout

As is the case for other types of CCDs, the first stage of amplification has to be integrated into the detector to take advantage of the low detector capacitance for low noise readout. For the p–n CCD again, a source follower has been built using a new type of junction field-effect transistor (SSJFET) operating on a fully depleted substrate. This type of electronics, which uses the same technological steps as already needed for detector production, will be described more fully in Chap. 9.

The readout topology (Fig. 6.36) used in a specific application of x-ray astronomy (Soltau et al. 1996) has been designed for high-speed readout. Each row of 200 pixels has its own readout node, so that parallel readout is possible that requires only 2 ms for the complete device.

Performance

For application in x-ray detection, the following properties are of fundamental importance:

- energy range of sensitivity;
- quantum efficiency;
- dynamic range: signal over background due to electronic noise and low-energy signal tails from partial charge collection;
- energy resolution;
- charge-transfer efficiency; and
- readout speed.

The range of sensitivity for x-ray energies at the upper limit is determined by the semiconductor properties and thickness, as high-energy photons will

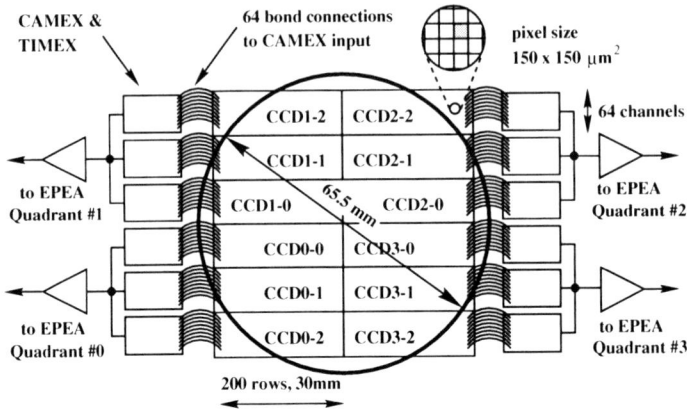

Fig. 6.36. Readout topology of the XMM fully depleted p–n CCD. The device consists of 12 functionally independent subunits of 3×1 cm², each containing 64 colums of 200 pixels and all produced on one single silicon wafer. A separate source follower amplifier for each column is integrated into the device. Readout is performed with specially developed integrated circuits (CAMEX) (see Sect 7.6) with 64 channels. (After Strüder et al. 1997, Fig. 3)

have little probability of interacting within the semiconductor. Compared with standard partially depleted CCDs, higher energies can be detected. The lower energy limit is determined by two requirements: the separation of the signal from the electronic noise and the charge-collection properties at the semiconductor radiation entrance window. The latter is important because of the low penetration depth of low-energy photons. In simple form, this charge loss near the surface may be described by a "dead layer" with a certain thickness. Such a description is, however, insufficient: it predicts either a full signal once the interaction occurs past the dead layer or no signal at all when it occurs within the dead layer, while in reality signals with lower amplitude are detected. This partial collection of charge leads to low-energy tails in the energy spectra, resulting in a background beneath the lower energy peaks.

A lower limit of the energy resolution is given by the statistics of the ionization process, which was discused in Chap. 4 and described by the introduction of the Fano factor (see equation (4.4.2)). Added to this are statistical and systematic fluctuations in charge collection and transfer, as well as electronic noise in charge amplification.

A particular problem in the CCDs described so far is their continuous sensitivity, which includes the time of transfer to the readout. Signals generated during the readout will thereby be assigned the wrong position. It is therefore desirable to have the exposure time – the time in which the signals are collected in a quiet CCD – relatively long with respect to the charge-transfer time. In the only p–n CCD which has been developed so far, this has been accomplished by providing a separate readout node to each column (Fig. 6.36), thus allowing the readout of all pixels in a row in parallel. This, together with the large pixel

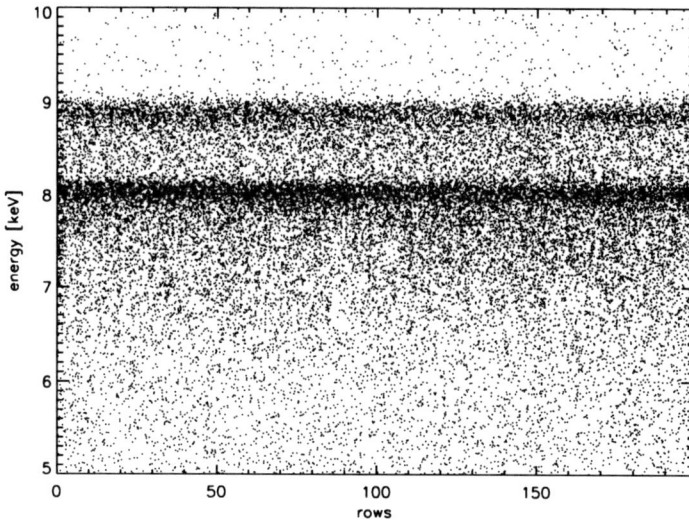

Fig. 6.37. Scatter plot of signal height versus pixel row of the XMM fully depleted p–n CCD, irradiated with intensive radiation from an x-ray tube using a copper target. The prominent bands are from the $CuK_\alpha = 8.05\,keV$ and $CuK_\beta = 8.90\,keV$ lines. (After Soltau et al. 1996, Fig. 6)

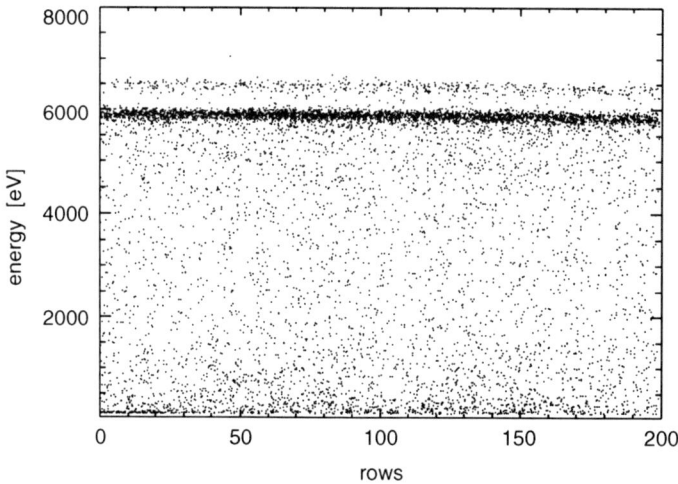

Fig. 6.38. Scatter plot of signal height versus pixel row of the XMM fully depleted p–n CCD irradiated with a ^{55}Fe source. Two well separated bands corresponding to the MnK_α (5.898 keV) and MnK_β (6.490 keV) are seen. The small slope of the bands from closest (first) to furthest (200th) readout row (3 cm transfer) is a measure of the transfer efficiency (above 99.9%). (After Strüder et al. 1997, Fig. 4)

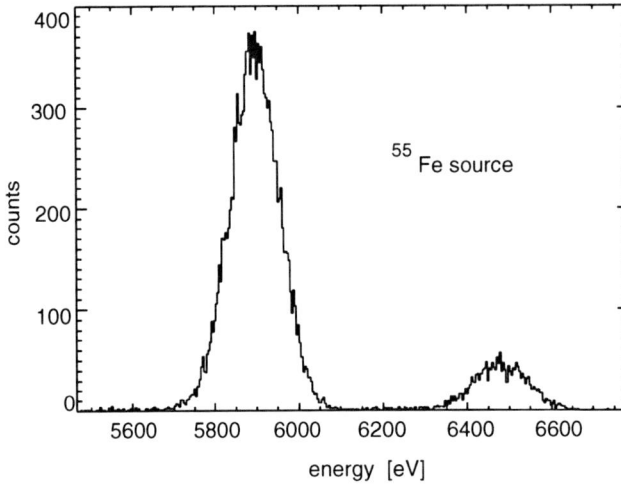

Fig. 6.39. Energy spectrum for a ^{55}Fe x-ray source measured with the XMM fully depleted p–n CCD. The energy resolution is 130 eV (FWHM). (After Strüder et al. 1997, Fig. 7)

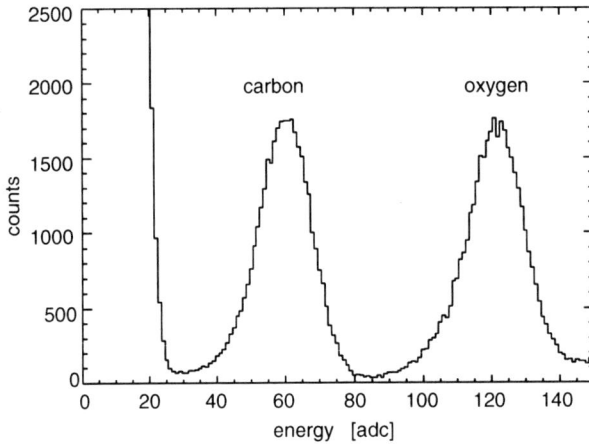

Fig. 6.40. Carbon ($E = 282$ eV) and oxygen ($E = 523$ eV) spectrum in a single-photon counting mode of the XMM fully depleted p–n CCD. The FWHM is about 75 eV for both lines and the quantum efficiency is above 85%. (After Strüder et al. 1997, Fig. 8)

size of $150 \times 150\,\mu\mathrm{m}^2$ matched to the resolution of the x-ray telescope, leads to a 2 ms readout time for the complete device.

In the following, some results of a $3 \times 1\,\mathrm{cm}^2$ section of a $36\,\mathrm{cm}^2$ device (Strüder et al. 1997) will be shown. Typical readout times are 2 ms and exposure times 20 ms. The scatter plot of signal height versus pixel row of the XMM fully depleted p–n CCD (Fig. 6.37) has been obtained with high-intensity irradiation from an x-ray tube with a copper target. A similar plot, with irradiation of the device with a ^{55}Fe source, is shown in Fig. 6.38. Two well separated bands, corresponding to the MnK$_\alpha$(5.9 keV) and MnK$_\beta$(6.5 keV), are seen in Fig. 6.38. The small slope of the bands from closest (first) to furthest (200th) readout row is a measure for the transfer efficiency (measured as above 99.9%), although this is dependent on the temperature ($\approx$ 140 K) and other operating conditions.

The energy resolution and the signal-over-background ratio can be judged from Fig. 6.39, which shows the energy spectrum obtained with an ^{55}Fe x-ray source. That the device is performing well for low-energy x-rays is demonstrated in Fig. 6.40. The carbon ($E = 277$ eV) and oxygen ($E = 525$ eV) peaks are well separated from each other and from the tails of the electronic noise peak. The good properties at low energies are due to the thin (and homogeneous) entrance window (Hartmann et al. 1997).

6.7 Summary

Using the basic understanding of the working principles of simple energy-sensitive detectors as presented in Chap. 5, the principles used for measurement of position in addition to energy have been explained.

The most straightforward extension is that of diode strip detectors, in which a large-area diode is split into strips and each of the strips is read out by its own electronics. In the most simple scheme, the position of the strip producing a signal provides the position information. In many circumstances the signal charge is distributed over more than one strip. This is due to diffusion during charge collection and to spatial extension of the primary charge for (for example) inclined ionizing tracks in a strip detector. Measurement of signal height in such a case allows interpolation of the position by comparison of the signal height, and it thus provides improvement in measurement precision.

Analog readout can also be used for the purpose of reduction of the number of readout channels, when leaving some strips floating between strips connected to the readout electronics. In these circumstances the charge collected on a floating strip is capacitively fed into the neighboring readout channels by the naturally present strip–strip capacitances, and the position can again be obtained by interpolation. It is important in this case that the a.c. floating strips are held on the same d.c. voltage as the readout strips.

Further developments of strip detectors are detectors with double-sided readout (Sect. 6.3), the integration of a.c. coupling into the detector structure, and the development of biassing structures that can be implemented with fairly simple technology (Sect. 6.4). Simplification of detector concepts and technology is important in view of proposed large-scale applications of strip detectors.

Completely new possibilities were brought about by the invention of the semiconductor drift detector (Sect. 6.5). In these devices, charge is collected in a potential valley inside the device and moved essentially parallel to the surface towards one or several collecting electrodes. This movement is accomplished by a drift field component parallel to the wafer surface. From the drift time to the collecting electrode the position can be determined. Drift detectors are of interest not only because of their position-measurement capability but perhaps even more because of the improvement in energy resolution compared with reverse-biased diodes of the same area. This is due to the radically reduced capacitive load that is seen by the output amplifier. Various drift devices can be built, providing linear or radial drift for example and – depending on the arrangement of charge collecting electrodes – one- or two-dimensional position measurement. For simple energy measurement, radial drift towards a single central collecting electrode is well suited. When a homogeneous entrance window is important, the single-sided drift diode described in Sect. 6.5.4 provides a solution. Further improvement of performance is provided by the integration of electronics into the detector, as will be described in Chaps. 8 and 9.

The working principles of MOS CCDs, which are standard devices in optical applications and have found their way also into nuclear radiation detection, are described in Sect. 6.6. Based on the drift-detector principle, fully depleted p–n CCDs have been developed. Their operating principle is very different from that of standard MOS CCDs, and for a variety of reasons they are better suited for nuclear radiation detection. In particular, they are fully depleted and therefore sensitive to radiation over the whole semiconductor volume.

Additional position-sensitive detectors will be described in Chaps. 8 and 9.

7 The Electronics of the Readout Function

As there is a close interplay between a detector and its electronics, both components have to be considered together when designing a detector for a specific application. In this chapter we will restrict ourselves to the analog front-end electronics, the part in which one tries to approach as close as possible the limits given by the laws of physics. Therefore, the problem of measuring charge will be addressed in some detail, after reviewing the working principles and properties of transistors. Special emphasis will be placed on noise minimization and on integrated electronics developed for the readout of semiconductor detectors.

7.1 Operating Principles of Transistors

Transistors are commonly classified into unipolar and bipolar, depending on whether only one or both types of charge carriers participate in the current flow. As a consequence of the difference in operating principles, their properties – and therefore their suitability for specific applications – differ greatly. Bipolar transistors are well suited for high-speed applications and for driving large currents. Unipolar transistors are common in moderate-speed low-noise applications (JFETs) and are most prominent in digital circuitry (MOSFETs).

7.1.1 Bipolar Transistors

The bipolar transistor consists essentially of two back-to-back p–n junctions with a thin common base (Fig. 7.1). As an example, a p–n–p transistor will be considered – the n–p–n transistor properties can be obtained by suitable changes of signs and carrier type. Under normal operating conditions the emitter–base junction is forward-biased, while the base–collector junction is reverse-biased. For this device to have good amplification properties, the following requirements have to be met:

- the doping concentration of the emitter has to be large with respect to the base;
- the base must be very thin; and
- the doping concentration of the collector should be small.

The significance of these conditions will become clear when the very basic model of the device is reviewed. Assuming, for simplicity, that we have an

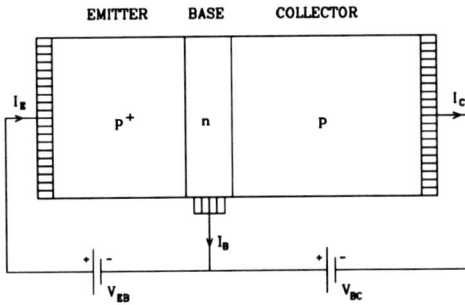

Fig. 7.1. Schematics of a bipolar transistor

abrupt change between the three homogeneously doped regions, we first look at the situation when no external voltages are applied to the device (Fig. 7.2). The same approximations – neglect of the very thin transition regions – that were made beforehand in the case of the single junction (see Sect. 3.1) lead to the charge density, electric field and energy band structure shown in the same figure.

With the doping concentrations $N_{A,E} \leq N_{D,B} \leq N_{A,C}$ for emitter, base and collector, respectively, the distance of the Fermi level to the closest band, either E_v or E_c is smallest in the emitter region and farthest in the collector region, while the distance to E_c in the base region lies in between. Without application

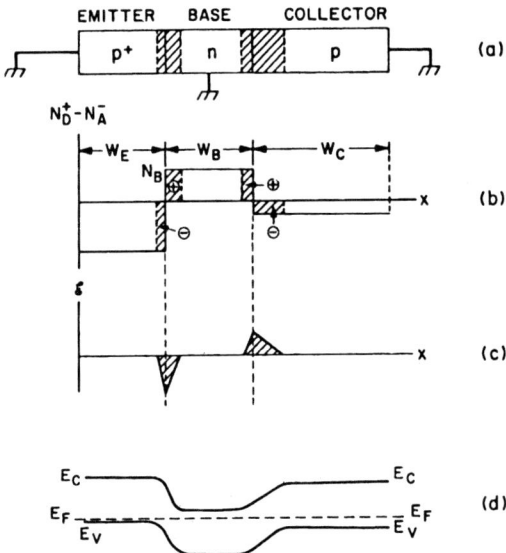

Fig. 7.2a–d. A p–n–p bipolar transistor in thermal equilibrium in the abrupt-change approximation: device structure and space-charge regions (**a**); net doping concentration $(N_D - N_A)$ and charge density (hatched) (**b**); electric field (**c**); and band structure (**d**) (After Sze 1985, p. 112 Fig. 3)

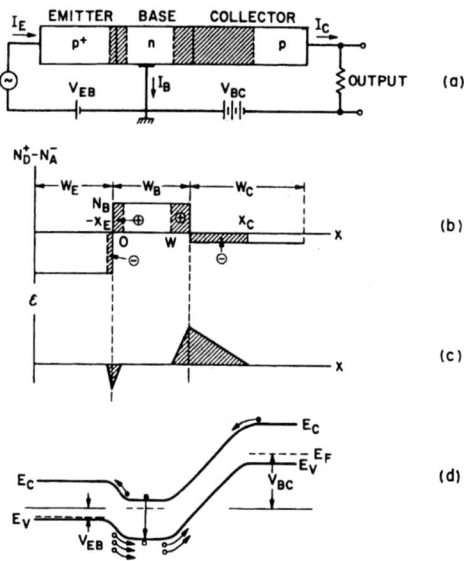

Fig. 7.3a–d. A p–n–p bipolar transistor in the active mode of operation: device structure and space-charge region (a); net-doping concentration and charge density (b); electric field (c); and energy band diagram (d). The direction and signs of currents and voltages chosen correspond to all positive values in the active mode of operation (After Sze 1985, p. 113 Fig. 4)

of an external voltage, the diffusion current at any place in the device will be balanced by the drift current individually for each type of carrier, and no net current will be flowing anywhere.

The situation in the normal or "active" mode of operation is shown in Fig. 7.3. Applying a forward bias to the emitter–base junction and a reverse bias to the collector–base junction results in a shrinking of the emitter and a widening of the collector space-charge regions. In the emitter–base junction the diffusion current will dominate the drift current both for holes moving from emitter to base and for electrons moving from base to emitter. Part of the holes will recombine with electrons in the base, but for a very thin base most holes will diffuse through the base and escape into the collector's space-charge region where the electric field drives them towards the collector.

In addition to this hole current, there will also be an electron current between collector and base, due to the diffusion of minority carrier electrons from the collector into the space-charge region. This (saturation) electron current will be largely independent of the operational condition of the device, as long as the base–collector junction is strongly reverse-biased, in which case it will not contribute to the amplification properties of the device.

Two basic parameters describe the performance of the transistor: the emitter efficiency γ and the base transport factor α_T. The emitter efficiency is the ratio between emitter hole current J_{Ep} and total emitter current $J_E \equiv$

$J_{Ep} + J_{En}$, and is given therefore by

$$\gamma = \frac{J_{Ep}}{J_{Ep} + J_{En}} = \frac{J_{Ep}}{J_E} . \tag{7.1.1}$$

The base transport factor is the ratio of hole current reaching the collector to the hole current injected from the emitter, and is given by

$$\alpha_T = \frac{J_{Cp}}{J_{Ep}} . \tag{7.1.2}$$

For a good transistor, both of these quantities are very close to unity, so that the common base current gain (compare Fig. 7.4) α_0, given by

$$\alpha_0 = \frac{J_{Cp}}{J_E} = \gamma \alpha_T , \tag{7.1.3}$$

also approaches unity.

Here one sees the reason for the requirements stated at the beginning of the section. The high emitter doping with respect to the base doping leads to a small electron current in the forward-biased emitter–base junction and consequently to γ close to unity. The thin base leads to a large base transport factor α_T and the low collector doping to a large width of the collector space-charge region and consequently to a small collector–base capacitance and a high breakdown voltage of the reversely biased collector–base junction.

The current–voltage characteristics can be obtained following the method and approximations already used for the diode (see Sect. 3.1). There, we determined the minority carrier concentration at the edges of the neutral regions from the externally applied voltages[25] and we calculated the minority carrier current at the edges of the neutral region after solving the continuity equation.

Using (3.1.6), we find directly the minority charge carrier concentration at the boundaries of emitter (n_E), collector (n_C) and two edges of base (p_{B_1} and p_{B_2}):

$$n_E = n_{E_0} \, e^{\frac{qV_{EB}}{kT}} \qquad\qquad n_C = n_{C_0} \, e^{\frac{qV_{CB}}{kT}} = n_{C_0} \, e^{-\frac{qV_{BC}}{kT}} \tag{7.1.4}$$

$$p_{B_1} = p_{B_0} \, e^{\frac{qV_{EB}}{kT}} \qquad\qquad p_{B_2} = p_{B_0} \, e^{\frac{qV_{CB}}{kT}} = p_{B_0} \, e^{-\frac{qV_{BC}}{kT}} .$$

In these equations, n_{E_0}, p_{B_0} and n_{C_0} are the minority carrier densities in the neutral emitter, base and collector regions respectively. We can also take over the electron current densities of emitter and collector from (3.1.7) and (3.1.13) – with the signs changed corresponding to the current and voltage directions indicated in Fig. 7.3. Thus:

$$J_{En} = \frac{qn_{E_0}D_n}{L_{n,E}} \left(e^{\frac{qV_{EB}}{kT}} - 1 \right) \tag{7.1.5}$$

$$J_{Cn} = -\frac{qn_{C_0}D_n}{L_{n,C}} \left(e^{\frac{qV_{CB}}{kT}} - 1 \right) = \frac{qn_{C_0}D_n}{L_{n,C}} \left(1 - e^{-\frac{qV_{BC}}{kT}} \right)$$

---

[25]Taking into account contact potentials between semiconductor and metal, the externally applied voltage equals the difference of the Fermi levels in the neutral semiconductor regions divided by the electron charge: $qV_{EB} = E_{F,B} - E_{F,E}$; $qV_{CB} = E_{F,B} - E_{F,C}$.

with $L_{n,E} = \sqrt{D_n \tau_{n,E}}$ and $L_{n,C} = \sqrt{D_n \tau_{n,C}}$, the minority carrier diffusion length in emitter and collector respectively. However, for the thin base we have to solve again the continuity equation (3.1.9) now written for holes:

$$D_p \frac{\partial^2 p_B}{\partial x^2} - \frac{p_B - p_{B_0}}{\tau_{pr}} = 0$$

with τ_{pr} the recombination lifetime and P_{B_0} the thermal equilibrium concentration of holes in the base. The solution to this equation is

$$p_B(x) = p_{B_0} + C_1 e^{x/L_p} + C_2 e^{-x/L_p} \qquad \text{with} \qquad L_p = \sqrt{D_p \tau_{pr}} \ ,$$

and with the new boundary conditions p_{B_1} and p_{B_2} at the edges of the neutral base region $x = 0$ and $x = W$, we obtain the solution for the hole current density and its gradient:

$$p_B(x) = p_{B_0} \left\{ 1 + \frac{1}{\sinh \frac{W}{L_p}} \right.$$

$$\left. \times \left[\left(e^{\frac{qV_{EB}}{kT}} - 1 \right) \sinh \frac{W-x}{L_p} + \left(e^{-\frac{qV_{BC}}{kT}} - 1 \right) \sinh \frac{x}{L_p} \right] \right\}$$

$$\frac{\partial p_B(x)}{\partial x} = p_{B_0} \frac{1}{L_p \sinh \frac{W}{L_p}}$$

$$\times \left[-\left(e^{\frac{qV_{EB}}{kT}} - 1 \right) \cosh \frac{W-x}{L_p} + \left(e^{-\frac{qV_{BC}}{kT}} - 1 \right) \cosh \frac{x}{L_p} \right] .$$

With (2.5.3) for the diffusion current density, we have

$$J_{Bp} = -qD_p \frac{\partial p_B}{\partial x}$$

and we can establish the hole current densities at the edges of the base, of the emitter and collector respectively, as:

$$J_{Ep} = qD_p \frac{p_{B_0}}{L_p \sinh \frac{W}{L_p}} \left[\left(e^{\frac{qV_{EB}}{kT}} - 1 \right) \cosh \frac{W}{L_p} + \left(1 - e^{-\frac{qV_{BC}}{kT}} \right) \right] \qquad (7.1.6)$$

$$J_{Cp} = qD_p \frac{p_{B_0}}{L_p \sinh \frac{W}{L_p}} \left[\left(e^{\frac{qV_{EB}}{kT}} - 1 \right) + \left(1 - e^{-\frac{qV_{BC}}{kT}} \right) \cosh \frac{W}{L_p} \right] .$$

Their difference then becomes

$$J_{Ep} - J_{Cp} = qD_p \frac{p_{B_0}}{L_p \sinh \frac{W}{L_p}} \left(\cosh \frac{W}{L_p} - 1 \right) \left(e^{\frac{qV_{EB}}{kT}} + e^{-\frac{qV_{BC}}{kT}} - 2 \right) .$$

With (7.1.5) and (7.1.6) we have expressed all current densities by the voltages applied to the emitter–base and collector–base junctions with the sign convention indicated in Fig. 7.3, which corresponds to all positive signs in the active operational condition.

For small base width $(W/L_p \ll 1)$ the equations simplify to

$$J_{Ep} \approx J_{Cp} \approx \frac{qD_p p_{B_0}}{W}\left[e^{\frac{qV_{EB}}{kT}} - e^{-\frac{qV_{BC}}{kT}}\right] \qquad (7.1.7)$$

$$J_{Ep} - J_{Cp} \approx \frac{qD_p p_{B_0} W}{2L_p^2}\left[e^{\frac{qV_{EB}}{kT}} + e^{-\frac{qV_{BC}}{kT}} - 2\right] \ ,$$

the electron current densities (equations (7.1.5)) not being affected.

For high reverse bias of the collector junction, all the current densities become independent of V_{BC}. Assuming in addition that the emitter junction is strongly forward-biased $(qV_{EB} \gg kT)$, we can obtain simple estimates for the quality parameters, namely the emitter efficiency γ and the base transfer factor α_T:

$$\gamma = \frac{J_{Ep}}{J_{Ep} + J_{En}} \approx \left(1 + \frac{D_n}{D_p}\frac{n_{E_0}}{p_{B_0}}\frac{W}{L_{n,E}}\right)^{-1} \approx 1 - \frac{D_n}{D_p}\frac{N_{DB}}{N_{AE}}\frac{W}{L_{n,E}} \qquad (7.1.8)$$

$$\alpha_T = \frac{J_{Cp}}{J_{Ep}} = 1 - \frac{J_{Ep} - J_{Cp}}{J_{Ep}} \approx 1 - \frac{W^2}{2L_{p,B}^2} \ .$$

The deviation of the common base current gain

$$\alpha_0 = \gamma\alpha_T \approx 1 - \frac{W^2}{2L_{p,B}^2} - \frac{D_n}{D_p}\frac{N_{DB}}{N_{AE}}\frac{W}{L_{n,E}} \qquad (7.1.9)$$

from unity, namely $1 - \alpha_0$, thus rises with the square of the ratio of the base width to the diffusion length for γ very close to 1.

Transistors are most frequently used in the common emitter configuration (Fig. 7.4). In this case the most relevant transistor parameters are the collector–base current gain, given by

$$\frac{\Delta I_C}{\Delta I_B} = \frac{\Delta I_C}{\Delta I_E - \Delta I_C} = \frac{\alpha_0}{1 - \alpha_0} \qquad (7.1.10)$$

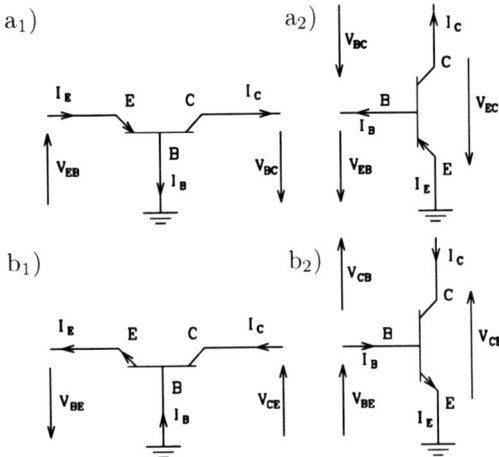

Fig. 7.4a,b. A p–n–p (a) and n–p–n (b) bipolar transistor: in common base (a_1, b_1) and common emitter operation (a_2, b_2)

and the transconductance $g_m = \frac{\Delta I_C}{\Delta V_{EB}}$. The transconductance is obtained by multiplying the collector hole current density (equation (7.1.7)) with the cross-sectional area A and taking the derivative with respect to V_{EB}. Notice that the electron current in the collector is constant in our approximation. With the above assumptions, we obtain the important expression for the transconductance of a bipolar transistor

$$g_m \approx A\frac{qD_p p_{B_0}}{W}\frac{q}{kT}e^{\frac{qV_{EB}}{kT}} \approx \frac{q}{kT}I_{Cp} \ . \tag{7.1.11}$$

The transconductance is thus given in good approximation by the ratio of transistor current and thermal voltage ($\frac{kT}{q} = 25\,\mathrm{mV}$ for silicon at room temperature). Note that the transconductance of the bipolar transistor equals the conductance of a forward-biased simple diode carrying the same bias current I. This is seen from (3.1.7) when taking the derivative with respect to V, the externally applied voltage.

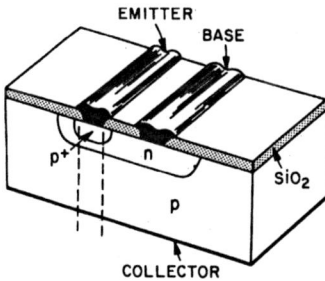

Fig. 7.5. A p–n–p transistor with the substrate acting as collector. (After Sze 1985, p. 110 Fig. 2)

A real device will deviate quite strongly from the idealistic picture presented so far. A very simple example is shown in Fig. 7.5 where the p substrate acts as collector. An n-type region acting as base is formed by diffusion or implantation. Within this region a shallow highly doped p-type region forms the emitter. Clearly, with such a production method the idealistic assumption of uniform doping is not fulfilled. This inaccuracy, and many other effects, which are even partially beneficial, may change the detail but will retain the basic features of the model.

Important effects that have to be considered in real devices include: breakdown (multiplication processes) when applying large collector–emitter voltages; series resistances in the emitter, base and collector regions, leading to a degradation of transistor parameters; and – in the case of high-frequency or low-noise applications – parasitic capacitances.

7.1.2 Junction Field Effect Transistors

Although the junction field effect transistor (JFET) is based on p-n junctions, only one type of charge carrier – the majority carrier – participates in the conduction process. Therefore it is called a unipolar device.

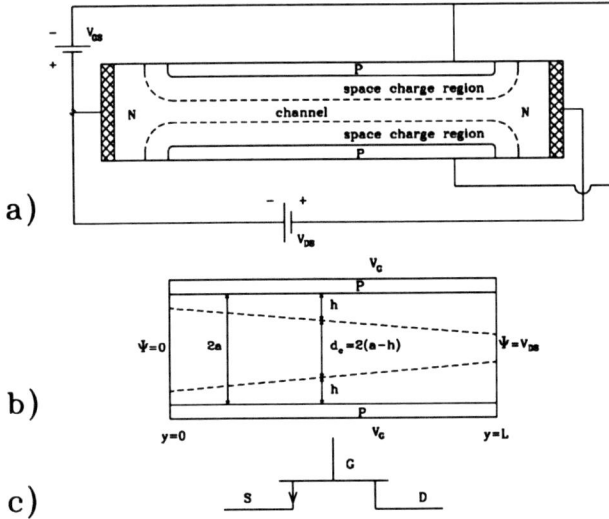

Fig. 7.6a–c. Principle of an n-channel Junction Field Effect transistor; operating at small drain-source voltage (a); channel region with potential drop along the channel (b); and device symbol (c)

The basic principle of the device is shown in Fig. 7.6. The highly doped source and drain regions are connected by a narrow channel made from the same conduction-type of material, with boundaries formed by two reversely biased p-n junctions. In normal operating conditions the drain of the n-channel device is positively biased with respect to the source, so as to cause an electron flow from source to drain, and the gate is negatively biased. The depth of the depletion layer – and therefore the channel depth – can be controlled by the gate voltage: for small drain voltage the depth of the depletion layer will be almost constant over the whole length of the channel and the device will therefore act like a resistor for which the resistance value can be adjusted by the gate voltage.

If the drain voltage is increased, however, the potential – and therefore also the depletion layer depth – will vary as a function of the channel position such that the channel depth decreases with the distance from the source. This will result in a nonlinearity in the current–voltage characteristics. If, with further increase of the drain voltage, the channel depth at the drain-side boundary reaches zero, the device is said to "operate in saturation", which means that the current is only weakly dependent on the drain voltage.

Current–Voltage Characteristics

We will now derive a simplistic model for the idealized device shown in Fig. 7.6, assuming very asymmetrically doped gate junctions ($N_A \gg N_D$) so that the space-charge region extends only into the channel but not into the highly doped gate. For simplicity we will refer all externally applied voltages with respect to the source contact. As nominal zero of the potential we define the undepleted region of the (homogeneously doped) channel at the source end.

The potential Ψ_c in the channel varies continuously from the source ($\Psi_c = 0$) to the drain end ($\Psi_c = V_D$). At an arbitrary point y in the channel, the depletion depth is given by (3.1.5) as

$$h(y) = \sqrt{\frac{2\epsilon\epsilon_0}{qN_D}(\Psi_c(y) - V_G + V_{bi})} \ . \tag{7.1.12}$$

The saturation or pinch-off condition is reached when the depth of the channel d_c, and therefore also the channel surface charge density $Q_c \equiv qN_A d_c$

$$d_c = 2(a - h) = 2\left[a - \sqrt{\frac{2\epsilon\epsilon_0}{qN_D}(\Psi_c(y) - V_G + V_{bi})}\right] \tag{7.1.13}$$

reach zero at the drain end. The pinch-off voltage V_p is defined as the total voltage $V_D - V_G + V_{bi}$ under this condition:

$$V_p = V_{D,sat} - V_G + V_{bi} = \frac{qN_D}{2\epsilon\epsilon_0}a^2 \ . \tag{7.1.14}$$

The surface charge density in the partially depleted channel is given by $Q_c = -qN_D d_c$ and for a transistor width W one obtains a current

$$I_D = Q_c \nu W = W Q_c \mu_n \frac{d\Psi_c}{dy} = W q N_D d_c \mu_n \frac{d\Psi_c}{dy} \tag{7.1.15}$$

$$= W q N_D \mu_n 2 \sqrt{\frac{2\epsilon\epsilon_0}{qN_D}} \left(\sqrt{V_p} - \sqrt{\Psi_c - V_G + V_{bi}}\right) \frac{d\Psi_c}{dy} \ .$$

Integrating (7.1.15) between source and drain end of the channel and dividing by the channel length L, we obtain

$$\frac{1}{L}\int_{y=0}^{L} I_D \, dy = I_D = \frac{W}{L}\mu_n \int_{\Psi_c=0}^{V_D} Q_c(\Psi_c) \, d\Psi_c \tag{7.1.16}$$

$$= \frac{W}{L} q N_D \mu_n 2 \sqrt{\frac{2\epsilon\epsilon_0}{qN_D}} \int_{\Psi_c=0}^{V_D} \left(\sqrt{V_p} - \sqrt{\Psi_c - V_G + V_{bi}}\right) d\Psi_c \ .$$

From the foregoing we obtain

$$I_D = I_p \left\{ \frac{V_D}{V_p} - \frac{2}{3}\left(\frac{V_D - V_G + V_{bi}}{V_p}\right)^{\frac{3}{2}} + \frac{2}{3}\left(\frac{-V_G + V_{bi}}{V_p}\right)^{\frac{3}{2}} \right\} , \tag{7.1.17}$$

with

$$I_{\mathrm{p}} = \frac{W}{L}\mu_n 2\sqrt{2\epsilon\epsilon_0 q N_{\mathrm{D}} V_{\mathrm{p}}^3} = \frac{W}{L}\mu_n \frac{q^2 N_{\mathrm{D}}^2 a^3}{\epsilon\epsilon_0}$$

and the pinch-off voltage V_{p} as given in (7.1.14).

The saturation voltage and current are obtained from (7.1.14) and (7.1.17) as

$$V_{\mathrm{D,sat}} = V_{\mathrm{p}} + V_{\mathrm{G}} - V_{\mathrm{bi}} = \frac{q N_{\mathrm{D}}}{2\epsilon\epsilon_0} a^2 + V_{\mathrm{G}} - V_{\mathrm{bi}} \tag{7.1.18}$$

$$I_{\mathrm{D,sat}} = I_{\mathrm{p}}\left[\frac{1}{3} + \frac{-V_{\mathrm{G}} + V_{\mathrm{bi}}}{V_{\mathrm{p}}} + \frac{2}{3}\left(\frac{-V_{\mathrm{G}} + V_{\mathrm{bi}}}{V_{\mathrm{p}}}\right)^{3/2}\right]. \tag{7.1.19}$$

Important parameters characterizing the transistor are the channel conductance g, the transconductance g_{m} and the gate-to-channel capacitance C_{G}. In the threshold region ($V_{\mathrm{D}} < V_{\mathrm{D,sat}}$) one obtains by simple differentiation of (7.1.16) or (7.1.17):

$$g = \frac{\partial I_{\mathrm{D}}}{\partial V_{\mathrm{D}}} = \frac{W}{L}\mu_n Q_{\mathrm{c}}(V_{\mathrm{D}}) = \frac{I_{\mathrm{p}}}{V_{\mathrm{p}}}\left[1 - \sqrt{\frac{V_{\mathrm{D}} - V_{\mathrm{G}} + V_{\mathrm{bi}}}{V_{\mathrm{p}}}}\right] \tag{7.1.20}$$

$$g_{\mathrm{m}} = \frac{\partial I_{\mathrm{D}}}{\partial V_{\mathrm{G}}} = \frac{I_{\mathrm{p}}}{V_{\mathrm{p}}}\left[\sqrt{\frac{V_{\mathrm{D}} - V_{\mathrm{G}} + V_{\mathrm{bi}}}{V_{\mathrm{p}}}} - \sqrt{\frac{-V_{\mathrm{G}} + V_{\mathrm{bi}}}{V_{\mathrm{p}}}}\right]. \tag{7.1.21}$$

In the saturation region the transconductance becomes

$$g_{\mathrm{m,sat}} = \frac{\partial I_{\mathrm{D,sat}}}{\partial V_{\mathrm{G}}} = \frac{I_{\mathrm{p}}}{V_{\mathrm{p}}}\left[1 - \sqrt{\frac{-V_{\mathrm{G}} + V_{\mathrm{bi}}}{V_{\mathrm{p}}}}\right]. \tag{7.1.22}$$

Notice that the channel conductance at threshold ($V_{\mathrm{DS}} = 0$) equals the transconductance in the saturation region. Furthermore, the current–voltage characteristics depend only on the ratio of the geometrical parameters $\frac{W}{L}$ but not on W or L separately. The conductance in the saturation region in our simplistic model is exactly zero, one of the most serious shortcomings of the model. In more refined models the shortening of the channel length with increasing V_{D} is taken into account. We then find an increase of g with a shortening of the channel length (keeping $\frac{W}{L}$ constant).

For a precise calculation of the gate capacitance, the position dependence of the depletion depth $h(y)$ should be known. As a first approximation we may assume a linear dependence and obtain

$$\frac{2}{C_{\mathrm{G}}} = \frac{1}{\epsilon_{\mathrm{s}}\epsilon_0}\frac{\langle h(y)\rangle}{WL}, \tag{7.1.23}$$

which below and above saturation yields, via (7.1.12):

$$C_G = WL\sqrt{2\epsilon_s\epsilon_0 qN_D}\,\frac{2}{\sqrt{-V_G + V_{bi}} + \sqrt{\sqrt{V_D - V_G + V_{bi}}}} \qquad (7.1.24)$$

$$C_{G,sat} = WL\sqrt{2\epsilon_s\epsilon_0 qN_D}\,\frac{2}{\sqrt{-V_G + V_{bi} + a}}\;. \qquad (7.1.25)$$

Example 7.1
Problem: *The idealized JFET considered for the derivation of the transistor characteristics was perfectly symmetrical with front- and back-gate kept at the same potential. Real devices are not symmetrical and often front- and back-gate can be supplied separately. Redo the derivation of the current–voltage characteristics for the device shown in Fig. 7.7 in which the back-gate is the low-doped p^- substrate (doping density N_A).*

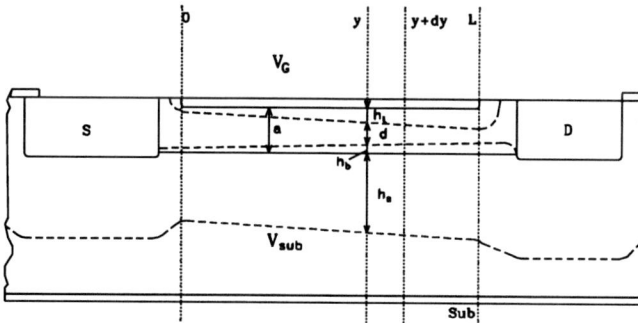

Fig. 7.7. Junction Field Effect transistor with substrate backgate

The highly doped n^+ source and drain regions are connected by the moderately n-doped channel (doping density N_D). The channel is confined by the shallow highly p^+-doped top gate and by the lowly doped n substrate.
Solution: For simplicity we will again assume abrupt changes between homogeneously doped semiconductor regions. In addition we use the graded channel approximation. This means that the changes in the direction parallel to the surface (y) are so small that we can use one-dimensional equations to relate charge density, electric field and potential:

$$\mathcal{E}_x = -\frac{\partial V}{\partial x} \qquad \rho = \epsilon\epsilon_0 \frac{\partial \mathcal{E}_x}{\partial x}\;.$$

The depletion width of the channel of the very asymmetric junction on the topside is given by (7.1.13), on the basis of (3.1.5), as

$$h_t = \sqrt{\frac{2\epsilon\epsilon_0}{qN_D}(\Psi_c - V_G + V_{bi,t})}\;, \qquad (7.1.26)$$

while for the bottom side we will assume the more general situation that extends into the channel as well as into the substrate. Then one has

$$h_{\mathrm{b}} = \sqrt{\frac{2\epsilon\epsilon_0}{q} \frac{N_{\mathrm{A}}}{N_{\mathrm{D}}(N_{\mathrm{A}} + N_{\mathrm{D}})} (\varPsi_{\mathrm{c}} - V_{\mathrm{sub}} + V_{\mathrm{bi,b}})} \qquad (7.1.27)$$

$$h_{\mathrm{s}} = \sqrt{\frac{2\epsilon\epsilon_0}{q} \frac{N_{\mathrm{D}}}{N_{\mathrm{A}}(N_{\mathrm{A}} + N_{\mathrm{D}})} (\varPsi_{\mathrm{c}} - V_{\mathrm{sub}} + V_{\mathrm{bi,b}})} \ ,$$

where V_{G} and V_{sub} are the externally applied gate and substrate voltages and $\varPsi_{\mathrm{c}}(y)$ is the potential in the channel at position y. The built-in voltages of the channel-to-top-gate diode $v_{\mathrm{bi,t}}$ and channel-to-substrate diode $v_{\mathrm{bi,b}}$ regions can be calculated from (3.1.1). The channel depth is then derived as

$$d(y) = a - h_{\mathrm{t}} - h_{\mathrm{b}} \qquad (7.1.28)$$

and the surface charge density in the channel as

$$\sigma(y) = -qN_{\mathrm{D}}d(y) \ .$$

The pinch-off condition is again obtained from the requirement that the channel depth at the drain end of the channel $(y = L)$ reaches zero:

$$d(L) = a - \sqrt{\frac{2\epsilon\epsilon_0}{qN_{\mathrm{D}}}(V_{\mathrm{D,sat}} - V_{\mathrm{G}} + V_{\mathrm{bi,t}})}$$

$$- \sqrt{\frac{2\epsilon\epsilon_0}{q} \frac{N_{\mathrm{A}}}{N_{\mathrm{D}}(N_{\mathrm{A}} + N_{\mathrm{D}})}(V_{\mathrm{D,sat}} - V_{\mathrm{sub}} + V_{\mathrm{bi,b}})} = 0 \ . \qquad (7.1.29)$$

It is left to the reader to solve this quadratic equation for the saturation voltage $V_{\mathrm{D,sat}}$, needed for calculation of the transistor properties in the saturation region.

With the drift velocity $v = \mu_n \frac{\partial \varPsi_{\mathrm{c}}(y)}{\partial y}$, the current of a transistor with channel width W becomes

$$I_{\mathrm{D}} = -W\sigma v = WqN_{\mathrm{D}}d(\varPsi_{\mathrm{c}}(y))\mu_n \frac{\partial \varPsi_{\mathrm{c}}(y)}{\partial y} \ . \qquad (7.1.30)$$

Integrating this equation between source and drain end of the channel, we obtain (again with $\varPsi_{\mathrm{c}}(0) = 0$ by definition) and using (7.1.26) and (7.1.27):

$$\int_{y=0}^{L} I_{\mathrm{D}}\,\mathrm{d}y = I_{\mathrm{D}}L = WqN_{\mathrm{D}}\mu_n \int_{0}^{V_{\mathrm{D}}} \left[a - h_{\mathrm{t}}(\varPsi_{\mathrm{c}}) - h_{\mathrm{b}}(\varPsi_{\mathrm{c}}) \right]\mathrm{d}\varPsi_{\mathrm{c}}$$

$$= WqN_{\mathrm{D}}\mu_n \int_{\varPsi_{\mathrm{c}}=0}^{V_{\mathrm{D}}} \Bigg\{ a - \sqrt{\frac{2\epsilon\epsilon_0}{qN_{\mathrm{D}}}} \Bigg[(\sqrt{\varPsi_{\mathrm{c}} - V_{\mathrm{G}} + V_{\mathrm{bi,t}}}$$

$$+ \sqrt{\frac{N_{\mathrm{A}}}{N_{\mathrm{A}} + N_{\mathrm{D}}}} \sqrt{\varPsi_{\mathrm{c}} - V_{\mathrm{sub}} + V_{\mathrm{bi,b}}} \Bigg] \Bigg\} \mathrm{d}\varPsi_{\mathrm{c}} \ .$$

Thus we have that

$$
I_D = \frac{W q N_D}{L} a \mu_n
$$

$$
\times \left\{ V_D - \sqrt{\frac{2 \epsilon \epsilon_0}{q N_D}} \left[(V_D - V_G + V_{bi,t})^{3/2} - (-V_G + V_{bi,t})^{3/2} \right. \right. \qquad (7.1.31)
$$

$$
\left. \left. + \sqrt{\frac{N_A}{N_A + N_D}} \left[(V_D - V_{sub} + V_{bi,b})^{3/2} - (-V_{sub} + V_{bi,b})^{3/2} \right] \right] \right\} .
$$

Remember that for an n-channel JFET in normal operating conditions $V_D > 0$, $V_G < 0$ and $V_{sub} < 0$. Transistor properties in the saturation region ($V_{D,sat}$, $I_{D,sat}$, $g_{m,sat}$) may be obtained from the condition of zero channel depth at the drain end (see (7.1.29)), in similar fashion to the symmetrical example considered previously.

7.1.3 Metal–Oxide–Semiconductor Field Effect Transistors

MOS transistors would more appropriately be called Conductor–Insulator–Semiconductor transistors since even in the silicon technology where this name originates the metal gate has been replaced by polysilicon in most cases, and sometimes insulating materials other than SiO_2 are used. It is the most important device for large-scale integrated circuits such as microprocessors and memories, and more recently it has also been used for high-density detector readout electronics. The electrical characteristics of a Metal–Oxide–Semiconductor field effect transistor (MOSFET) are similar to those of the JFET.

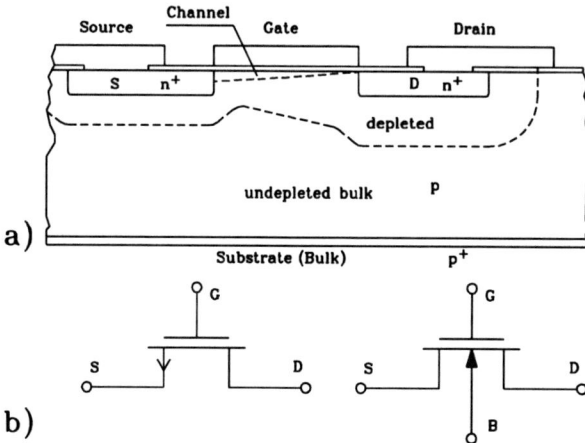

Fig. 7.8a,b. An n-channel MOSFET: Cross-section (a) and device symbol (b). The separation of the space-charge region from the channel immediately below the gate and from the undepleted bulk is indicated by the *dashed lines*

An n-channel MOSFET device (Fig. 7.8) consists of two very asymmetrically doped n^+-p junctions joined by a MOS structure. At negative or zero gate voltage – using the convention that externally applied voltages are taken with respect to the source contact – source and drain are insulated from each other. At sufficiently positive gate voltage, an inversion layer will form below the gate at the semiconductor–insulator interface. Therefore drain and source will be connected by a conductive layer. Its conductivity can be controlled by the gate voltage and – to a lesser extent – by the substrate voltage.

The derivation of the current–voltage characteristics is similar to (and slightly simpler than) that for the JFET. Again, we use the graded channel approximation, which allows us to use the one-dimensional relationships between charge density, electric field and potential.

The MOS structure has already been cosidered in Sect. 3.3. There, the thermal equilibrium condition has been studied and the potential at the semiconductor–oxide interface has been determined from the semiconductor substrate potential and the voltage applied between metal and substrate contact only. When the applied voltage exceeded a certain value, the depletion depth did not increase further, because an inversion layer was formed by electron–hole pairs created thermally in the depletion layer. These restricting properties do not apply in the case of the MOSFET, where the inversion layer is connected to source and drain and its voltage therefore can be controlled by these electrodes. Electrons created in the space-charge region below the channel are therefore collected in the channel and transported to the drain of the transistor.

Current–Voltage Characteristics of an n-Channel MOSFET

Figure 7.9 shows the channel region of an n-channel MOSFET and indicates the symbols used in the following derivation. As was done already in Sect. 3.3, we again introduce the flat-band voltage V_{FB}, namely the voltage applied between gate and substrate contact of a MOS structure at which the electric field at the boundary of the semiconductor is zero.

Since it is somewhat difficult to keep track of the contact potentials involved in the device, we will first consider the situation in which source, drain and substrate contacts of the n-channel MOSFET are connected and the gate is held at the flat-band voltage $V_G = V_{FB}$ with respect to this common contact. As has been shown in Sect. 3.3, the potential within the semiconductor bulk is constant and the electric field is zero up to the oxide boundary under this condition. We define the semiconductor bulk potential in this condition as nominally zero.

If the gate voltage is increased sufficiently, an inversion layer forms just below the oxide–semiconductor interface. This inversion layer – the "channel" – will form an electrical connection between source and drain. At the onset of strong inversion, its potential will be

$$\Psi_s = 2\Psi_B \ , \tag{7.1.32}$$

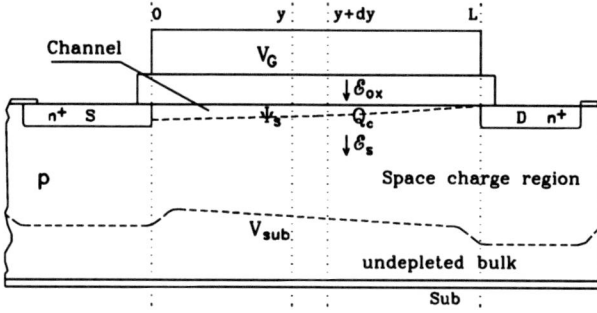

Fig. 7.9. Channel region of an n-channel enhancement-type MOSFET. By use of the graded channel approximation, a small element of the channel with length dy is considered in an approximate one-dimensional fashion. $\mathcal{E}_{ox}$ and E_s are the electrical field strengths in the oxide and the silicon, Ψ_s is the potential at the Si$-$SiO$_2$ interface and Q_c is the surface charge density of the channel region (integrated over the depth of the channel)

$q\Psi_B$ being the distance of the intrinsic level E_i from the Fermi level E_F of the substrate, which depends on the doping concentration according to (3.3.1). The potential of the inversion layer will remain almost constant with further increase of the gate voltage and also with a change of the substrate potential in the negative direction; this is because its potential remains fixed by the source (and drain) to which it is electrically connected.

In order to relate the voltage drop along the channel with the channel current, we now calculate the channel's surface-charge density Q_c as a function of the surface potential Ψ_s, and the externally applied voltages (with respect to the source contact) of gate V_G and substrate V_{sub}. The depth of the space-charge region below the channel may be calculated from the potential difference between channel Ψ_s and substrate V_{sub} according to

$$\Psi_s - V_{sub} = \frac{qN_A}{\epsilon\epsilon_0}\frac{d_s^2}{2}$$

$$d_s = \sqrt{\frac{2\epsilon\epsilon_0}{qN_A}(\Psi_s - V_{sub})} \ . \tag{7.1.33}$$

The electric field just below the channel is given as

$$\mathcal{E}_s = \frac{qN_A}{\epsilon\epsilon_0}d_s = \sqrt{\frac{2qN_A}{\epsilon\epsilon_0}(\Psi_s - V_{sub})} \ , \tag{7.1.34}$$

while above the interface the change in the electric field from the flat-band condition is given by

$$\mathcal{E}_{ox} = (V_G - V_{FB} - \Psi_s)/d_{ox} \ . \tag{7.1.35}$$

The channel's surface-charge density may thus be found as

$$Q_c = -\epsilon_{ox}\epsilon_0\mathcal{E}_{ox} + \epsilon\epsilon_0\mathcal{E}_s$$

$$= -\left[C_{ox}(V_G - V_{FB} - \Psi_s) - \sqrt{2qN_A\epsilon\epsilon_0(\Psi_s - V_{sub})}\right] \ , \tag{7.1.36}$$

with $\qquad C_{\mathrm{ox}} = \dfrac{\epsilon_{\mathrm{ox}} \epsilon_0}{d_{\mathrm{ox}}}$,

the oxide capacitance per unit area.

As the channel potential Ψ_s equals $2\Psi_B$ (at least close to the source for $V_D \neq 0$), the gate voltage at which the channel starts to form, the threshold voltage V_T, is obtained from (7.1.36) with $Q_c = 0$ and $\Psi_s = 2\Psi_B$. We thus obtain

$$V_T = V_{FB} + 2\Psi_B + \frac{1}{C_{\mathrm{ox}}} \sqrt{2\epsilon_0 q N_A (2\Psi_B - V_{\mathrm{sub}})} \ . \tag{7.1.37}$$

Putting now a (small) positive voltage V_D between drain and source contacts, the channel potential will vary between $2\Psi_B$ and $V_D + 2\Psi_B$ from the source $(y = 0)$ to the drain $(y = L)$ end of the channel respectively, so that we can parameterize the channel potential as

$$\Psi_s(y) = V_c(y) + 2\Psi_B$$

with $V_c(0) = 0$, $V_c(L) = V_D$.

In the graded channel approximation, (7.1.36) is assumed to hold for any position in the channel, and we may write down the drain current I_D of a transistor of width W as

$$I_D = -\nu(y) Q_c(y) W = \mu_n \mathcal{E}_y(y) Q_c(y) W = -W \mu_n Q_c(y) \frac{\partial V_c}{\partial y}$$

$$= W \mu_n \frac{dV_c(y)}{dy} \Big[C_{\mathrm{ox}} (V_G - V_{FB} - V_c(y) - 2\Psi_B)$$

$$- \sqrt{2\epsilon_0 q N_A (V_c(y) + 2\Psi_B - V_{\mathrm{sub}})} \Big] \ , \tag{7.1.38}$$

ν being the drift velocity given in (2.5.1).

Integrating (7.1.38) between source and drain ends of the channel and dividing by the channel length L leads to

$$I_D = \frac{1}{L} \int_{y=0}^{L} I_D \, dy = -\frac{W}{L} \mu_n \int_{V_c=0}^{V_D} Q_c(V_c) \, dV_c \tag{7.1.39}$$

$$= \frac{W}{L} \mu_n C_{\mathrm{ox}} \int_{V_c=0}^{V_D} \Big\{ V_G - V_{FB} - 2\Psi_B - V_c$$

$$- \frac{\sqrt{2\epsilon_0 q N_A}}{C_{\mathrm{ox}}} (V_c + 2\Psi_B - V_{\mathrm{sub}})^{\frac{1}{2}} \Big\} \, dV_c$$

$$= \frac{W}{L} \mu_n C_{\mathrm{ox}} \Big\{ \Big(V_G - V_{FB} - 2\Psi_B - \frac{V_D}{2} \Big) V_D$$

$$- \frac{2}{3} \frac{\sqrt{2\epsilon_0 q N_A}}{C_{\mathrm{ox}}} \Big[\big(V_D + 2\Psi_B - V_{\mathrm{sub}} \big)^{\frac{3}{2}} - \big(2\Psi_B - V_{\mathrm{sub}} \big)^{\frac{3}{2}} \Big] \Big\} \ . \tag{7.1.40}$$

Considering the dependence of the drain current I_D at fixed gate (V_G) and substrate voltage (V_{sub}) on the drain voltage, it is clear that the current increases linearly at first and then reaches a maximum (it saturates) when the surface charge density at the drain end of the channel reaches zero. We again use (7.1.36) to obtain the saturation voltage $V_{D,sat}$ by setting $\Psi_s = V_D + 2\Psi_B$ and $Q_c = 0$ and obtain after solving a quadratic equation

$$V_{D,sat} = V_G - V_{FB} - 2\Psi_B + \frac{1}{C_{ox}^2}\epsilon\epsilon_0 q N_A$$

$$\times \left[1 - \sqrt{1 + 2\frac{C_{ox}^2}{\epsilon\epsilon_0 q N_A}(V_G - V_{FB} - V_{sub})}\right] . \tag{7.1.41}$$

The sign of the square root has been chosen from the requirement that an increase in the substrate voltage V_{sub} leads to a rise of saturation voltage $V_{D,sat}$. Inserting (7.1.41) into (7.1.40), we obtain the saturation current, which in our simple model is assumed to stay constant for $V_D \geq V_{D,sat}$. It is given by

$$I_{D,sat} = \frac{W}{L}\mu_n C_{ox}\left\{\left(V_G - V_{FB} - 2\Psi_B - \frac{V_{D,sat}}{2}\right)V_{D,sat}\right. \tag{7.1.42}$$

$$\left. - \frac{2}{3}\frac{\sqrt{2\epsilon\epsilon_0 q N_A}}{C_{ox}}\left[(V_{D,sat} + 2\Psi_B - V_{sub})^{\frac{3}{2}} - (2\Psi_B - V_{sub})^{\frac{3}{2}}\right]\right\} .$$

We now derive the important transistor parameters, transconductance $g_m \equiv \partial I_D/\partial V_G$ and output conductance $g \equiv \partial I_D/\partial V_D$, as well as the parameter describing the influence of the substrate on the current $g_{sub} \equiv \partial I_D/\partial V_{sub}$. For the threshold region $V_D < V_{D,sat}$ one finds from (7.1.39) or (7.1.40) that

$$g = \frac{\partial I_D}{\partial V_D} = -\frac{W}{L}\mu_n Q_c(V_D) = \frac{W}{L}\mu_n C_{ox}\left\{V_G - V_{FB} - V_D - 2\Psi_B\right.$$

$$\left. - \frac{\sqrt{2\epsilon\epsilon_0 q N_A}}{C_{ox}}\sqrt{V_D - V_{sub} + 2\Psi_B}\right\} \tag{7.1.43}$$

$$g_m = \frac{\partial I_D}{\partial V_G} = \frac{W}{L}\mu_n C_{ox} V_D \tag{7.1.44}$$

$$g_{sub} = \frac{\partial I_D}{\partial V_{sub}} = \frac{W}{L}\mu_n \sqrt{2\epsilon\epsilon_0 q N_A}\left[\sqrt{V_D - V_{sub} + 2\Psi_B}\right.$$

$$\left. - \sqrt{-V_{sub} + 2\Psi_B}\right] . \tag{7.1.45}$$

For the saturation region $(V_D > V_{D,sat})$ we cannot calculate the output conductance as it is zero in our simple model – slightly more sophisticated models introduce a shortening of the channel length, resulting in a better description of the saturation region. The transconductance can, however, be calculated from (7.1.42) and (7.1.41) respectively with (7.1.39):

$$g_{m,sat} = \frac{\partial I_{D,sat}}{\partial V_G} = -\frac{W}{L}\mu_n\left[Q_c(V_{D,sat})\frac{\partial V_{D,sat}}{\partial V_G} + \int_{V_c=0}^{V_{D,sat}}\frac{\partial Q_c}{\partial V_G}dV_c\right]$$

$$= \frac{W}{L}\mu_n C_{ox} V_{D,sat} . \tag{7.1.46}$$

Here we have made use of the fact that the surface charge density at the drain end of the channel in saturation is zero and the derivative of Q_c with respect to the gate voltage V_G is $-C_{ox}$.

Similarly, we find the steering of the drain current I_D by the substrate voltage V_{sub} given by

$$
\begin{aligned}
g_{sub,sat} &= \frac{\partial I_{D,sat}}{\partial V_{sub}} = -\frac{W}{L}\mu_n \left[Q_c(V_{D,sat})\frac{\partial V_{D,sat}}{\partial V_{sub}} + \int_{V_c=0}^{V_{D,sat}} \frac{\partial Q_c}{\partial V_{sub}}\, dV_c \right] \\
&= \frac{W}{L}\mu_n C_{ox} \frac{\sqrt{2\epsilon\epsilon_0 q N_A}}{C_{ox}} \left[\sqrt{V_{D,sat} - V_{sub} + 2\Psi_B} \right. \\
&\qquad \left. - \sqrt{-V_{sub} + 2\Psi_B} \right] .
\end{aligned}
\tag{7.1.47}
$$

Output conductance at $V_D = 0$ and transconductance in the saturation region are related, as can be derived from (7.1.43), (7.1.46) and (7.1.47):

$$
\begin{aligned}
g_{m,sat} + g_{sub,sat} &= -\frac{W}{L}\mu_n \int_{V_c=0}^{V_{D,sat}} \left[\frac{\partial Q_c}{\partial V_G} + \frac{\partial Q_c}{\partial V_{sub}} \right] dV_c \\
&= \frac{W}{L}\mu_n \int_{V_c=0}^{V_{D,sat}} \frac{\partial Q_c}{\partial V_c}\, dV_c = -\frac{W}{L}\mu_n Q_c(0) = g(V_D = 0) .
\end{aligned}
\tag{7.1.48}
$$

Replacing $\frac{\partial Q_c}{\partial V_G} + \frac{\partial Q_c}{\partial V_{sub}}$ by $-\frac{\partial Q_c}{\partial V_c}$ is either verified explicitly or inferred by the fact that a small change in the channel potential has the same effect on the channel surface charge as an opposite change of gate and substrate potential simultaneously.

Current–Voltage Characteristics of a p-channel MOSFET

The p-channel transistor characteristics can be calculated in complete analogy to the n-channel set, with appropriate sign changes. A hole inversion layer has to be formed on the n-type bulk of doping concentration N_D. Therefore the channel potential at the onset of strong inversion is negative:

$$
\Psi_s = -2\Psi_B ,
\tag{7.1.49}
$$

$q\Psi_B$ being the absolute distance of the intrinsic level E_i from the Fermi level E_F of the substrate, which depends on the doping concentration according to (3.3.1).

Before proceeding, we need to make some remarks on sign convention. We take I_D as positive when it flows into the drain. The drain current of a p-channel device in its normal operation mode will therefore be negative: it will increase in magnitude with gate voltage or substrate voltage being changed in the negative direction. The substrate will be at source potential or biased positively with respect to it, which, compared to n-channel devices, reflects in sign changes under square roots connected with substrate biasing. In the following, the formulas derived for the n-channel device will be rewritten for the p-channel case, as follows:

$$V_{\text{sub}} - \Psi_{\text{s}} = \frac{qN_{\text{D}}}{\epsilon\epsilon_0}\frac{d_{\text{s}}^2}{2} \tag{7.1.50}$$

$$d_{\text{s}} = \sqrt{\frac{2\epsilon\epsilon_0}{qN_{\text{D}}}(V_{\text{sub}} - \Psi_{\text{s}})} \tag{7.1.51}$$

$$\mathcal{E}_{\text{s}} = -\frac{qN_{\text{D}}}{\epsilon\epsilon_0}d_{\text{s}} = -\sqrt{\frac{2qN_{\text{D}}}{\epsilon\epsilon_0}(V_{\text{sub}} - \Psi_{\text{s}})} \tag{7.1.52}$$

$$\mathcal{E}_{\text{ox}} = (V_{\text{G}} - V_{\text{FB}} - \Psi_{\text{s}})/d_{\text{ox}} \ . \tag{7.1.53}$$

The channel surface charge density thus may be found as

$$\begin{aligned}
Q_{\text{c}} &= -\epsilon_{\text{ox}}\epsilon_0\mathcal{E}_{\text{ox}} + \epsilon\epsilon_0\mathcal{E}_{\text{s}} \\
&= -\left[C_{\text{ox}}(V_{\text{G}} - V_{\text{FB}} - \Psi_{\text{s}}) + \sqrt{2qN_{\text{D}}\epsilon\epsilon_0(V_{\text{sub}} - \Psi_{\text{s}})}\right] \ ,
\end{aligned} \tag{7.1.54}$$

leading to the threshold gate voltage with $Q_{\text{c}} = 0$ and $\psi_{\text{s}} = -2\psi_{\text{B}}$:

$$V_{\text{T}} = V_{\text{FB}} - 2\Psi_{\text{B}} - \frac{1}{C_{\text{ox}}}\sqrt{2\epsilon\epsilon_0 qN_{\text{D}}(V_{\text{sub}} + 2\Psi_{\text{B}})} \ . \tag{7.1.55}$$

Parameterizing the channel potential as

$$\Psi_{\text{s}}(y) = V_{\text{c}}(y) - 2\Psi_{\text{B}}$$

with $V_{\text{c}}(0) = 0$, $V_{\text{c}}(L) = V_{\text{D}}$, the drain current of a transistor of width W becomes

$$\begin{aligned}
I_{\text{D}} &= -\nu(y)Q_{\text{c}}(y)W = -\mu_p\mathcal{E}_y(y)Q_{\text{c}}(y)W = W\mu_p Q_{\text{c}}(y)\frac{\partial V_{\text{c}}}{\partial y} \\
&= -W\mu_p\frac{dV_{\text{c}}(y)}{dy}\Bigg[C_{\text{ox}}(V_{\text{G}} - V_{\text{FB}} - V_{\text{c}}(y) + 2\Psi_{\text{B}}) \\
&\quad + \sqrt{2\epsilon\epsilon_0 qN_{\text{D}}(V_{\text{sub}} - V_{\text{c}}(y) + 2\Psi_{\text{B}})}\Bigg] \ .
\end{aligned} \tag{7.1.56}$$

Integrating (7.1.56) between source and drain ends of the channel and dividing by the channel length L leads to

$$\begin{aligned}
I_{\text{D}} &= \frac{1}{L}\int_{y=0}^{L} I_{\text{D}}\,dy = \frac{W}{L}\mu_p\int_{V_{\text{c}}=0}^{V_{\text{D}}} Q_{\text{c}}(V_{\text{c}})\,dV_{\text{c}} \\
&= -\frac{W}{L}\mu_p C_{\text{ox}}\int_{V_{\text{c}}=0}^{V_{\text{D}}}\Bigg\{V_{\text{G}} - V_{\text{FB}} + 2\Psi_{\text{B}} - V_{\text{c}} \\
&\quad + \frac{\sqrt{2\epsilon\epsilon_0 qN_{\text{D}}}}{C_{\text{ox}}}(V_{\text{sub}} - V_{\text{c}} + 2\Psi_{\text{B}})^{\frac{1}{2}}\Bigg\}\,dV_{\text{c}} \\
&= -\frac{W}{L}\mu_p C_{\text{ox}}\Bigg\{\left(V_{\text{G}} - V_{\text{FB}} + 2\Psi_{\text{B}} - \frac{V_{\text{D}}}{2}\right)V_{\text{D}} \\
&\quad - \frac{2}{3}\frac{\sqrt{2\epsilon\epsilon_0 qN_{\text{D}}}}{C_{\text{ox}}}\left[(V_{\text{sub}} - V_{\text{D}} + 2\Psi_{\text{B}})^{\frac{3}{2}} - (V_{\text{sub}} + 2\Psi_{\text{B}})^{\frac{3}{2}}\right]\Bigg\} \ .
\end{aligned} \tag{7.1.58}$$

We can then use (7.1.54) to obtain the saturation voltage $V_{D,sat}$ by setting $\Psi_s = V_D - 2\Psi_B$ and $Q_c = 0$, and after solving a quadratic equation we obtain

$$
\begin{aligned}
V_{D,sat} =& V_G - V_{FB} + 2\Psi_B \\
&+ \frac{1}{C_{ox}^2} \epsilon\epsilon_0 q N_D \left[1 + \sqrt{1 + 2\frac{C_{ox}^2}{\epsilon\epsilon_0 q N_D}(V_{sub} - V_G + V_{FB})} \right] , \quad (7.1.59)
\end{aligned}
$$

with which the saturation current becomes

$$
\begin{aligned}
I_{D,sat} =& -\frac{W}{L} \mu_p C_{ox} \Bigg\{ \left(V_G - V_{FB} + 2\Psi_B - \frac{V_{D,sat}}{2} \right) V_{D,sat} \quad (7.1.60) \\
&- \frac{2}{3} \frac{\sqrt{2\epsilon\epsilon_0 q N_D}}{C_{ox}} \left[(V_{sub} - V_{D,sat} + 2\Psi_B)^{\frac{3}{2}} - (V_{sub} + 2\Psi_B)^{\frac{3}{2}} \right] \Bigg\} .
\end{aligned}
$$

For the threshold region $V_D > V_{D,sat}$, we find from (7.1.56) that

$$
\begin{aligned}
g \equiv \frac{\partial I_D}{\partial V_D} = \frac{W}{L} \mu_p Q_c(V_D) =& -\frac{W}{L} \mu_p C_{ox} \Bigg\{ V_G - V_{FB} - V_D + 2\Psi_B \\
&+ \frac{\sqrt{2\epsilon\epsilon_0 q N_A}}{C_{ox}} \sqrt{V_{sub} - V_D + 2\Psi_B} \Bigg\} \quad (7.1.61)
\end{aligned}
$$

$$
g_m \equiv \frac{\partial I_D}{\partial V_G} = -\frac{W}{L} \mu_p C_{ox} V_D \quad (7.1.62)
$$

$$
\begin{aligned}
g_{sub} \equiv& \frac{\partial I_D}{\partial V_{sub}} \\
=& \frac{W}{L} \mu_p \sqrt{2\epsilon\epsilon_0 q N_D} \left[\sqrt{V_{sub} - V_D + 2\Psi_B} - \sqrt{V_{sub} + 2\Psi_B} \right] . \quad (7.1.63)
\end{aligned}
$$

The transconductance in the saturation region is calculated from (7.1.60) and (7.1.59), respectively from (7.1.57) with $Q_c(V_{D,sat}) = 0$ and $\frac{\partial Q_c}{\partial V_G} = -C_{ox}$, to give

$$
\begin{aligned}
g_{m,sat} =& \frac{\partial I_{D,sat}}{\partial V_G} = \frac{W}{L} \mu_p \left[Q_c(V_{D,sat}) \frac{\partial V_{D,sat}}{\partial V_G} + \int_{V_c=0}^{V_{D,sat}} \frac{\partial Q_c}{\partial V_G} dV_c \right] \\
=& -\frac{W}{L} \mu_p C_{ox} V_{D,sat} \quad (7.1.64)
\end{aligned}
$$

$$
\begin{aligned}
g_{sub,sat} =& \frac{\partial I_{D,sat}}{\partial V_{sub}} = \frac{W}{L} \mu_p \left[Q_c(V_{D,sat}) \frac{\partial V_{D,sat}}{\partial V_{sub}} + \int_{V_c=0}^{V_{D,sat}} \frac{\partial Q_c}{\partial V_{sub}} dV_c \right] \quad (7.1.65) \\
=& \frac{W}{L} \mu_p C_{ox} \frac{\sqrt{2\epsilon\epsilon_0 q N_A}}{C_{ox}} \left[\sqrt{V_{sub} - V_{D,sat} + 2\Psi_B} - \sqrt{V_{sub} + 2\Psi_B} \right] .
\end{aligned}
$$

Again, one finds the relation between output conductance at $V_D = 0$ and transconductance in the saturation region from (7.1.61), (7.1.64) and (7.1.65), yielding

$$
\begin{aligned}
g_{m,\text{sat}} + g_{\text{sub,sat}} &= \frac{W}{L}\mu_p \int_{V_c=0}^{V_{D,\text{sat}}} \left[\frac{\partial Q_c}{\partial V_G} + \frac{\partial Q_c}{\partial V_{\text{sub}}}\right] dV_c \\
&= -\frac{W}{L}\mu_p \int_{V_c=0}^{V_{D,\text{sat}}} \frac{\partial Q_c}{\partial V_c}\, dV_c = \frac{W}{L}\mu_p Q_c(0) = g(V_D = 0) \ .
\end{aligned}
\tag{7.1.66}
$$

Simplification of MOSFET Current–Voltage Characteristics

For practical MOSFET transistors, steering from the bulk is much less strong than from the gate. Furthermore the bulk voltage is usually held constant (with respect to source or drain), so that the bulk effect can in good approximation be taken into account in the threshold voltage and the more subtle variation of depletion depth along the channel can be ignored.

We then have for *n-channel MOSFETs*, via (7.1.40) and (7.1.37)

$$
I_D = \frac{W}{L}\mu_n C_{\text{ox}} \left\{ \left(V_{G,\text{eff}} - \frac{V_D}{2}\right) V_D \right\} \ ,
\tag{7.1.67}
$$

$$
\begin{aligned}
V_{G,\text{eff}} &= V_G - V_T \\
&= V_G - \left[V_{\text{FB}} + 2\Psi_B + \frac{1}{C_{\text{ox}}}\sqrt{2\epsilon\epsilon_0 q N_A (2\Psi_B - V_{\text{sub}})} \right] \ .
\end{aligned}
\tag{7.1.68}
$$

The saturation voltage follows from the condition

$$
\frac{\partial I_D}{\partial V_D} = 0 \ ,
$$

leading to

$$
V_{D,\text{sat}} = V_{G,\text{eff}}
\tag{7.1.69}
$$

$$
I_{D,\text{sat}} = I_D(V_{D,\text{sat}}) = \frac{W}{L}\mu_n C_{\text{ox}} \frac{V_{D,\text{sat}}^2}{2}
\tag{7.1.70}
$$

$$
g_{m,\text{sat}} = \frac{\partial I_{D,\text{sat}}}{\partial V_{G,\text{eff}}} = \frac{W}{L}\mu_n C_{\text{ox}} V_{D,\text{sat}} = \sqrt{2\frac{W}{L}\mu_n C_{\text{ox}} I_{D,\text{sat}}} \ .
\tag{7.1.71}
$$

For *p-channel MOSFETs*, we find similarly from (7.1.58) and (7.1.55):

$$
I_D = -\frac{W}{L}\mu_p C_{\text{ox}} \left\{ \left(V_{G,\text{eff}} - \frac{V_D}{2}\right) V_D \right\} \ ,
\tag{7.1.72}
$$

$$
\begin{aligned}
V_{G,\text{eff}} &= V_G - V_T \\
&= V_G - \left[V_{\text{FB}} - 2\Psi_B - \frac{1}{C_{\text{ox}}}\sqrt{2\epsilon\epsilon_0 q N_A (V_{\text{sub}} + 2\Psi_B)} \right] \ .
\end{aligned}
\tag{7.1.73}
$$

In this case the saturation voltage once more follows from the condition

$$\frac{\partial I_D}{\partial V_D} = 0 \; ,$$

leading to

$$V_{D,\text{sat}} = V_{G,\text{eff}} \tag{7.1.74}$$

$$I_{D,\text{sat}} = I_D(V_{D,\text{sat}}) = -\frac{W}{L}\mu_p C_{\text{ox}}\frac{V_{D,\text{sat}}^2}{2} \tag{7.1.75}$$

$$g_{m,\text{sat}} = \frac{\partial I_{D,\text{sat}}}{\partial V_{G,\text{eff}}} = -\frac{W}{L}\mu_p C_{\text{ox}} V_{D,\text{sat}} = \sqrt{2\frac{W}{L}\mu_p C_{\text{ox}}(-I_{D,\text{sat}})} \; . \tag{7.1.76}$$

Example 7.2
Problem: *Changing suddenly the gate voltage of a unipolar (e.g. MOS) transistor does not change abruptly the drain current. This is due to gate-channel capacitances that have to be charged up or, in other words, the channel charge density has to be changed by having a transition period with unequal drain and source currents. Estimate the time it takes to change from one stationary state to another for the transistor, operating in the saturation region, after a sudden change of the gate voltage, keeping source and drain at fixed potential during the transition.*
Solution: *As a measure for the transition time, the ratio between the change in total channel charge and the change in current between the two steady states will be taken. The simplified relationship for an n-channel MOS transistor, ignoring bulk defects, will be taken. We also will express the saturation current* $I_{D,\text{sat}}$ *in terms of the channel's surface-charge density at the source end,* $Q_c(0)$:

$$V_{D,\text{sat}} = \frac{Q_c(0)}{C_{\text{ox}}}$$

$$I_{D,\text{sat}} = \frac{W}{L}\mu_n C_{\text{ox}}\frac{V_{D,\text{sat}}^2}{2} = \frac{W}{L}\mu_n \frac{Q_c^2(0)}{C_{\text{ox}}} \; .$$

The total charge in the channel is

$$Q_t = W \int_0^L Q_c(y)\, dy = \frac{2}{3} W L Q_c(0)$$

with the factor 2/3 being due to the linear relationship between channel surface charge density and channel potential and the requirement of $Q_c(L) = 0$.
Taking the ratio

$$\tau = \frac{\partial Q_t/\partial Q_c(0)}{\partial I_{D,\text{sat}}/\partial Q_c(0)} = \frac{1}{3}\frac{C_{\text{ox}}}{\mu_n}\frac{L^2}{Q_c(0)} = \frac{1}{3}\sqrt{\frac{C_{\text{ox}} W L^3}{\mu_n I_{D,\text{sat}}}}$$

and some reasonable values $l = 5\,\mu\text{m}$, $W = 20\,\mu\text{m}$, $d_{\text{ox}} = 40\,\text{nm}$ and $I_{\text{D,sat}} = 10\,\mu\text{A}$, we have:

$$C_{\text{ox}} = \frac{\epsilon\epsilon_0}{d_{\text{ox}}} = \frac{3.84 \cdot 8.854 \times 10^{-14}\,\text{F/cm}}{40 \times 10^{-7}\,\text{cm}} = 8.5 \times 10^{-8}\,\text{F/cm}^2$$

$$\tau = \frac{1}{3}\sqrt{\frac{8.5 \times 10^{-8}\,\text{F/cm}^2 \cdot 20 \cdot 5^3 \times 10^{-16}\,\text{cm}^4}{1450\,\text{cm}^2/\text{Vs} \cdot 10^{-5}\,\text{A}}} = 1.2 \cdot 10^{-9}\,\text{s} \ .$$

It is also instructive to express the transition time in terms of the transconductance instead of the current. We obtain with (7.1.71)

$$\tau = \frac{\sqrt{2}}{3}\frac{C_{\text{ox}}WL}{g_{\text{m,sat}}} \ .$$

Note that $C_{\text{G}} = C_{\text{ox}}WL$ is the total gate-channel capacitance.

7.1.4 Threshold Behavior of Unipolar Transistors

The enhancement type of MOSFET transistor's characteristics so far have been developed for the condition of strong inversion. For JFETs and depletion-type MOSFETs, the condition of equality between majority carrier concentration and doping density in the undepleted region of the channel has been assumed and the abrupt change approximation of the majority carrier density from the neutral region to zero in the space-charge region has been used. As a result of these assumptions, the current is zero below the gate threshold voltage and starts to rise linearly above the threshold.

Clearly these assumptions are not valid close to threshold. In MOS transistors a *weak inversion* layer forms already when the gate voltage is kept below the threshold value and the JFET channel is still conducting, when the majority carrier density drops below the channel doping density. Operation in these conditions is common in detector front-end electronics, when low noise is required under the constraint of limited power consumption.

Predicting quantitatively the transistor characteristics close to and below the threshold is rather difficult and requires numerical simulation. We can, however, consider the situation in which the carrier density in the channel is so low that its influence on the potential can be neglected. In such a case the potential along the channel direction will be uniform and carrier transport in the channel will be by diffusion only. Making the somewhat arbitrary assumption that the channel can be treated as a tube with cross-sectional area A, with carrier density depending on the longitudinal coordinate y only, we have reduced the problem to a one-dimensional case (Fig. 7.10). The current is determined by the carrier densities at the channel ends:

$$I_{\text{D}} = -qAD_p\frac{\partial p}{\partial y} = -qAD_p\frac{p(0) - p(L)}{L} \ . \tag{7.1.77}$$

Source   **Channel**   **Drain**

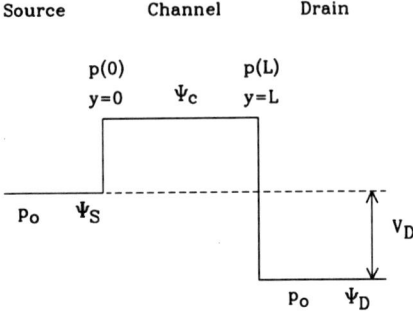

Fig. 7.10. Potential in the subthreshold region of a p-channel unipolar transistor

Similarly to previous treatments, such as in case of the diode, the carrier densities at the channel ends are assumed to be exponentially dependent on the voltage difference between source and channel, or drain and channel respectively:

$$p(0) = p_0 \exp\left(-\frac{\psi_c - \psi_S}{V_T}\right) \tag{7.1.78}$$

$$p(L) = p_0 \exp\left(-\frac{\psi_c - \psi_D}{V_T}\right) = p(0)e^{\frac{V_D}{V_T}}$$

with p_0 as the carrier density, ψ_S and ψ_D being the potentials in the neutral source and drain regions respectively, and $V_T = \frac{kT}{q}$ the thermal voltage. With $V_D = \psi_D - \psi_S$, (7.1.77) can be rewritten as

$$I_D = -qAD_p p(0)\frac{1 - e^{\frac{V_D}{V_T}}}{L} = \text{const.} \; e^{-\frac{\psi_c}{V_T}}\frac{1 - e^{\frac{V_D}{V_T}}}{L} \; . \tag{7.1.79}$$

Here we have included everything within the constant except the quantities depending on the applied voltages, such as V_D and ψ_c. This is reasonable, as we will not make an attempt to determine the effective channel cross-section A but will rather look at the dependence of the transconductance $g_m = \frac{\partial I_D}{\partial V_G}$ as a function of the transistor current. In our approximation, the channel potential differs from the gate potential by a constant amount. This is true for a JFET with front and back gate connected, and it is a good approximation for MOSFETs.

Taking the derivative of (7.1.79) with respect to the channel potential, we find:

$$g_m = \frac{\partial I_D}{\partial V_G} \approx \frac{\partial I_D}{\partial \psi_c} = \frac{|I_D|}{V_T} \; . \tag{7.1.80}$$

The transconductance is given by the ratio of drain current to thermal voltage. The same relationship was found for bipolar transistors in normal operational mode.

Finding the same optimum relation between transconductance and transistor current as for bipolar devices, one may be wondering about possible differences. They are found in device speed. Changing the gate voltage suddenly will not lead to an instantaneous change of drain current; it will take some time until a new equilibrium in the charge carrier density in the channel is reached. One can investigate the transition between the two stationary states by solving the time-dependent diffusion equation for the carrier density in the channel.

It is much simpler, however, to get an estimate for the transition time by considering the two steady states and taking the ratio between change in total channel charge and change in current, as was done in Example 7.2. Assuming that the transistor is operated in the saturation region, the charge density at the drain end of the channel is zero, and then the total charge in the channel is

$$Q_T = \frac{1}{2}qALp(0) \tag{7.1.81}$$

because of the linear relationship between charge density and position, while the current according to (7.1.79) is given by

$$I_D = -qAD_p p(0)\frac{1}{L} \tag{7.1.82}$$

and the ratio between channel charge and current is given by

$$\tau = -\frac{Q_T}{I_D} = \frac{L^2}{2D_p} \quad . \tag{7.1.83}$$

This ratio is thus seen to be dependent only on the square of the channel length.

Example 7.3
Problem: *Find the carrier transit time, namely the average time it takes for a charge carrier to cross the channel, and show that it equals the transition time between two steady states from the derivation above. Find numerical values for n- and p-channel transistors with 5 μm gate length.*
Solution: The velocity at lateral position y of the channel is

$$v(y) = \frac{-I_D}{qAp(y)} \quad , \qquad \text{with} \qquad p(y) = p(0)\left(1 - \frac{y}{L}\right) \quad . \tag{7.1.84}$$

The time dt to move a distance dy is given by $dt = dy/v(y)$. Integrating over the channel length, the transition time becomes

$$\begin{aligned}
\tau &= \int_0^L \frac{1}{v(y)}\,dy = -\frac{qA}{I_D}\int_0^L p(y)\,dy \\
&= -\frac{qA}{I_D}\int_0^L p(0)\left(1 - \frac{y}{L}\right)dy = -\frac{qA}{I_D}\frac{p(0)L}{2} = \frac{L^2}{2D_p} \quad .
\end{aligned} \tag{7.1.85}$$

For 5 μm-long transistors, the transit times are then

$$\tau_p = \frac{L^2}{2D_p} = \frac{L^2}{2V_T\mu_p} = \frac{(5 \times 10^{-4}\,\text{cm})^2}{2 \cdot 0.0259\,\text{V} \times 505\,\text{cm}^2/\text{Vs}} = 9.6\,\text{ns}$$

$$\tau_n = \frac{L^2}{2D_n} = \frac{L^2}{2V_T\mu_n} = \frac{(5 \times 10^{-4}\,\text{cm})^2}{2 \cdot 0.0259\,\text{V} \cdot 1450\,\text{cm}^2/\text{Vs}} = 3.3\,\text{ns} \quad.$$

7.1.5 The Different Types of JFETs and MOSFETs

In the models presented so far, we have only considered one channel type at a time and have assumed constant doping concentration in distinct regions of the devices. It is straightforward to introduce the proper type and sign changes in order to convert from n-channel to p-channel devices and vice versa. The assumption of a constant doping concentration is rarely fulfilled. However, qualitatively the behavior remains similar, as it does if position-dependent doping in the channel is assumed. In fact the graded channel approximation can be used almost unchanged so long as doping depends on channel depth only. It is then possible to obtain, similarly to (7.1.20) and (7.1.43) for the output conductance in the threshold region:

$$g = \frac{\partial I_D}{\partial V_D} = \left| \frac{W}{L} \mu Q_c(L) \right| \tag{7.1.86}$$

and for the sum of transconductances of front and back gate (substrate for MOS transistors):

$$(g_{m,1} + g_{m,2})_{\text{sat}} = \left(\frac{\partial I_D}{\partial V_G} + \frac{\partial I_D}{\partial V_{BG}} \right)_{\text{sat}}$$

$$= g(V_D = 0) = \left| \frac{W}{L} \mu Q_c(0) \right| \quad. \tag{7.1.87}$$

As well as the graded channel approximation, which is inadequate for very short channel lengths, several other approximations also have their limitations. The channel of the MOS transistor already forms below the threshold gate voltage. The device is then said to work in "weak inversion". This operational mode is often used in low-noise low-power applications.

For use in circuits, the distinction between enhancement and depletion devices is important. Enhancement transistors do not draw current when the gate is connected to the source. Junction field effect transistors are always of the depletion type. The type of MOS transistors depends on many parameters, the most important of which is the material comprising the gate insulator. Silicon oxide always has positive charges, so that simple p-channel MOS transistors are enhancement devices.

The threshold voltage can be changed in either direction by suitable doping close to the surface. Similar to JFETs, one may also construct a channel by doping. In normal operation this channel is depleted from the top (MOS) gate, so that the current flow is moved from the surface into the semiconductor. These kinds of devices are MOS n- or p-channel depletion-type transistors.

The schematics of standard JFET and MOS transistors is shown in Fig. 7.11.

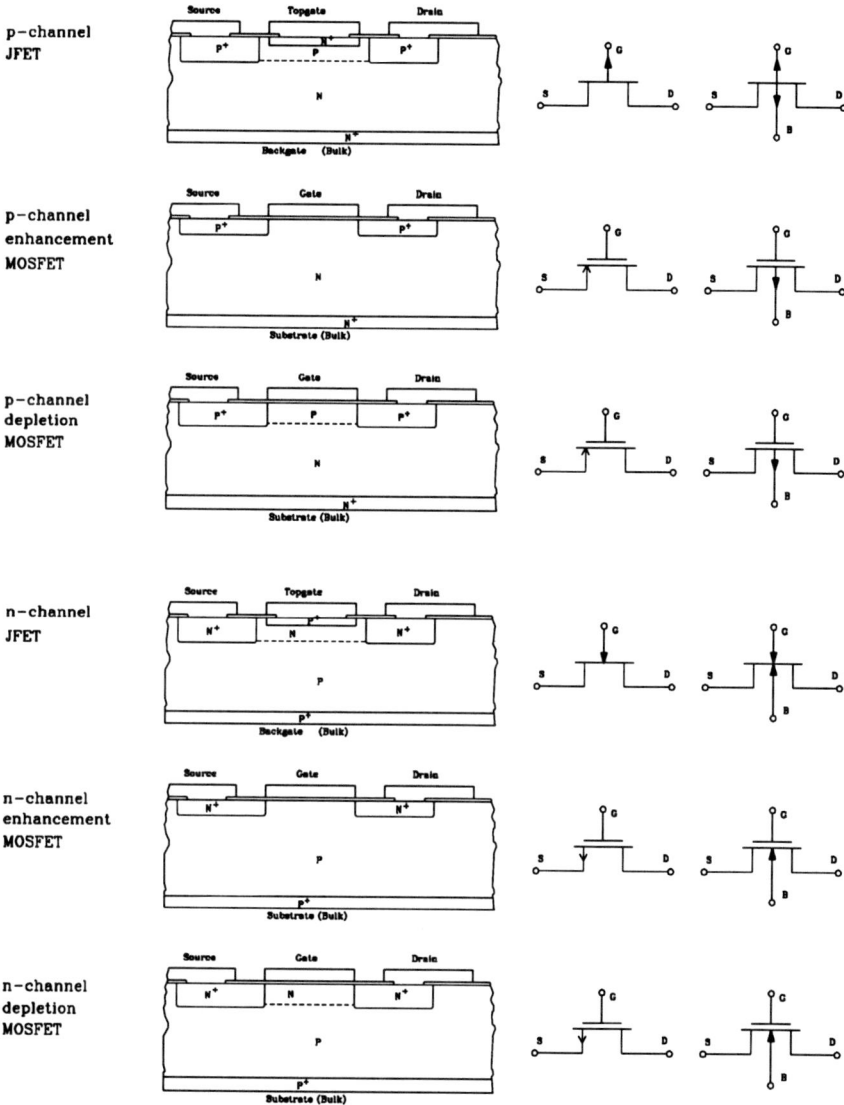

Fig. 7.11. Various types of JFETs and MOSFETs: cross-section along the channel (*left*) and commonly used device symbols. The rightmost symbols contain also the substrate contact, the *arrow* indicating the diode between substrate and channel that is normally reversely biased

7.2 Noise Sources

There exist three different sources of noise, present in most electronic elements: thermal noise, low-frequency voltage noise, and shot noise.

Noise can be considered as macroscopic manifestations of microscopic random processes. It is therefore derived from considerations of microscopic processes. In some cases it is, however, also possible to obtain results from macroscopic (thermodynamic) considerations only. This is the case for thermal noise, where a system in thermal equilibrium can be considered.

7.2.1 Thermal Noise

This noise is theoretically very well understood and can be calculated by thermodynamics. The basic reason for this noise is the thermal fluctuations of the electron distribution in a conductor. Consider, for example, a resistor of resistance R. Even without having a current flowing, it is possible to measure a noise voltage U_n between the two terminals. The spectral density of the noise power, according to thermodynamic calculation, is given by

$$\frac{d\overline{U_n^2}}{df} = 4kTR \tag{7.2.1}$$

with f the frequency (taking positive values only), k the Boltzmann constant and T the absolute temperature.[26]

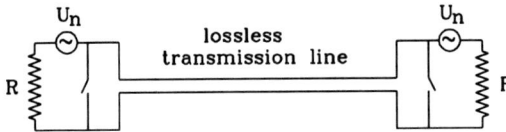

Fig. 7.12. Gedankenexperiment for the derivation of the thermal noise spectrum of a resistor, proposed by Nyquist in 1928

Equation (7.2.1) can be derived by considering a system consisting of a delay line of impedance R that is properly terminated at both ends, as shown in Fig. 7.12 (Nyquist 1928). The noise voltage created by one resistor will travel through the delay line towards the other resistor, where the signal will be converted to heat and vice versa. Shortening the delay line suddenly at both ends will lead to a reflection of the travelling waves at the ends of the delay line, such that standing waves will be present in the delay line. Counting the number of possible states in a given frequency interval and multiplying it by the

---

[26]The power spectrum in (7.2.1) is called "physical" or "unilateral", with the frequencies assuming positive values. The "mathematical" or "bilateral" power spectrum, with frequencies runing from $-\infty$ to $+\infty$, is $\frac{d\overline{U_n^2}}{df} = 2kTR$.

occupation probability according to Boltzmann statistics, and realizing that the waves will have been originally produced by the resistors, we arrive at above formula given in (7.2.1).[27] As the thermal noise spectrum is independent of the frequency, thermal noise is often referred to as "white noise".

A physical resistor, according to (7.2.1), can be described as an idealized noiseless resistor, with a noise voltage source U_n in series or a noise current source I_n in parallel with the spectral densities, given as

$$\frac{\mathrm{d}\overline{U_n^2}}{\mathrm{d}f} = 4kTR \qquad\qquad \frac{\mathrm{d}\overline{I_n^2}}{\mathrm{d}f} = \frac{4kT}{R} \quad .$$

7.2.2 Low-Frequency Voltage Noise

This type of noise is seen in most electronic devices. The noise power spectrum in most cases has an approximate $1/f$ dependence, shown by:

$$\frac{\mathrm{d}U_n^2}{\mathrm{d}f} \approx \frac{A_n}{f^\alpha} \quad, \qquad \text{with} \quad \alpha \approx 1 \quad . \tag{7.2.2}$$

The physical source of this noise is not unique, and may be due to widely differing mechanisms for different types of electronic elements. Also, for the same type of elements the strength of the noise may depend on details in the technological production process. Furthermore, the $1/f$ dependence is only a rough approximation.

Because of the importance of low-noise signal measurement, a separate section (Sect. 7.2.4) will be devoted to a discussion of the physical origin of low-frequency noise in field effect transistors.

7.2.3 Shot Noise

This noise is a consequence of the discrete nature of electric charge and represents the statistical fluctuation in the number of charge carriers making up a charge $Q = Nq$. Considering a "constant" current I, one expects a charge $\Delta Q = I\Delta t$ crossing a boundary along the current path in a (short) time interval Δt. This corresponds to $\Delta N = \frac{I\Delta t}{q}$ electrons on average, with a statistical fluctuation $\delta\Delta N = \sqrt{\Delta N}$. Thus the mean-square variation of the current measured over a short time interval is

$$\langle \delta I^2 \rangle = \frac{q^2(\delta\Delta N)^2}{\Delta t^2} = \frac{q^2\Delta N}{\Delta t^2} = \frac{qI}{\Delta t} \quad .$$

---

[27] Notice that in this thermodynamic derivation the elementary charge q and the Planck constant h do not feature. If quantum effects are included, (7.2.1) changes to $\frac{\mathrm{d}U_n^2}{\mathrm{d}f} = 2Rhf \coth\frac{hf}{kT}$, which reduces to (7.2.1) for $h \to 0$.

It is also possible to derive from these considerations a frequency spectrum of the noise current, leading to

$$\frac{\mathrm{d}\langle i_n^2 \rangle}{\mathrm{d}f} = 2Iq \ . \tag{7.2.3}$$

In deriving this equation, the implicit assumption is made that the probability of an electron passing the boundary is independent of the status of all other electrons. This means that the change of the potential distribution caused by the electrons that have already crossed the boundary does not influence the probability of another electron crossing the boundary. In general this condition will not be fulfilled. It is, however, fulfilled to a very good approximation in some important cases. One well known example is the thermionic emission of electrons out of a hot metal, in which case the electrons have to cross a potential barrier whose height remains almost unchanged as long as the electrons are moved away from the vicinity of the surface sufficiently fast. Another important example is the reverse-biased detector.

Notice that shot noise requires the presence of a current flow caused by an external power source, while thermal noise is present in devices even without application of external power.

7.2.4 Noise in Transistors

The noise properties of the various transistor types treated so far are rather different, and this is due to their different working principles.

Thermal noise is important in all types of transistors. Shot noise is important in the case of bipolar transistors, with their base currents typically only two orders of magnitude below the emitter current so that the statistical fluctuations of the base current cannot be neglected in comparison to the signal current. In Junction Field Effect transistors, shot noise plays a minor role and in MOSFETs it is completely absent. $1/f$ noise is very prominent in MOSFETs but much less significant in JFETs and bipolar transistors.

Thermal Noise
Thermal noise in field effect transistors is due to the resistance of the channel. However, one cannot simply calculate the resistance of the channel and use the resistor's thermal noise formula given in (7.2.1): a somewhat more sophisticated analysis is necessary, which takes into account the interaction of the channel with the gate.

In Fig. 7.13 the channel has been split into many short regions, each with correspondingly small resistances and noise sources. In order to investigate the influence of these noise sources on the drain current, we will first consider the influence of a single voltage change δV_n at a position y in the channel. We split the transistor (Fig. 7.13) into two parts, I and II, with channel length y and $L-y$ respectively. The current in these two parts is the same, and we are looking for the current change ΔI when the source voltage of the second part differs by δV_n from the drain voltage of the first. The first transistor operates in the threshold

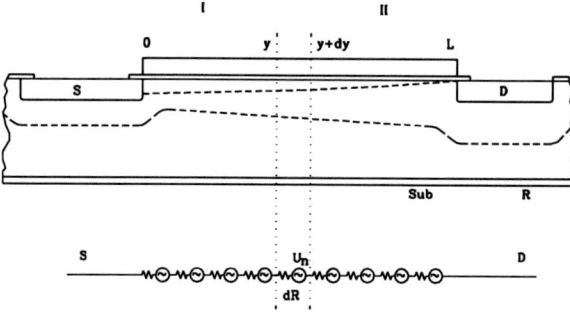

Fig. 7.13. Derivation of noise in a MOS transistor. The channel is split into many short pieces of resistance dR, and corresponding noise voltages are obtained from (7.2.1)

region while the second one is in the saturation region. The introduction of the (noise) voltage δV_n will lead to a change of the drain voltage of the first transistor of ΔV_c and of the source voltage of the second by $\Delta V_c + \delta V_n$. Thus the change of the drain current is obtained via (7.1.86) and (7.1.87) from the requirement of current equality in the two transistor regions, and is given by

$$\Delta I_D = g_I \Delta V_c = -(g_{m1} + g_{m2})_{II}(\Delta V_c + \delta V_n)$$
$$= \frac{W}{y}\mu Q_c(y)\Delta V_c = -\frac{W}{L-y}\mu Q_c(y)(\Delta V_c + \delta V_n)$$

$$\Delta V_c = -\frac{y}{L}\delta V_n \tag{7.2.4}$$

and so

$$\Delta I_D = g_I \Delta V_c = -\frac{W}{y}\mu Q_c(y)\frac{y}{L}\delta V_n = -\frac{W}{L}\mu Q_c(y)\delta V_n \quad . \tag{7.2.5}$$

We will now add up the effect of noise voltages generated in regions dy of the channel of length L. The resistance of this piece of channel is

$$dR = \frac{dy}{\mu Q_c(y)W} \quad , \tag{7.2.6}$$

and the corresponding noise spectrum is

$$\frac{d\langle \delta V_c^2 \rangle}{df} = 4kT\,dR = \frac{4kT}{W\mu Q_c(y)}\,dy \quad , \tag{7.2.7}$$

so that the noise spectrum for the transistor becomes

$$\frac{d\langle \Delta I_D^2 \rangle}{df} = \int_{y=0}^{L} \left[\frac{W}{L}\mu Q_c(y)\right]^2 \frac{4kT}{W\mu Q_c(y)}\,dy$$
$$= 4kT\frac{W}{L}\mu\frac{1}{L}\int_{y=0}^{L} Q_c(y)\,dy = 4kT\frac{W}{L}\mu\langle Q_c\rangle \quad . \tag{7.2.8}$$

With (7.1.87) this may be rewritten as

$$\frac{\mathrm{d}\langle \Delta I_\mathrm{D}^2 \rangle}{\mathrm{d}f} = 4kT(g_{m1} + g_{m2})\frac{\langle Q_c \rangle}{Q_c(y=0)} \ . \tag{7.2.9}$$

The noise frequency spectrum of a transistor is that of a resistor with resistance equal to the inverse of the transconductance (front plus back), modified by the ratio of average channel surface charge density and channel charge density at the source end.

The modification factor $\langle Q_c \rangle / Q_c(y=0)$ is dependent on technical details such as the channel doping profile. For a MOS enhancement transistor for which a linear relationship between channel surface charge density and channel potential (ignoring the substrate effect) exists, we obtain, taking into account the voltage drop along the channel (see (7.2.6)):

$$Q_c(y) = Q_c(0)\sqrt{1 - \frac{y}{L}} \ , \tag{7.2.10}$$

and for the correction factor:

$$\frac{\langle Q_c \rangle}{Q_c(y=0)} = \frac{1}{L}\int_{y=0}^{L}\sqrt{1 - \frac{y}{L}}\,\mathrm{d}y = \frac{2}{3} \ . \tag{7.2.11}$$

It is customary to characterize the noise behavior of a transistor by a noise voltage source positioned at the gate. The white channel noise of the transistor current (see (7.2.9)) can be represented by a white noise voltage source at the gate, given by

$$\frac{\mathrm{d}\langle v_n^2 \rangle}{\mathrm{d}f} = 4kT\frac{2}{3}\frac{1}{g_m} \ , \tag{7.2.12}$$

with the assumption that source and backgate are connected and that g_m represents the sum of the transconductances of the two gates.

Low-Frequency Noise

The physical reasons for low-frequency noise in unipolar transistors are crystal defects, which cause trapping of charge carriers. Movable charge carriers become bound on local trapping centers and are released with some time delay. While being trapped, the presence of the temporarily fixed charge induces charge in the channel und thus modulates the transistor current.

Considering now an individual trap in the semiconductor bulk region of an n-channel JFET transistor, as indicated in Fig. 7.14, the amplitude of modulation of the transistor current will be constant, because it depends only on the position of the trap with respect to the channel. The average time it takes until a trap changes from the empty state to the filled state will be dependent on the electron capture cross-section and the electron concentration in the vicinity of the trap. The time constant for re-emission of the electron from the trap is strongly dependent on the depth of the trap and on the temperature. The random filling and re-emptying of traps therefore causes a random

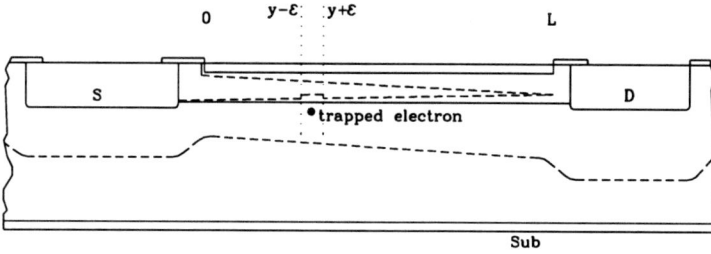

Fig. 7.14. Low-frequency noise: the effect of a single electron trap on an n-channel JFET transistor. The electron trapped at position y decreases the channel surface density in its vicinity and thereby decreases the transistor current

change over time of the drain current between two constant levels. It therefore was named "random telegraph signal" (RTS) by Kandiah, who was able to observe experimentally this effect of single traps in transistors (Kandiah 1981). The systematic study of random telegraph signals, especially their dependence on the temperature, gives insight into the trapping mechanisms and allows optimization of devices with respect to their noise behavior (Kandiah 1983 and 1986; Kandiah et al. 1989).

Some examples of random telegraph signals observed by Kandiah in small-sized JFET and MOS transistors are shown in Fig. 7.15. The drain current of a JFET is shown in the top part of the figure. A single trap with a characteristic time of 1 ms is present. In the bottom part of the figure the temperature dependence of the drain current of a MOSFET is shown. One notices the large effect of a temperature change of only 14 K.

In the following we will try to obtain a quantitative understanding of the effects by first investigating the amplitude of the random telegraph signal, then considering the time needed for changes in the charge state of the trap, and finally deriving the noise frequency spectrum due to a single trap.

Looking at the amplitude of the random telegraph signal, we again use the graded channel approximation, as has been done for all transistors considered so far. We assume that in the vicinity of position y along the channel a charge $q \cdot m$ is induced in the channel.[28] We distribute this charge over the width W of the channel, and over a length $y_t - \epsilon$ to $y_t + \epsilon$ so that the surface channel density has a smooth distribution $Q_c(y)$ in the regions $0 < y < (y_t - \epsilon)$ and $(y_t + \epsilon) < y < L$, and $Q_c(y) + \Delta Q_c$ for $(y_t - \epsilon) < y < (y_t + \epsilon)$ with $\Delta Q_c = \frac{q \cdot m}{2\epsilon W}$. Except in the vicinity of the trap, the potential is assumed not to be directly affected by the trapped charge, and the channel surface charge density $Q_c(V_c)$ is dependent only on the channel voltage Ψ_c.

In determining the current $I - \Delta I$ for a given drain voltage V_D, we follow exactly the same procedure as was done earlier for all transistor types. We express the drain current by the channel surface charge density Q_c and the

---

[28] The factor $m \leq 1$ has been introduced to provide for the situation where only part of the trapped charge is induced in the channel, with the rest induced in other conductors such as the bulk, source and drain.

drift velocity ν, integrate over the channel length L, divide by L, and obtain:

$$
\begin{aligned}
I - \Delta I = \frac{1}{L} \Bigg\{ & \int_{y=0}^{y_t-\epsilon} -W Q_c(y)\nu(y)\,dy \\
& + \int_{y=y_t-\epsilon}^{y_t+\epsilon} -W[Q_c(y) + \Delta Q_c]\nu(y)\,dy \\
& + \int_{y=y_t+\epsilon}^{L} -W Q_c(y)\nu(y)\,dy \Bigg\} \\
= & \frac{W}{L} \int_{y=0}^{L} -Q_c(\Psi_c)\mu\frac{\partial \Psi_c}{\partial y}\,dy - \frac{W}{L}\Delta Q_c \int_{y=y_t-\epsilon}^{y_t+\epsilon} \nu(y)\,dy \\
= & \frac{W}{L}\mu \int_{\Psi_c=0}^{V_D} Q_c(\Psi_c)\,d\Psi_c - \frac{qm}{L}\nu(y_t) \ .
\end{aligned}
$$

As the first term can be identified with the current, obtained with a drain voltage V_D without the presence of a trapped charge (compare (7.1.16)), the change in the drain current due to the trapped charge is obtained as

$$
\Delta I = \frac{qm}{L}\nu(y_t) \ . \tag{7.2.13}
$$

As the drift velocity in the channel increases from source to drain end, the amplitude of the RTS is strongest for traps close to the drain end. For traps in the space-charge region beyond the drain end of the channel, RTS amplitudes decrease again.[29]

Considering now the mechanism of trapping and detrapping, we attempt to estimate the average time it takes for a single unoccupied trap to capture an electron and for a filled trap to release an electron. We consider the situation where the trap is located sufficiently above the intrinsic level E_i and where electrons are the only charge carrriers present in the vicinity of the trap, so that we can ignore hole capture and hole emission. (A generalization of the treatment to include holes is straightforward.) In these circumstances the probability for capturing an electron is proportional to the electron concentration n and the electron (thermal) velocity ν_{th}, so that we can write, for the average time it takes to capture an electron:

$$
t_1 = \frac{1}{n\nu_{th}\sigma_n} \ , \tag{7.2.14}
$$

with σ_n representing the electron capture cross-section.

The probability of electron emission from the filled trap will be a strong function of the position within the bandgap but not of the electron density.

---

[29]Kandiah derives (7.2.13) in a different way using Ramo's theorem, which considers the currents induced in electrodes when the velocity of a moving charge is changed. He concludes that for a transistor operating in saturation, the RTS signal amplitude should be maximal and constant for traps located in the pinch-off region (near the drain) and be given by the same formula, with the drift velocity replaced by the saturation velocity. This conclusion is, however, contradicted by computer simulations (Longoni et al. 1995).

XSO1

T = 166K V_D = 2V I_D = 17μA

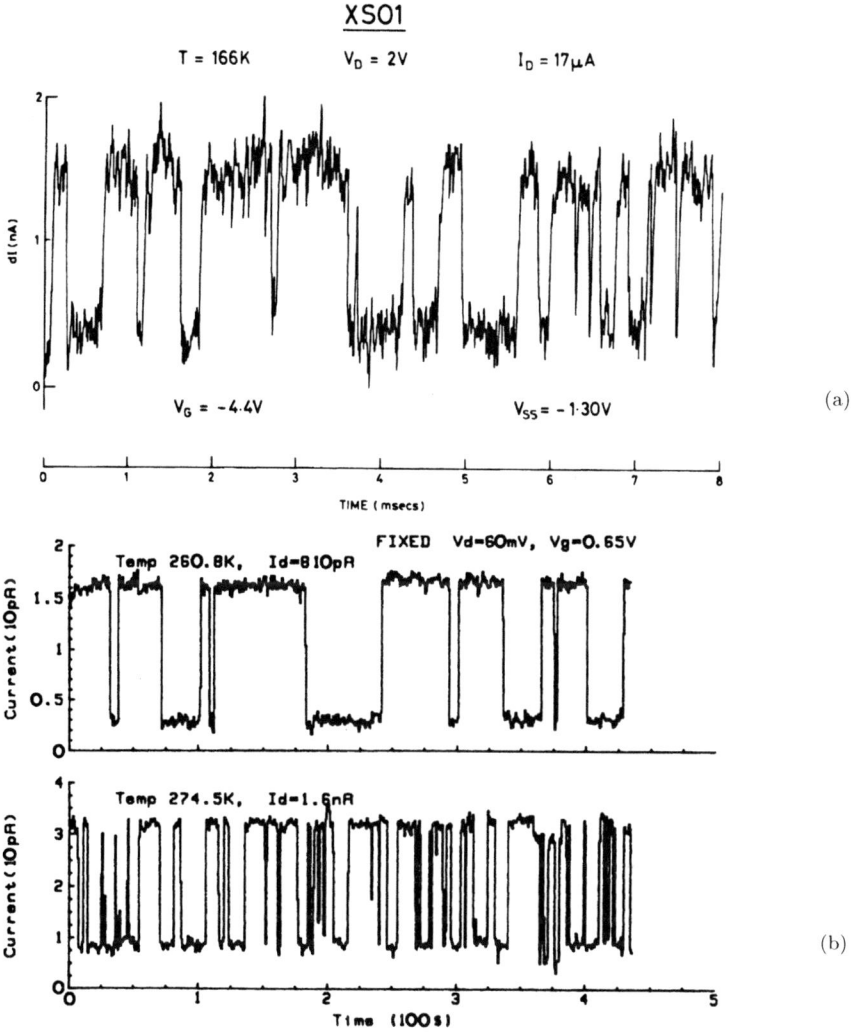

(a)

V_G = −4.4V V_{SS} = −1·30V

TIME (msecs)

FIXED Vd=6OmV, Vg=O. 65V

Temp 26O.8K,   Id=81OpR

Temp 274.5K,   Id=1.6nR

(b)

Time (1OO s)

Fig. 7.15a,b. Random telegraph signals due to bulk traps observed in a small-sized JFET (a) and a MOS transistor operated at two different temperatures (b). For a single defect the transistor current changes between two levels whenever the trap is being filled or emptied, thus changing its charge state. The amplitude of the change depends on the position of the trap; the characteristic times for filling and re-empting depend in addition on the nature of the trap and on temperature. (After Kandiah et al. 1989, Figs. 7 and 5)

In order to find this probability, we will consider the trap in the condition of thermal equilibrium i.e. the device without application of external voltages. For this situation we can calculate the probability P_{occ} that the trap is occupied with an electron from the Fermi–Dirac function [(2.3.1)] and compare it with the probability obtained from average electron emission time, t_2, and capture time in thermal equilibrium, t_1'. We thus have

$$P_{\text{occ}} = \frac{1}{1 + e^{\frac{E_t - E_F}{kT}}} = \frac{t_2}{t_1' + t_2} \quad,$$

resulting in

$$t_2 = \frac{t_1'}{e^{\frac{E_t - E_F}{kT}}} = \frac{e^{\frac{E_F - E_t}{kT}}}{n_{\text{th}} \nu_{\text{th}} \sigma_n} = \frac{e^{\frac{E_F - E_t}{kT}}}{N_c e^{-\frac{E_c - E_F}{kT}} \nu_{\text{th}} \sigma_n} = \frac{e^{\frac{E_c - E_t}{kT}}}{N_c \nu_{\text{th}} \sigma_n} \quad. \tag{7.2.15}$$

Here we have used (2.3.4) to express the thermal equilibrium electron concentration n_{th} in terms of the effective density of states in the conduction band N_c and the Fermi level E_F.

Next, we consider the noise frequency spectrum of the drain current, corresponding to the random telegraph signal with amplitude $a = \frac{qm}{L}\nu(y_t)$ (see (7.2.13)) and random high and low times t_1 and t_2. The energy spectrum of this noise signal $\frac{d\langle i_n^2 \rangle}{df}$ is obtained by applying Fourier analysis to the RTS amplitude containing the high and low times, which are statistically distributed according to exponential time distributions with mean time t_1 and t_2. Writing the current time dependence as $i(t)$, we have the Fourier transform

$$I(\omega) = \int_{-\infty}^{\infty} i(t) e^{-jwt} \, dt \quad, \tag{7.2.16}$$

while the current can be re-expressed by the Fourier transform as

$$i(t) = \frac{1}{2\pi} \int_{-\infty}^{\infty} I(\omega) e^{jwt} \, d\omega \quad. \tag{7.2.17}$$

Multiplying the complex conjugate of (7.2.17) with $i(t)$ and integrating over t, we can easily verify that the time integral over the square of the current $i^2(t)$ may be expressed as the frequency integral of the square of the Fourier transform (Parseval's Theorem), thus:

$$\int_{-\infty}^{\infty} |i(t)|^2 \, dt = \int_{-\infty}^{\infty} \frac{1}{2\pi} |I(\omega)|^2 \, d\omega \quad,$$

so that we can identify the absolute square of the Fourier transform as the spectral "current energy" density

$$S_i(\omega) = \frac{di^2}{df} = 2|I(\omega)|^2 \quad, \tag{7.2.18}$$

the factor two being due to taking positive frequencies only, i.e. using the unilateral spectral density. Averaging this expression over many current shapes with statistically distributed low and high times, one arrives at the spectral energy density for a random telegraph signal of

$$\frac{\mathrm{d}\langle i_n^2 \rangle}{\mathrm{d}f} = \frac{i_{n0}^2}{1 + \omega^2 t_0^2} \quad , \tag{7.2.19}$$

with the characteristic time t_0 and the low-frequency current power density i_{n0}^2 of the Lorentzian spectrum given as

$$\frac{1}{t_0} = \frac{1}{t_1} + \frac{1}{t_2} \quad , \qquad i_{n0}^2 = \frac{(2at_1 t_2)^2}{(t_1 + t_2)^3} \quad , \tag{7.2.20}$$

and with a the RTS amplitude estimated in (7.2.13).

For a single trap, the current noise power density is finite and constant at low frequencies and falls with $1/f^2$ at high frequencies. The corner frequency is at $f_0 = \frac{1}{2\pi t_0}$. The frequency spectrum of a single trap does not resemble $1/f$ behavior. This is also observed experimentally in small-sized JFETS that contain a very small number of defects.

The $1/f$ behavior, as observed in MOSFETs, can however be explained by a superposition of many traps with different time constants. Such is the case for traps within the oxide close to the oxide–silicon interface. The charge state of those traps can be changed by capture and emission of electrons or holes tunneling through a thin barrier of oxide from and towards the silicon. As the probability for tunneling has an exponential dependence on the product of barrier height and depth, one arrives at a continuum of characteristic times for the traps in the oxide under the assumption that the space distribution of the traps is continuous near the oxide–silicon interface.

For practical purposes, $1/f$ noise can be described by a gate noise voltage with spectral power density parameterized as

$$\frac{\mathrm{d}\langle v_n^2 \rangle}{\mathrm{d}f} = \frac{A_f}{f} = \frac{K_\mathrm{F}}{WLC_\mathrm{ox}^2} \frac{1}{f} \quad , \tag{7.2.21}$$

with K_F a parameter that is characteristic for the specific device production process. Equation (7.2.21) is very useful for scaling noise properties with transistor geometry. Note that $1/f$ gate noise voltage is independent of the transistor current. Typical values for K_F of MOS enhancement transistors are $10^{-32}\,\mathrm{C}^2/\mathrm{cm}^2$ (p-channel) and 3–$4 \times 10^{31}\,\mathrm{C}^2/\mathrm{cm}^2$ (n-channel) (Sansen 1987) This parameterization is, however, only approximate. Much lower noise can be obtained in JFETs and depletion-type MOSFETs when the channel does not touch the oxide.

It is beyond the scope of this book to discuss in detail the various models for low-frequency noise. The book is, rather, intended to give a physical understanding of noise mechanisms and to refer the reader to the literature for further studies.

7.3 The Measurement of Charge

The standard problem in the readout of a semiconductor detector is the low-noise measurement of the signal charge, usually under severe constraints such as the requirements of high-speed operation, low power consumption, restricted space and/or high radiation levels.

In this section the general problems of charge measurement will be addressed, while specific solutions for the electronics will be considered later.

7.3.1 The Charge-Sensitive Amplifier

The charge-sensitive amplifier (CSA), invented by Emilio Gatti (Cottini et al. 1956) and represented in Fig. 7.16, consists of an inverting amplifying circuit which – in the ideal case – delivers an output voltage proportional to the input ($U_{\text{out}} = -A U_{\text{in}}$) and a feedback capacitor C_f. In addition, a high-resistance feedback or a switch is needed in the feedback loop, in order to bring the circuit into its operating condition. C_D represents the capacitive load of the detector at the input, C_i the capacitive load to earth present in the amplifier, usually dominated by the gate capacitance of the input transistor.

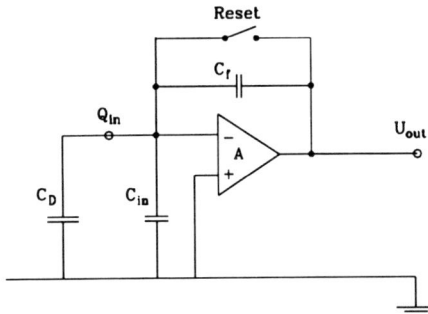

Fig. 7.16. Principle of a Charge Sensitive Amplifier. The inverting amplifier has a capacitive feedback. The reset switch is only used for bringing the system into its operating condition, and is often replaced by a high-ohmic resistor

Putting a charge Q_{in} at the input will result in an output voltage change of

$$U_{\text{out}} = -\frac{Q_{\text{in}}}{C_f + \frac{C_D + C_i + C_f}{A}} \rightarrow -\frac{Q_{\text{in}}}{C_f} , \qquad (7.3.1)$$

which for large amplification is given by the ratio of signal charge over feedback capacitance, indicating the fact that the charge has been transferred completely from the detector to the feedback capacitor. For low frequencies the input impedance of the CSA will be represented by a capacitance of the value

$$C_{\text{eff}} = (A + 1)C_f + C_i . \qquad (7.3.2)$$

A high value of $C_{\text{eff}} > C_{\text{D}}$, i.e. a low input impedance, is important because when C_{eff} is only of the same order of magnitude as the detector capacitance C_{D} the charge would be incompletely transferred to the electronics. This would result in a loss of sensitivity and possibly crosstalk within the detector to neighboring channels.

7.3.2 Noise in a Charge-Sensitive Amplifier

A further important consideration in applying a charge-sensitive amplifier for the readout of a detector concerns noise produced in the detector's amplifier system. The contribution of the amplifier to the noise is due to the noise generated by the electronic components of the circuit. For a properly designed amplifier, its noise contribution will be dominated by the noise generated in the input transistor(s). In line with the discussion in Sect. 7.2.4, it must be remembered that the noise of MOSFETs can be represented by a noise voltage source put in series with the gate of the input transistor. For bipolar transistors – having a significant base current – a current source representing the fluctuation of this base current is added to the circuit diagram, representing the noise properties of the transistor. With the transistor at the input of the circuit, the noise voltage and noise current sources are positioned in series and parallel to the input. Consequently one speaks of "serial" and "parallel" noise.

Serial Noise

It is customary to describe the noise properties of the amplifier by a single voltage source U_n representing the effects of all sources transferred to the input.[30] The presence of this noise voltage will result in an output voltage even if there is no signal charge present. For an evaluation of the corresponding noise charge, it is easiest to consider the charge necessary to compensate for the effect of the noise voltage, such that the output voltage remains at zero. The value can be immediately read from Fig. 7.17:

$$-Q_n = U_n(C_{\text{D}} + C_{\text{in}} + C_{\text{f}}) = C_{\text{T}} U_n \ , \tag{7.3.3}$$

with C_{T} the total "cold" input capacitance.

Parallel Noise

A realistic detector cannot be presented as a simple capacitor as was done in Fig. 7.17. It will draw some leakage current, which has statistical fluctuations. These fluctuations can be represented as a noise current source i_n feeding into the amplifier input. If this noise is at a thermodynamic minimum (shot noise), the current power frequency spectrum is given by (7.2.3).

Parallel noise can also be generated by the amplifier itself. Having as input the base of a bipolar transistor, for instance, the base current contributes to the noise in the same way as the detector leakage current.

---

[30]This is possible for, for example, MOS devices, in which the input gate leakage current is negligible. For bipolar devices this "serial" noise is also presented in the same way, but a "parallel" noise source representing the gate leakage current has to be added.

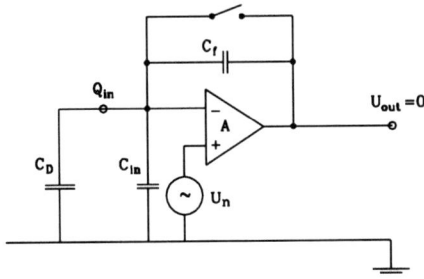

Fig. 7.17. The effect of amplifier noise in a Charge Sensitive Amplifier

Frequency Dependence

In the description of a CSA, the assumption of an ideal amplifier has been made, meaning that the output voltage is exactly proportional to the input voltage ($U_{out} = A_0 U_{in}$). A_0 is called the open loop gain. As is well known, this will only be a good approximation for low-frequency signals. At higher frequencies – sinusoidal signals are assumed – the amplification will decrease and the output signal will not stay in phase with the input voltage. The frequency at which the amplification drops by $\sqrt{2}$ is called the "corner" frequency. A phase change of the amplifier close to $180°$ and amplification above unity will lead to oscillations, when coupled back to the input.

We will leave the subject with these sketchy remarks alone and refer the reader to the electronics literature for further studies. The only remark to be added refers to the equivalent noise source at the input. In transforming the various noise sources that are distributed in the amplifier to the input, one should strictly speaking take into account the frequency-dependent amplification properties, which will vary with the position of the noise source in the circuit.

7.3.3 Filtering and Shaping

The signal produced by the amplifier will usually not be used directly; it will be further amplified and shaped. The aim of these procedures is to optimize the ratio of signal to noise and to reduce the interference between subsequent signals. We will only consider a few very simple cases, the simplest being an idealized charge-sensitive amplifier followed by an RCCR filter. For a more elaborate treatment, the reader is referred to the literature (e.g. Gatti and Manfredi 1986).

The arrangement of a CSA followed by an RCCR filter is shown in Fig. 7.18. The output of the CSA is a voltage step given by (7.3.1) for very high amplification as $\frac{Q}{C_f}$. The shaper does an RC integration followed by a CR differentiation. This procedure results in a signal peak, which for the same integration and differentiation time constant $\tau = R_1 C_1 = R_2 C_2$ has the shape and peak value given by

a)

b)

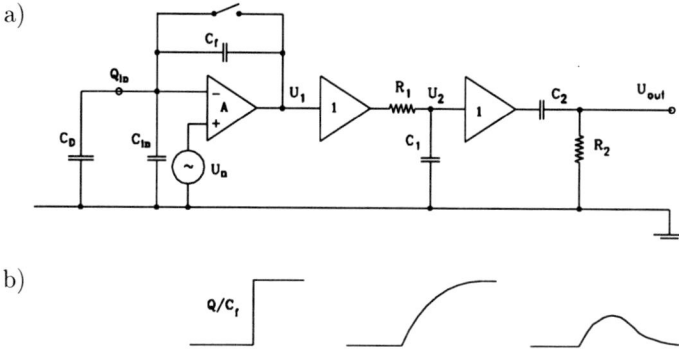

Fig. 7.18a,b. Noise filtering and signal shaping in a RCCR filter following a charge-sensitive amplifier (a). The two unity gain amplifiers have been introduced in order to completely decouple the functions of the CSA, the integration (RC) and the differentiation (CR) stages. The signal form is indicated for each stage (b)

$$U_{\text{out}} = \frac{Q}{C_f}\frac{t}{\tau}e^{-\frac{t}{\tau}} \ , \qquad U_{\text{peak}} = \frac{Q}{C_f}e^{-1} \ . \tag{7.3.4}$$

The height of this peak is a measure of the signal charge. Superimposed on the signal is the noise voltage, and we are interested in the signal-to-noise ratio, which is defined as the ratio of the height of the peak value to the root-mean-square value of the noise voltage measured at the same point in the circuit. In order to find the noise voltage at the output point, each noise source in the circuit has to be traced to the output and the resulting voltages added in quadrature.

Doing so, one finds the important result that, for white (thermal) serial noise, the ratio of noise to signal (N/S) decreases with the square root of the shaping time constant τ, while for 1/f noise this ratio stays constant. Parallel noise, given as a time integral over current fluctuations, increases with the square root of the shaping time. We thus have:

$$N/S_{\text{white, series}} \propto \frac{1}{\sqrt{\tau}} \tag{7.3.5}$$

$$N/S_{1/f, \text{series}} \propto 1 \tag{7.3.6}$$

$$N/S_{\text{white, parallel}} \propto \sqrt{\tau} \ . \tag{7.3.7}$$

More sophisticated continuous time filtering methods use (for example) Gaussian shape filtering, which can be approximated to by several RC integration and differentiation steps in sequence. Especially important in integrated electronics are the techniques in which the output signal is sampled several times and mathematical manipulation of the samples is performed. This can be done either after the measurement by numerical processing or directly via the local readout electronics. In the latter case, it is usually achieved by using switched capacitor techniques for analog algebraic manipulations. Common to both methods, however, is the need to sample the signal at fixed (or, at least,

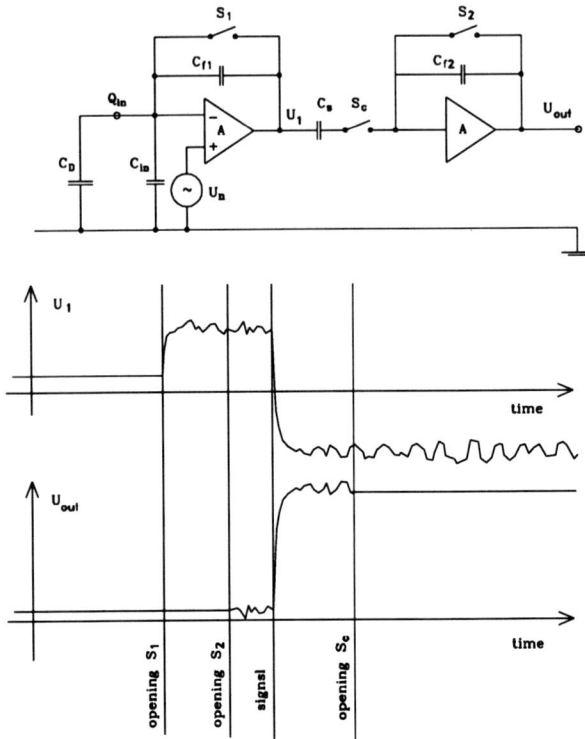

Fig. 7.19. Double-correlated sampling of the output of a charge-sensitive amplifier (CSA)

known) times with respect to its generation. With dense enough sampling, the arrival time of the signal can also be extracted from the data.

In the following narrative and in Sect. 7.6, examples of using switched capacitor techniques for filtering the output of a CSA will be given. The circuit shown in Fig. 7.19 is used for double-correlated sampling of the CSA output signal. Advance knowledge of signal arrival time is required. Such advance knowledge is quite common in accelerator experiments, in which bunches of accelerated particles arrive at fixed time intervals.

The circuit consists of two sequential charge-sensitive amplifiers connected by a coupling capacitor C_s and switch S_c. Initially all switches are closed. Thus both CSAs have reset their input and output voltages to proper working conditions and a possible offset voltage between CSA1 and CSA2 is stored on capacitor C_s. Ahead of the expected signal, the following operations are performed in sequence:

- opening switch S_1 at time t_1, resulting in an unwanted charge injection into the input of CSA1 and therefore an output voltage change that will be stored on capacitance C_s;

- opening of reset switch S_2. Any voltage change on the output of CSA1 (e.g. signal or noise) is also seen in the output of CSA2, amplified by the ratio C_s/C_{f2};
- signal charge Q_s generation at time t_3 changes the output of CSA1 by $\Delta U_1 = Q_s/C_{f1}$ and the output voltage by $\Delta U_{\text{out}} = Q_s \frac{C_s}{C_{f1}C_{f2}}$; and
- opening of switch S_c at time t_4 inhibits further change of output voltage. The difference of output voltage of CSA1 (amplified by C_{f2}/C_s) between times t_2 and t_4 remains present at the output of the circuit.

Considering filtering properties, one notices that the full signal step is measured. The result of a sinusoidal signal sampled at a time difference $\Delta t = \tau = t_4 - t_2$ will give a difference signal that will not only depend on the frequency but also on the phase. Squaring, averaging over the phase, and integrating over the noise frequency spectrum will give the mean-square noise signal to which the true signal has to be compared.

The procedure leads to the introduction of a weighting function, with which the noise frequency spectrum has to be convoluted before integration. For double-correlated sampling, this weighting function has the form

$$W(f) = 1 - \cos(2\pi f \tau) \ . \tag{7.3.8}$$

This function tends to zero for low frequencies and thus provides suppression of very low-frequency noise. However, it enhances higher frequencies by up to a factor of two before it goes to zero again at frequency $f = 1/\tau$ (and multiples thereof).

Double-correlated sampling does not suppress white noise, as the periodic weighting function extends to infinity. This has to be done by limiting the bandwidth of the charge-sensitive amplifier. As was the case for RCCR filtering, the N/S ratio for $1/f$ noise is almost independent of τ. Only for noise with faster-than-linear drop with frequency does double-correlated sampling give better results than RCCR filtering. Its virtue is in the suppression of switching noise from the CSA1 and in the easy implementation in integrated electronics circuits.

More sophisticated schemes of switched capacitor filtering, taking several samples (sometimes with different weight), have also been implemented.

7.4 Basic Electronic Circuits

The basic circuits described in this section have been selected in view of their application in integrated circuit technology. Since only the most basic properties are given, the reader is referred to the general literature (e.g. Gatti and Manfredi 1986; Gray and Meyer 1993) for a thorough treatment of the subject.

Properties of the circuits often are investigated using small-signal analysis: small deviations of voltages (u) and currents (i) from the static operational values. Such an analysis is not only appropriate for small signals but also for the investigation of noise properties.

In small signal notation a "noiseless" transistor is described as

$$i_D = \frac{\partial I_D}{\partial V_G} u_{GS} + \frac{\partial I_D}{\partial V_{DS}} u_{DS} = g_m u_{GS} + g u_{DS} \ . \tag{7.4.1}$$

7.4.1 Current Sources and Mirrors

Current sources are widely used elements in integrated circuit technology. They provide a current that is "independent" (within limits) of the applied voltage and that can be used, for example, as a "load" in an amplifying circuit. An often used feature is also current mirroring and scaling, so that a fixed ratio of currents in different branches of an electronic circuit can be enforced.

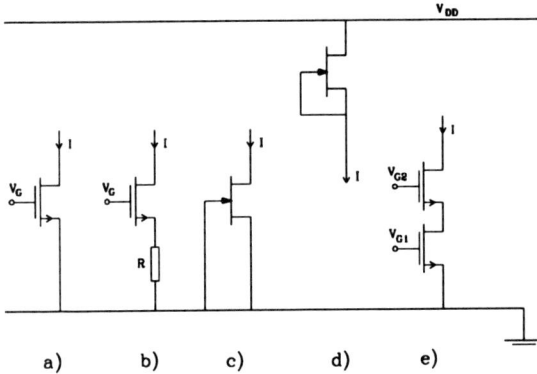

Fig. 7.20a–e. Various types of current sources, as described in the text

The simplest example of a current source of moderate quality is a transistor with fixed gate voltage working in the saturation region (Fig. 7.20a), having an internal conductance $\frac{\partial I}{\partial V} = g$. The quality of this source may be improved by putting a resistor into the source (Fig. 7.20b), thereby decreasing the conductance to $\frac{\partial I}{\partial V} = \frac{g}{1+(g_m+g)R}$. This result is easily derived using small-signal analysis with (7.4.1) and $i = u_R/R$ so that

$$i_{out} = -g_m u_R + g(u_{out} - u_R) = u_R/R \ .$$

In the case of a depletion-type transistor, e.g. a JFET, the constant gate voltage may be the source voltage – source and gate may be connected (Fig. 7.20c). Then the source may also be separated from the supply voltage and we have a current source that can be set at a floating potential (Fig. 7.20d). A drastically improved source can be built with two transistors (Fig. 7.20e). The internal conductance of this "cascode" circuit source is $\frac{\partial I}{\partial V} = \frac{g_1 g_2}{g_{m_2}+g_1+g_2} \approx g_1 \frac{g_2}{g_{m_2}}$.

The simplest and most common example of a NMOS current mirror is shown in Fig. 7.21a. The voltage of the gate connected to the drain of enhancement transistor T_1 will adjust itself in such a way that the transistor current equals

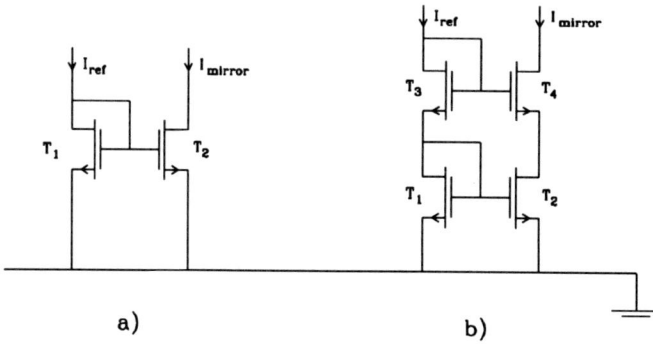

Fig. 7.21a,b. Various types of NMOS current mirrors: simple version (a) and cascode version (b)

the externally supplied current I_{ref}. As the gate voltages of T_1 and T_2 are equal, the ratio of the currents will be given by the ratios of transistor geometries

$$\frac{I_{mirror}}{I_{ref}} \approx \frac{W_2/L_2}{W_1/L_1} \tag{7.4.2}$$

provided that the transistors are equal in all other properties except geometries. The relation is only approximate, since the drain voltage may be rather different for the moderate quality current source T_2. A drastic improvement in the "stiffness" of the mirroring, i.e. the independence of the mirrored current on the applied voltage, is obtained with the "cascode" version shown in Fig. 7.21b.

7.4.2 Inverters

The inverters shown in Fig. 7.22 consist of an input transistor and a load, which can be a resistor (a), a current source (b), which might be realized by a depletion-type transistor (c) or an active load (d), as is usually the case in complementary technologies such as CMOS.

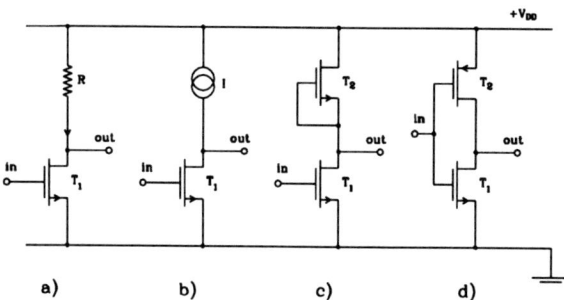

Fig. 7.22a–d. Various types of inverters, as described in the text

The inverter is the most common element in digital electronics, where only two extreme situations are of interest:

- input low, input transistor closed and output at positive supply voltage; and
- input high, input transistor open and output low.

While, in the examples a) to c) for low input, the power consumption of the circuits is zero, finite power consumption is present for high input. In contrast, the CMOS inverter shown in Fig. 7.22d with its active load, a p-channel transistor also steered by the input signal, has zero standing current for both input states; power is needed only in the transition between the two states. This is the basic reason for the low power consumption of digital CMOS circuits.

Inverters may also be used as amplifiers for analog applications. In this case the input voltage has to be adjusted to such a value that the output is not driven into low or high saturation. A common method consists of connecting input and output of the amplifier with a resistor or with a switch, which is closed from time to time. Then the input and output voltage at the operation point of the amplifier will adjust to the same level somewhere between the supply voltages in such a way that the current in the input transistor and the (active) load are equal.

We will calculate the low-frequency open loop gain (the ratio between output and input voltage with open output) and the output conductance for this situation. Taking as load the ideal current source of Fig. 7.22b, we find, using (7.4.1):

$$i = g_m u_{in} + g u_{out} = 0 \;\rightarrow\; \frac{u_{out}}{u_{in}} = -\frac{g_m}{g} \;. \tag{7.4.3}$$

For a resistive load (Fig. 7.22a) this procedure gives

$$i = g_m u_{in} + g u_{out} = -\frac{u_{out}}{R} \;\rightarrow\; \frac{u_{out}}{u_{in}} = -\frac{g_m}{g + 1/R} \;, \tag{7.4.4}$$

while for the CMOS inverter (Fig. 7.22d) with the active load we find

$$i = g_{m1} u_{in} + g_1 u_{out} = -g_{m2} u_{in} - g_2 u_{out} \;\rightarrow\; \frac{u_{out}}{u_{in}} = -\frac{g_{m1} + g_{m2}}{g_1 + g_2} \;. \tag{7.4.5}$$

In a similar way, the output impedance r_{out} and the output conductance $g_{out} \equiv 1/r_{out}$ can be be obtained by varying the output voltage and keeping the input voltage constant. One finds for the same three cases (Fig. 7.22b, a and d respectively):

$$i_{out} = g u_{out} \;\rightarrow\; g_{out} = \frac{1}{r_{out}} = g \qquad\qquad \text{(b)} \quad (7.4.6)$$

$$i_{out} = g u_{out} + \frac{u_{out}}{R} \;\rightarrow\; g_{out} = \frac{1}{r_{out}} = g + \frac{1}{R} \qquad\qquad \text{(a)} \quad (7.4.7)$$

$$i_{out} = g_1 u_{out} + g_2 u_{out} \;\rightarrow\; g_{out} = \frac{1}{r_{out}} = g_1 + g_2 \;. \qquad\qquad \text{(d)} \quad (7.4.8)$$

So far, we have only considered the low-frequency behavior of the inverters. At high frequencies, capacitive loads to the nodes of the circuit play a role. These capacitive loads may be connected externally, for instance by a detector connected to the input, by the input of a second stage amplifier connected to the output, or by stray capacitances of cables or connections. Furthermore, already the transistors have built-in capacitors, e.g. from gate to channel, in relation to both source and drain. Considering a sine-wave input signal, this will lead to a frequency-dependent amplification and phase shift of the output signal.

Alternatively one may consider the effect on the pulse form. A voltage step at the input will not result in a perfect step of the output voltage but rather in a smooth transition from one level to another. The time constant for this transition can be estimated by considering the situation of the circuit immediately after application of the input voltage step, as well as the final situation.

Example 7.4
Problem: *Estimate the rise time at the output of a charge-sensitive amplifier (CSA) built of a CMOS inverter with feedback capacitor when connected to a detector with capacitance C_D (Fig. 7.23). The capacitive load at the output representing the following stage is C_L.*

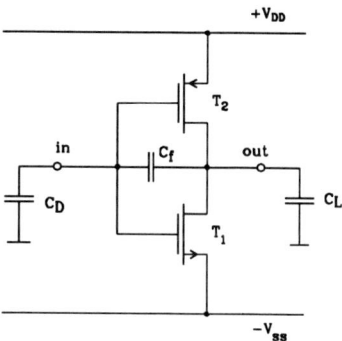

Fig. 7.23. CMOS inverter used as a charge-sensitive amplifier (CSA)

Solution: *The charge Q generated in the detector will in the first instance be split between the capacitors at the input C_D and C_f (the output being held fixed by the capacitive load $C_L \gg C_f$. Immediately after depositing a charge Q at the detector, the input voltage rises by $\frac{Q}{C_D+C_f} \approx \frac{Q}{C_D}$, thus changing the currents in the two transistors initially by $\Delta I = i_1 - i_2 = (g_{m1} + g_{m2})\frac{Q}{C_D}$. The asymptotic output voltage change $u_{out} = \frac{Q}{C_f}$ corresponds to a charge $\Delta Q_{out} = u_{out}(C_L + C_f) = Q(\frac{C_L}{C_f} + 1)$ to be stored at the output. Dividing this* charge by the initial current difference in the two transistors, one obtains for the time constant τ

$$\tau \approx \frac{\Delta Q_{\text{out}}}{\Delta I} = \frac{C_D}{g_{m1} + g_{m2}} \left(\frac{C_L}{C_f} + 1 \right) .$$

Considering the role of parasitic capacitances in this special circuit, one immediately recognizes that the gate–source capacitances have to be added to the detector capacitance and the gate–drain capacitances to the feedback capacitance. One may even do without a separate feedback capacitor but may just use the drain gate capacitances for the feedback. The effective input capacitance in this case is the Miller capacitance, the product of gate–drain capacitance and amplification (see (7.4.5)):

$$C_M = \frac{g_{m1} + g_{m2}}{g_1 + g_2} (C_{\text{gd1}} + C_{\text{gd2}}) .$$

7.4.3 Source Followers

Frequently, one needs to drive a low-impedance load, be it a large capacitance or a cable. This cannot be done from a high-impedance output; it is simply done by the addition of a source follower, which provides a low-impedance output. This circuit element consists of a transistor and a load in the source branch (Fig. 7.24). This load may be a resistor or a current source.

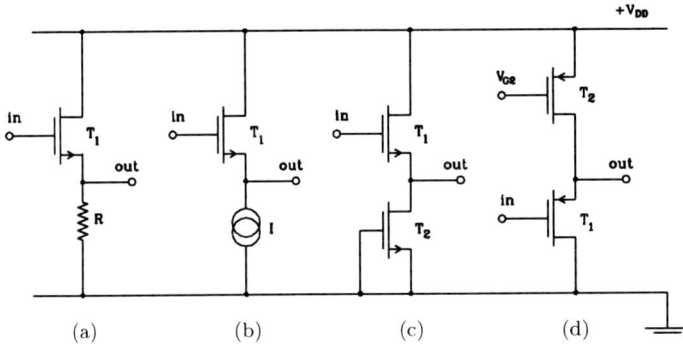

Fig. 7.24a–d. Several examples of source followers

The amplification of the source follower is close to unity. For the version with the resistive load (Fig. 7.24a), one obtains the amplification

$$\frac{u_{\text{out}}}{u_{\text{in}}} = \frac{g_m}{g_m + g + 1/R}$$

and the output conductance

$$g_{\text{out}} = \frac{1}{R_{\text{out}}} = g_m + 1/R \approx g_m .$$

The rise-time for a capacitive load C_L is

$$\tau = R_{\text{out}} C_L \approx \frac{C_L}{g_m} \quad .$$

7.4.4 Cascode Amplifiers

As has been shown in Sect. 7.4.2, the amplification of a simple inverter circuit is limited to the ratio of transconductance over channel conductance in the saturation region of a transistor. In order to obtain higher amplification, two or several inverters may be put in series. This, however, is dangerous if a feedback from output to input is required, as is the case in many applications. If the phase shift of the output signal is close to $180°$ at a frequency at which the amplification is larger than one, the circuit will oscillate.

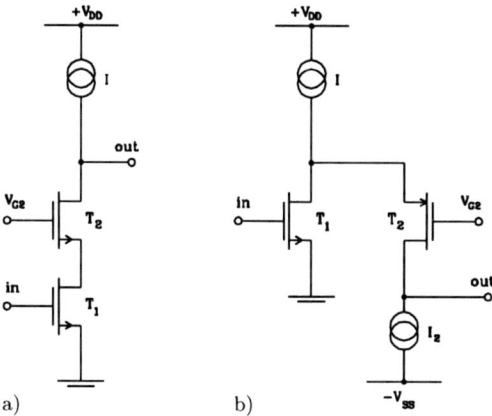

Fig. 7.25a,b. Cascode amplifiers: simple (a) and folded (b)

Cascode amplifiers provide high gain while suppressing the phase-shift problem to a large extent. Two examples of cascode amplifiers are shown in Fig. 7.25. The cascode transistor T_2 has its gate at fixed potential and thus reduces the change in the drain voltage of the input transistor T_1, due to the output voltage change, by a factor $\frac{g_{m2}}{g_2}$. The low-frequency gain is obtained as

$$\frac{u_{\text{out}}}{u_{\text{in}}} = \frac{g_{m1}}{g_1} \frac{g_{m2}}{g_2} \quad .$$

Phase changes in the output signals are due to the capacitive loads at the cascode node (drain of transistor T_1) and at the output node. The reason that cascode amplifiers are less prone to oscillations than the sequence of two simple amplifiers is due to the possibility of setting the onset of phase shifts in the cascode node far above that occurring in the output node. The amplification then drops below unity before the phase shift becomes sufficiently large to cause oscillations.

7.4.5 Differential Amplifiers

In a differential amplifier, the difference between two input voltages is amplified. In the ideal case the output voltage is independent of the average value
of the two input signals. Such amplifiers are well suited for amplification of
small signals, such as those that are transmitted through a (twisted) pair of
cables. Signals picked up the same way in both lines (the common mode) are
suppressed. Differential amplifiers are also frequently used for analog signal
processing in electronic networks.

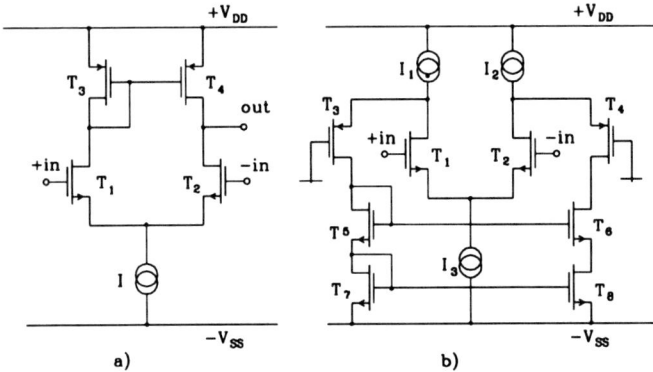

Fig. 7.26a,b. Differential amplifiers: simple (a) and with folded cascode (b)

A simple example is shown in Fig. 7.26a. Two n-channel input transistors
are connected with their common source to a current source. The source current
is split symmetrically between the two branches by the two transistors T_3 and
T_4 which, having their gates connected, form a current mirror. The output is
taken from the load of one of the two branches.

Figure 26b gives an example that is somewhat more complicated. It consists of two input transistors (T_1, T_2), two cascode transistors (T_3, T_4), three
current sources (I_1, I_2, I_3) and a cascode current mirror consisting of four transistors $(T_5$ to $T_8)$. The current source I_3 draws less current than the sum of the
two equal current sources I_1 and I_2, so that the difference goes into the two
folded cascode branches with cascode transistors T_3 and T_4. Putting a small
voltage difference at the input will in the first instance result in unequal currents in input transistors T_1 and T_2. As these current changes are fed directly
into the cascode branches and then mirrored onto the output node, the output
node (and also other nodes in the circuit) will change their potential – thereby
charging the (parasitic) capacitances not shown in the figure – until current
conservation is restored at each node. For an infinite resistive load (open output) and ideal current mirror $(T_5$ to $T_8)$ the original currents are restored. The
output voltage range of the circuit includes the input voltage. This is important when a d.c. feedback is needed for keeping the circuit in proper operational
condition.

7.5 Integrated Circuit Technologies

The development of position-sensitive semiconductor detectors, and in particular silicon strip detectors with strip pitch as low as $20\,\mu$m, have required major developments in the electronics of the readout, as a very large number of signal channels with high spatial density have to be read out. Therefore at a rather early stage the development was started of integrated electronics adapted to the specific needs of the readout of silicon detectors. The specific needs were the low-noise analog readout, combined with high density and low power consumption. Originally the circuits were based largely on existing electronics technologies; only later did the requirement for very low noise circuits and for radiation hardness lead to extensions and modifications of the standard MOS technology. A particular example is the development of electronics with a technology that could easily be integrated into the detector fabrication process (Sect. 8.2).

In this current section we will describe and compare some integrated circuit technologies. A specific circuit for strip detector readout will be discussed in the next section (Sect. 7.6).

Compared with discrete electronics, integrated electronics is much more restricted in the sense that only a limited number of device types are available. This is due to the fact that all devices have to be built on a single semiconductor substrate, without making the fabrication process too complex. Therefore usually only two types of transistors are available, because amplifying elements and the properties of these transistors can only be varied by changing the geometry, not by other parameters (for instance, doping concentrations). As passive components, only capacitors and resistors can be chosen, while inductances are practically not realizable. In addition, there exist severe restrictions on the ranges in which these passive components can be realized in an integrated circuit.

The most common integrated circuit technologies are based on MOS transistors as amplifying elements. The older NMOS technology is based on n-channel enhancement and depletion transistors, the CMOS technology on n-channel and p-channel enhancement transistors. Besides the relative simplicity in fabrication, low power consumption in digital applications is a strong point for MOS applications, especially CMOS technology. Its drawback is the limitation in speed; for high speed applications and for driving large currents, bipolar technology is superior.

In the following, the basic structures and production sequences of a few illustrative technology examples will be shown.

7.5.1 NMOS Technologies

We start with an example of NMOS technology. The elements that can be realized in basic NMOS technology are n-channel enhancement and depletion transistors, as well as capacitors and resistors. In order to keep the process reasonably simple, there is generally an aim by technologists to implement the passive elements with the same technological steps as used already in the more complicated transistor structures. Interconnections between devices have to be made with low-resistivity lines, which therefore are constructed of metal – usually aluminum. This metal can and has been used also for the gates of the transistors. More recently, however, there has been a significant switch to heavily doped polysilicon gates. The reason for this change has been the possibility of self-aligning the gate with respect to source and drain of the transistors – avoidance of overlap between gate and source or drain – thus minimizing parasitic capacitances. The introduction of polysilicon provides also a second connection layer, although one of much reduced conductivity. In the meantime, technologies using several metal-connection layers have become available.

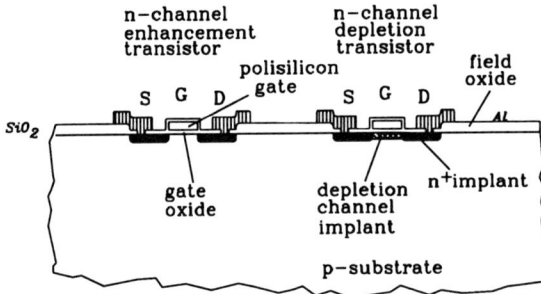

Fig. 7.27. NMOS technology: enhancement and depletion transistors

In order to reduce parasitic capacitances of the connecting lines between different devices and to avoid the appearance of parasitic electronic structures, the regions between the individual electronic elements are covered by a thick oxide layer, the "field oxide", on top of which the connections are located. A much thinner insulator (oxide) forms the gate barrier. A cross-section of the active elements used in one version of this technology is shown in Fig. 7.27. The substrate is lightly doped p-type silicon. Starting with the transistors, source and drain are heavily n-doped, and the thin gate oxide is covered by a polysilicon gate, which is exactly aligned with source and drain. Source and drain are connected to the metal layer by a small contact hole in the thin oxide. The connection of the oxide-covered polysilicon gate, with the metal outside the transistor area, is not shown in the figure. The depletion-type transistor has an additional n-type channel doping below the gate.

There are two simple ways of making a resistor: using a strip of polysilicon or making it out of the n^+ source implant that remains insulated from the bulk material due to the reversely biased junction.

Capacitances can be formed from polysilicon and n^+ implant separated by the thin oxide or between polysilicon and metal layer.

Not shown in Fig. 7.27 is a p-type surface doping, which is needed for compensation for the positive oxide and interface charges (see Sect. 3.3.2).

7.5.2 CMOS Technologies

Complementary Metal–Oxide–Semiconductor (CMOS) electronics uses in its simplest form only n- and p-channel enhancement transistors as active devices. These complementary transistors are insulated from each other by putting one of them into a well with opposite doping to the original bulk material. Such a well is usually produced by diffusion or by implantation and following drive-in diffusion.

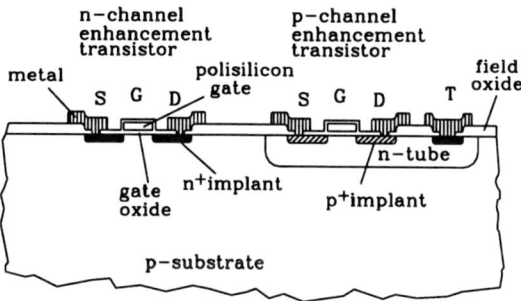

Fig. 7.28. CMOS technology: n- and p-channel enhancement transistors

The cross-section through the active devices of a CMOS version based on p-type substrate is shown in Fig. 7.28. Compared with the NMOS technology described above, the n-channel depletion transistor has been replaced by a p-channel enhancement transistor sitting in an n-doped well. This well has to be put at defined potential (as it controls the transistor threshold voltage), in the most common case to the positive supply voltage.

The properties of n- and p-channel devices will not be perfectly complementary, as already mobility of electrons and holes are different by a factor of three. Furthermore, bulk and well doping concentrations are significantly different, and the well potential can be chosen individually for each p-type transistor, while all n-channel transistors are embedded in the same bulk at or below the negative supply voltage. These differences have to be considered when trying to match n- and p-type transistors, so that for electrically similar behavior the geometry of complementary transistors will be different.

A special problem with CMOS electronics is "latchup". A self-sustaining discharge can be triggered by, for example, electrical pickup in a n–p–n–p thyristor-type structure built from real and parasitic structures. Sufficiently large distances between structures are required in order to avoid such problems. Sometimes additional guard structures are implemented.

7.5.3 Bipolar Technologies

Compared with MOS technologies, bipolar electronics requires higher techno-logical complexity. This is due to the necessity of building multilayer vertical structures, in which a very thin base layer has to be interleaved between emitter and collector layers, as shown in Fig. 7.29 for the example of a bipolar n–p–n transistor. A highly doped n^+ buried layer makes the connection of the buried n-type collector to the sideways displaced contact at the surface. This buried layer is produced by phosphorous diffusion into the original p-type wafer. An epitaxial n-type layer (see Chap. 10) is grown on top of the wafer and structured before growing the oxide, which laterally insulates the transistor structure. The base is formed by a flat boron implantation and the emitter's and collector's external contacts are constructed as a flat arsenic implantation. Most of the space is used up for connections; the proper transistor region is the small region below the emitter.

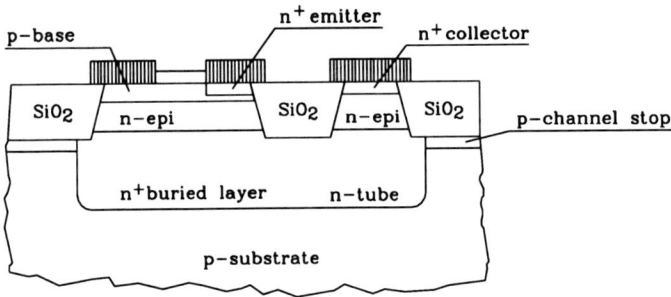

Fig. 7.29. Simplified cross-section of a bipolar n–p–n transistor

7.5.4 SOI Technologies

Silicon substrates are usually single crystals, with a thickness of several hundred microns whereby only a very thin top layer of a few microns is electrically active. The rest of the (undepleted) bulk acts as a fairly bad conductor, whose influence is largely detrimental because it produces crosstalk between different parts of a circuit through the so-called bulk effect, namely the steering of transistor current by the substrate. Furthermore, space between adjacent circuit elements is needed in order to protect against unwanted parasitic circuit elements, as well as latchup.

In Silicon-On-Insulator (SOI) technologies, a thin silicon layer on top of an insulator is created by, for example, high energy, high intensity implantation of oxygen followed by recrystallization of the top silicon layer. This top layer is then structured so that each transistor is built on its own island of silicon.

In some cases the silicon layer thickness is increased by epitaxial growth techniques, so that devices requiring a larger depth, such as JFETs, can also be implemented.

SOI technology allows a larger packing density of electronics than is possible with bulk techniques. Furthermore parasitic capacitances of interconnections are reduced because the insulator is much thicker than the field oxide in MOS or CMOS techniques. A potential problem is due to the undefined conditions on the buried $Si-SiO_2$ interface and on the side edges of the silicon islands. These problems can, however, be avoided by proper design of the circuit elements.

7.5.5 Mixed Technologies

The desire for making use of complementary features of devices on the same electronic circuit has led to the development of sophisticated combined technologies. BICMOS combines the advantages of low power consumption provided by CMOS and the large current-driving capabilities and speed of bipolar transistors. Recently, technologies implementing in addition JFET transistors have been developed. JFET transistors are of interest because of their good $1/f$ noise behavior and their intrinsic radiation-hardness. Some of these technologies are available in an SOI version.

7.6 Integrated Circuits for Strip Detectors

The development of integrated detector readout electronics was initiated by the simultaneous requirements of high density, low power and low noise for use with silicon strip detectors in the tight spatial environment of elementary particle physics collider experiments.

This development was started in Germany (Hofmann et al. 1984; Buttler et al. 1988) and first implemented in an experiment in the USA (Walker et al. 1984). Meanwhile, many circuits have been developed for this purpose, the basic principle of essentially all of them being:

- parallel amplification using a charge-sensitive amplifier at each input;
- parallel signal filtering combined with second-stage amplification and parallel storing within capacitive hold circuits; and
- serial readout through one single output channel.

The various circuits partially have special properties, such as: sparse readout (Kleinfelder et al. 1988), in which strips without signal are suppressed; slow continuous signal filtering, which simultaneously delays the signal until the arrival of an external signal (Beuville et al. 1990); and very low serial noise with extremely slow filtering (Toker et al. 1994).

The first of these developments, containing as it does most of the features of later projects, will be taken as an example. The basic functional principle of a single channel is shown in Fig. 7.30. It consists of two charge-sensitive amplifiers, each of them followed by a source follower, and four sets of capacitors

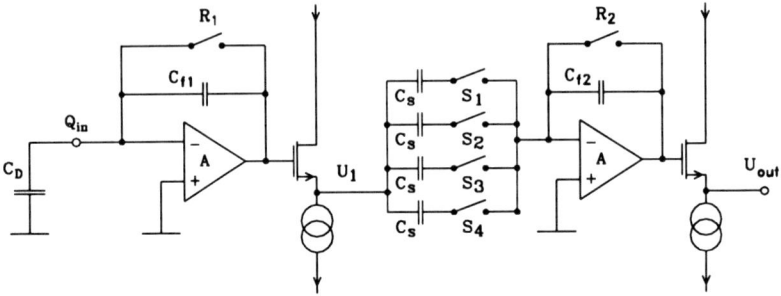

Fig. 7.30. Single channel readout schematics of the CAMEX64 strip detector readout circuit

and switches that connect the output of the first amplifier with the input of the second amplifier. The circuit is rather similar to the one shown in Fig. 7.19 and described in Sect. 7.3.3, but the essential difference is the fourfold multiplication of the capacitive coupling between the amplifiers. In this way it is possible to perform fourfold double-correlated sampling at times that are shifted relative to each other. This procedure provides a good approximation to trapezoidal shaping, which means averaging the output over time intervals before and after signal arrival and taking the difference between the averaged samples.

The switching sequence that performs this function is the following:

- Close R_1 and R_2. The charge on the feedback capacitances C_{f1} and C_{f2} is cleared.
- Open R_1: Some (unwanted) charge will be injected into the input by the switching procedure, producing an offset in U_1.
- Close and open in sequence S_1 to S_4. The U_1 offset values at the four times t_1 to t_4 will be stored on the four capacitors C_s.
- Open switch R_2. A small offset voltage appears at the output.
- Deposit signal charge Q_{sig} at input. U_1 changes by an amount of $\Delta U_2 = Q_{sig}/C_f$.
- Close and open S_1 to S_4 in sequence at times t'_1 to t'_4. A charge $C_i\Delta U_{2i}$ is inserted into the second amplifier at each sample. The total output voltage is $4C_s\Delta U_2 = Q_{sig}4\frac{C_s}{C_{f1}C_{f2}}$.

The complete chip, containing 64 channels, also comprises additional electronics, as shown in Fig. 7.31. Three test inputs allow deposition of a defined charge through test capacitors. Digital steering signals are regenerated by comparators. The decoder switches one signal at a time on the single output line where a driving circuit for the external load is attached. A circuit diagram valid for both amplifiers is shown in Fig. 7.32.

While the circuits mentioned so far are mainly designed for moderate speed of applications in low-radiation environments, recently a large amount of effort has been put into high speed operation and radiation hardness. This effort is due to planned use in elementary particle experiments at high luminosity accelerators, in particular at the Large Hadron Collider (LHC) at CERN,

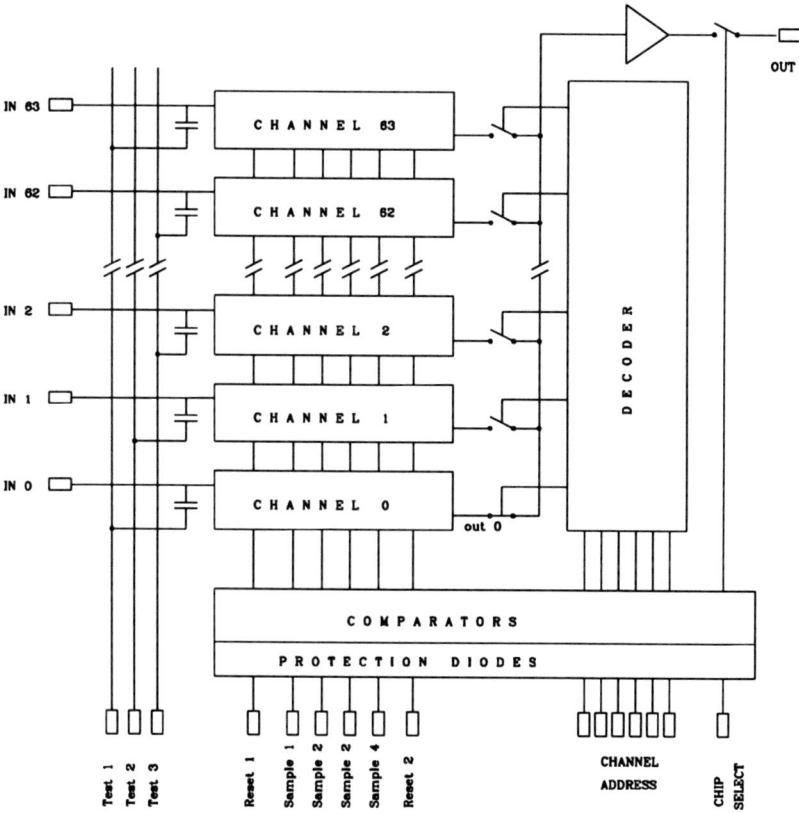

Fig. 7.31. Block diagram of the CAMEX64 strip-detector readout chip

Geneva, Switzerland. In these applications, the time difference between consecutive crossings of particle bunches (25 ns) is much shorter than the time it takes to decide whether or not the data of a particular event needs to be kept (trigger delay $\approx 2\,\mu s$). Therefore it is necessary not only to have fast amplifiers but in addition to store the recorded information for roughly one hundred crossings of particle bunches. As signal shaping with the MOS devices presently available is somewhat above the needed requirement of 20 ns, two different approaches are being taken for amplification. The straightforward method uses bipolar transistors in the amplifier (Anghinolfi et al. 1997). Alternatively, in the "deconvolution method" (Gadomski et al. 1992) a moderate speed (75 ns) CMOS amplifier is used and the output is sampled in 25 ns time intervals. The equivalent of 25 ns shaping is reconstructed via analog manipulation of the sampled signals. Both approaches use "circular" buffers into which the sampled data or the reconstructed signals are written in analog form. In some approaches a discrimination step precedes the storing and the information is kept in binary form only. Delayed reading of selected data out of the circular

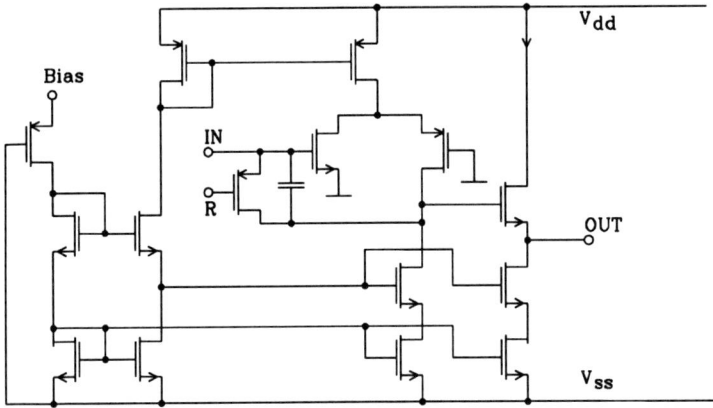

Fig. 7.32. Circuit diagram of the amplifier, including source follower and biasing circuit, of the CAMEX64 strip-detector readout chip

buffer is done in parallel with data recording. The devices are produced using a commercial radiation-hard technology. One of these technologies (see Dentan et al. 1996) has been developed in close relation to applications in particle physics.

7.7 Integrated Circuits for Pixel Detectors

Hybrid pixel detectors, as will be described more fully in Chap. 8, are constructed from separate detector and electronics chips. The detector consists of a two-dimensional array of diodes, and the electronics of a geometrically matching array of electronics circuits. The connection between the two chips is done by the so-called "flip chip" bonding technique. The chips are mounted face to face. Tiny bumps of indium, gold, solder or conducting glue provide the electrical connection between the detector and its electronics. Depending on the application, the electronics has to provide several functions for each individual cell, and may include such things as:

- signal charge amplification;
- noise filtering and signal storage for later readout;
- applying a signal threshold and setting a marker for the readout cycle; and
- the restoration of the pixel electronics after readout or, in the absence of a readout request, after a finite delay.

These functions have to be implemented in an area which is smaller than the pixel size as additional space is needed for readout control and signal bussing.

A variety of architectures have been proposed for the readout. In essentially all of them sparse readout, namely the skipping over empty pixels, is foreseen. Most of them are "column-based stuctures", where each column of pixels is treated separately and the result of the readout stored in an end-of-column buffer that can be read out asynchronously to the data capture. The advantage of the column structure with respect to an "x-y" structure is in the tolerance of malfunctionings: an error in one pixel will usually affect only one column rather than the whole device.

In some concepts only the information in the cell which had a signal above a preset threshold is kept; in others the signal height is also stored. The signal height can be measured by restoring the input of a charge-sensitive amplifier with a constant current and measuring the time it takes until the output voltage goes back below a preset threshold.

The development of pixel electronics is still in its early stages. One development has made it so far into real application (Heijne et al. 1994 and 1996). In this digital scheme, crossing a threshold sets a bit in each cell with a signal. This bit is sent through a delay line with fixed delay, matching the expected trigger delay. If a trigger coincides with the delayed bit, the address of the pixel cell is recorded and sent to the readout buffer.

In similar fashion to strip detectors, a large amount effort is now being put into developments aimed at high speed and radiation hardness for the new colliders. Rather sophisticated logic has to be implemented in the readout to associate the delayed trigger with the signals of the correct time slot. In addition, the radiation exposure is approximately one order of magnitude higher than for strip detectors at their respective positions in the experiments.

7.8 Noise in Strip Detectors – Front-End Systems

While the basic physics of electronics noise was discussed in Sect. 7.2, noise in strip detectors was only peripherally touched on in Chap. 6. A thorough analysis of noise is fairly complex as it involves noise sources in the basic detector, in the front-end amplifier, and in the biasing network supplying the detector's reverse-bias voltage and the sink for the detector leakage current. In addition, strips cannot be treated separately from each other as noise generated in one strip is also partially seen in neighboring strips. This noise correlation is due to the built-in capacitances between neighboring strips.

We will analyze, as an example, a capacitively coupled strip detector. We first assume that every strip is read out individually, and we then consider the somewhat more complicated case of capacitive charge division. A symbolic circuit diagram for the latter case, with one charge-division strip between the readout strips, is shown in Fig. 7.33.

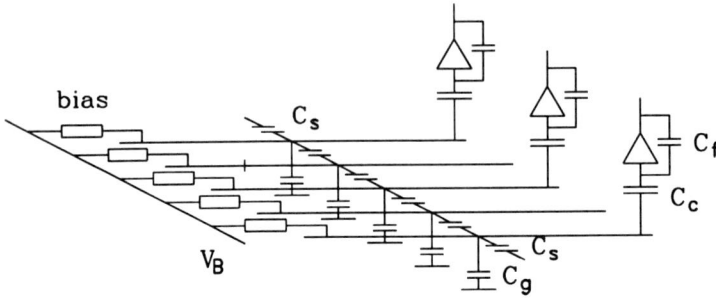

Fig. 7.33. Circuit diagram of a capacitively coupled strip detector with integrated biasing structure and capacitive charge division using one charge-division strip in between adjacent readout strips. A charge-sensitive amplifier is connected to each readout strip. C_c is the coupling capacitor implemented onto the detector by the strip implant–oxide–metal strip structure; C_s and C_g are the naturally occuring strip-to-strip and strip-to-earth (backplane) capacitances respectively. As the coupling capacitance C_c is much larger than the strip-to-strip capacitance C_s, capacitance between metal and neighboring strip can be lumped together with capacitance between strip implant and neighbor strip to a single capacitance C_s. Capacitances between non-neighboring strips are ignored in this consideration. Each strip is connected to a (usually built-in) biasing circuit that provides the bias voltage and removes the charge collected at the strip

7.8.1 Biasing Circuits

Before turning to the questions of noise in the example we shall examine, we will have to consider in more detail the biasing methods. In its simplest form the biasing circuit is a simple resistor, frequently realized as a polysilicon structure. Considerable simplification in detector-processing technology is obtained with a punch-through structure located at the end of the strip, such as is shown in Fig. 7.34.

The working mechanism of a punch-through structure depends on the potential that is established on the outer oxide surface. For bare oxide this potential will adjust to a value close to (and between) the potentials of neighboring electrodes, which in the top part of the figure are the metal strip (usually connected to an amplifier input) and the bias contact. For differing potentials of the neighboring electrodes, there will be a surface voltage gradient on top of the bare oxide, the potential distribution depending on the usually uncontrolled surface resistivity. Depending on the order of magnitude of the surface resistivity, it can take a very long time – minutes or even days – until an equilibrium distribution is reached.

A structure in which the surface potential can be controlled is shown in Fig. 7.34b. Here the oxide surface is covered by a metal electrode to which an external independent voltage can be applied. This MOSFET structure (using the thick "field oxide" as the gate insulator) is often referred to as a FOXFET structure.

Depending on the applied voltages, this structure can be operated in the punch-through mode or in the surface-channel mode. In the punch-through

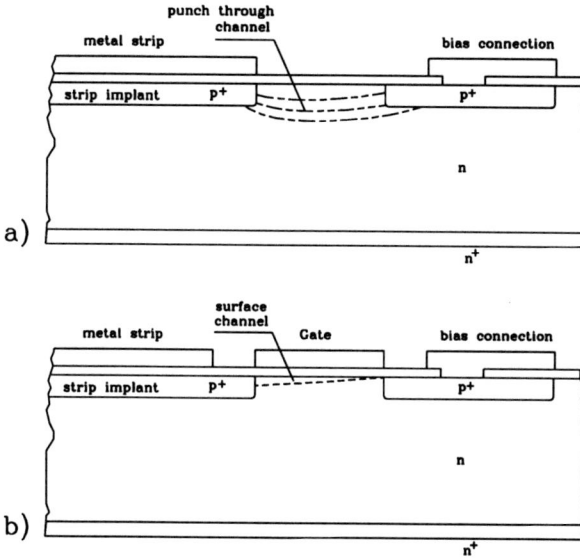

Fig. 7.34a,b. Punch-through biasing structure with bare oxide (a) and oxide covered by a metal gate (b). The outer oxide surface voltage adjusts to such a value that the punch-through current flows at finite depth across a potential barrier from the strip towards the bias contact. If the surface voltage can be externally controlled, as is the case in (b), the depth of the channel can be controlled. It is then also possible to adjust to zero depth, thus arriving at the function of a MOSFET operating for small leakage currents in weak inversion

mode the surface channel is blocked and for high enough potential difference between bias contact and strip implant a hole current will flow from the strip through the bulk towards the bias electrode. The strip implant voltage will adjust itself so as to equalize the leakage current collected by the strip with the bias current. The surface-channel operational mode is equivalent to that of a MOSFET, with the strip implant representing the source and the bias implant the drain. One can then adjust the gate voltage in such a way as to operate the transistor in saturation ($|V_D - V_G| > |V_{D,sat} - V_{G,eff}|$) or in the threshold region ($|V_D - V_G| < |V_{D,sat} - V_{G,eff}|$). The latter situation is similar to having a bias resistor formed by the channel charge layer. The voltage drop along the channel will be small and a rather long channel geometry ($\geq 100\,\mu m$) has to be chosen in order to produce bias resistors with sufficiently high resistance.

As the working principles of transistors have been dealt in considerable detail already, we turn now to the punch-through mechanism before turning to noise considerations.

The Punch-Through Mechanism
We consider a situation in which positive charge carriers from one highly doped p region move to another highly doped p region at lower potential across a potential barrier. This is the case in, for instance, the punch-through structure

of Fig. 7.34, where a curved punch-through channel in the bulk is indicated. For reasons of clarity and simplification, a one-dimensional approximation of a channel with cross-section A and channel length L will be analyzed.

The channel potential as a function of the position y along the channel is shown in Fig. 7.35. As we are restricting our analysis to situations in which the current stays small (this is the case for not too small a barrier height), we can expect that, for sufficiently large distance from the barrier peak, both to the right and left, the hole density is close to the thermal equilibrium distribution taken with respect to the corresponding p^+ region, and that any movement across the peak is by diffusion only. The situation is similar to that of a transistor in weak inversion, which was dealt with in Sect. 7.1.4.

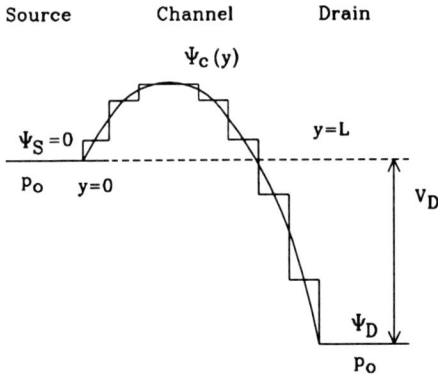

Fig. 7.35. Approximation of the channel potential by a step function

We will generalize the formalism used in that section in such a way as to treat the whole channel region in a uniform manner. The continually changing potential is approximated by a step function ψ_i in interval Δy_i (Fig. 7.35) and the charge transport within each interval is assumed to be due to diffusion only.

Calling p_i the hole concentration at the left side of interval i, and p'_i the similar concentration at the right, we have from (7.1.77):

$$I = -qAD_p \frac{p_i - p'_i}{\Delta y_i} \tag{7.8.1}$$

$$p'_i = p_i - \alpha \Delta y_i \ , \qquad \text{with} \qquad \alpha = -\frac{I}{qAD_p} \ . \tag{7.8.2}$$

(Note that I is negative for a p-channel device if the sign convention from transistors is kept.)

The relationship between charge densities before and after interval boundaries is assumed to follow quasi-equilibrium conditions:

$$p_i = p'_{i-1} \exp\left(-\frac{\psi_i - \psi_{i-1}}{V_T}\right) \ , \tag{7.8.3}$$

where $V_T = \frac{kT}{q}$ the thermal voltage, and ψ_i the potential in the interval i. Successive application of these relations, starting from the source side at $x = 0$ with potential ψ_S and hole density p_0 towards the drain with ψ_D at $x = L$, leads to

$$p_1 = p_0 \exp\left(-\frac{\psi_1 - \psi_S}{V_T}\right)$$

$$p_1' = p_0 \exp\left(-\frac{\psi_1 - \psi_S}{V_T}\right) - \alpha \Delta y_1$$

$$p_2 = p_0 \exp\left(-\frac{\psi_2 - \psi_S}{V_T}\right) - \alpha \Delta y_1 \exp\left(-\frac{\psi_2 - \psi_1}{V_T}\right)$$

$$p_2' = p_0 \exp\left(-\frac{\psi_2 - \psi_S}{V_T}\right) - \alpha \left[\Delta y_1 \exp\left(-\frac{\psi_2 - \psi_1}{V_T}\right) - \Delta y_2\right]$$

$$p_i' = p_0 \exp\left(-\frac{\psi_i - \psi_S}{V_T}\right) - \alpha \sum_{k=1}^{i} \Delta y_k \exp\left(-\frac{\psi_i - \psi_k}{V_T}\right). \tag{7.8.4}$$

Making the transition to infinitesimally small intervals, this relation is rewritten as

$$p(y) = p_0 \exp\left(-\frac{\psi(y) - \psi_S}{V_T}\right) - \alpha \int_{\eta=0}^{y} \exp\left(-\frac{\psi(y) - \psi(\eta)}{V_T}\right) d\eta. \tag{7.8.5}$$

The value of α, and the punch-through current $I \equiv -qAD_p\alpha$ can be found from the requirement that the hole concentration at the drain is given by $p(L) = p_0$:

$$\alpha = p_0 \frac{\exp\left(-\frac{\psi_D - \psi_S}{V_T}\right) - 1}{\int_{\eta=0}^{L} \exp\left(-\frac{\psi_D - \psi(\eta)}{V_T}\right) d\eta} \tag{7.8.6}$$

$$I = -qAD_p p_0 \frac{1 - \exp\left(-\frac{\psi_S - \psi_D}{V_T}\right)}{\int_{\eta=0}^{L} \exp\left(-\frac{\psi_S - \psi(\eta)}{V_T}\right) d\eta}. \tag{7.8.7}$$

The integral in the denominator is dominated by the region around the potential peak. The influence of the drain voltage on the current is determined not by the term in the numerator but indirectly by the change in the potential barrier height, which enters exponentially into the integral in the denominator.

Note that this analysis is well suited for determining the current that is determined by diffusion across the barrier, but not for finding the charge-transit time through the total length of the channel. This is because in the region beyond the barrier excess carriers will be moved by drift in the electric field, which has been ignored in our treatment.

Example 7.5

Problem: *Find the punch-through current as a function of the barrier height $V_B = \psi_{\max} - \psi_S$ for a parabolic form of the potential barrier, the barrier maximum being at a distance y_B from the source, and the channel length L being large with respect to the source–barrier distance.*

Solution: *Setting the source potential to zero, the potential as a function of position is given as*

$$\psi(y) = V_B \left[1 - \frac{(y - y_B)^2}{y_B^2}\right] .$$

As $L \gg y_B$, $\psi_S - \psi_D \gg V_T$ and the exponential term in the numerator of (7.8.7) can be ignored. It remains therefore to evaluate the integral in the denominator:

$$\int_{y=0}^{L} \exp\left(\frac{\psi(y)}{V_T}\right) dy$$

$$= \int_{y=0}^{L} \exp\left(\frac{V_B\left[1 - \frac{(y - y_B)^2}{y_B^2}\right]}{V_T}\right) dy$$

$$= \sqrt{V_T/V_B}\, y_B \exp\left(\frac{V_B}{V_T}\right) \int_{-\sqrt{V_B/V_T}}^{(L/y_B - 1)\sqrt{V_B/V_T}} e^{-\eta^2} d\eta$$

$$\approx \sqrt{V_T/V_B}\, y_B \exp\left(\frac{V_B}{V_T}\right) \int_{-\sqrt{V_B/V_T}}^{\infty} e^{-\eta^2} d\eta$$

$$= \sqrt{V_T/V_B}\, y_B \exp\left(\frac{V_B}{V_T}\right) \sqrt{\pi}\mathrm{erf}(\sqrt{V_B/V_T}) .$$

Here we have made the parameter transformation $\frac{y - y_B}{y_B}\sqrt{\frac{V_B}{V_T}} = \eta$ and have extended the integration interval from L to ∞.

Substituting this expression into (7.8.7), we obtain, with the additional assumption $\psi_S - \psi_D \gg V_T$:

$$I = -qAD_p p_0 \frac{1}{y_B} \frac{\sqrt{V_B/V_T}}{\sqrt{\pi}\mathrm{erf}(\sqrt{V_B/V_T})} \exp\left(-\frac{V_B}{V_T}\right) .$$

The current is roughly exponentially dependent on the ratio between barrier height V_B and thermal voltage $V_T \equiv kT/q$. It is inversely proportional to the distance of the parabolic barrier maximum from the source y_B.

7.8.2 Noise in Biasing Circuits

Biasing circuits by their very nature generate parallel noise because the fluctuations of the bias current are fed into the amplifier input in exactly the same way as is the case for the detector leakage current. The noise charge is given by these current fluctuations integrated over time, with some weighting function that reflects the filtering properties of the amplifier.

An equivalent circuit diagram for a capacitively coupled detector connected to a charge-sensitive amplifier is shown in Fig. 7.36. Noise sources are the amplifier serial noise voltage u_n, which represents the noise sources within the amplifier, the shot noise of the detector leakage current $i_{D,n}$, and the noise

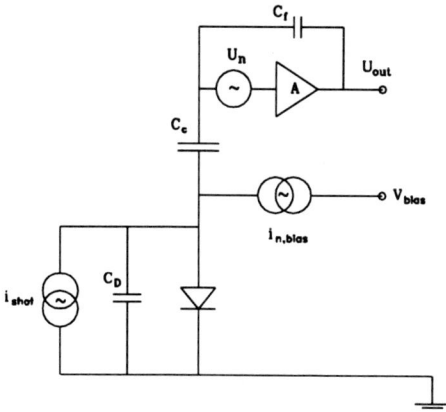

Fig. 7.36. Equivalent circuit diagram for a capacitively coupled detector connected to a charge-sensitive amplifier. The detector is represented by a revesely biased diode in parallel with a capacitor and a (shot) noise current source

current of the biasing circuit $i_{b,n}$. While the amplifier (serial) noise voltage produces an apparent output signal corresponding to a charge which is dependent on the (detector) capacitance (see Sect. 7.3.3), the signal of the parallel noise current sources $i_{D,n}$ and $i_{b,n}$ is the time integral of the current flowing into the amplifier. While the detector leakage current can be essentially shot noise, exhibiting a white current noise spectrum, both the serial amplifier noise as well as the bias current noise may have a noticeable low-frequency noise component, which for simplicity we will assume is described by a $1/f$ noise power density spectrum.

Even without a full analysis we can draw some conclusions about the dependence of the noise on the signal shaping time. Referring back to Sect. 7.3.3, we have found that white parallel noise scales with the square root of the filtering time constant $\sqrt{\tau}$ while white serial noise scales with $1/\sqrt{\tau}$ (see (7.3.5) to (7.3.7)). This difference of a factor τ reflects the time integration. $1/f$ noise therefore can also be scaled from $1/f$-series noise, leading to a noise dependence of

$$N/S_{\text{white, serial}} \propto \frac{1}{\sqrt{\tau}} \tag{7.8.8}$$

$$N/S_{1/f,\text{serial}} \propto 1$$

$$N/S_{\text{white, parallel}} \propto \sqrt{\tau}$$

$$N/S_{1/f,\text{parallel}} \propto \tau \ .$$

In the following the noise mechanisms and their noise power spectra will be discussed. For low-frequency noise that is due to trapping and detrapping (see Sect. 7.2.4), the considerations will be restricted to discussion of the RTS amplitudes due to single traps and will leave aside the capture and emission probabilities and any averaging over many traps.

Resistive Biasing

In the ideal – and often almost realistic – case, a resistor has negligible low-frequency $(1/f)$ noise and can therefore be represented by an idealized resistor with a parallel noise current source of noise spectrum given by

$$\frac{d\langle i_n^2 \rangle}{df} = 4kT/R \ . \tag{7.8.9}$$

Note the independence of this spectrum on the bias current.

Example 7.6

Problem: *The frequency spectrum of the biasing resistor has the same form as that of a shot noise (7.2.3). Find the bias resistance which at $T = 300\,\text{K}$ produces the same noise as a $1\,\mu\text{A}$ detector leakage current.*
Solution: *Setting (7.8.9) and (2.2.3) equal one has*

$$\frac{4kT}{R} = 2Iq \qquad R = \frac{4kT}{2q}\frac{1}{I} = 2V_{\text{T}}/I \ .$$

This leads at room temperature and for a $1\mu\,\text{A}$ current to
$R = 2 \times 0.0259\,\text{V}/10^6\,\text{A} = 52\,\text{k}\Omega.$

Biasing by a Field Effect Transistor

The noise properties of a field effect transistor operating in strong inversion have been derived in Sect. 7.2.4. Thermal noise – due to channel resistance – has been found to depend on the transconductance g_{m} (see (7.2.9)) according to:

$$\frac{d\langle i_n^2 \rangle}{df} = 4kTg_{\text{m}}\frac{\langle Q_{\text{c}} \rangle}{Q_{\text{c}}(y = 0)} \ , \tag{7.8.10}$$

similar to a resistor with resistance $\frac{1}{g_{\text{m}}}$ modified by the ratio between average channel surface charge density $\langle Q_{\text{c}} \rangle$ and density at the source end of the channel. This factor is close to one in threshold mode of operation and drops to $2/3$ in the saturation region. For a MOSFET in strong inversion, g_{m} is given by (7.1.76).

It is interesting to consider the result obtained when the same equation (7.8.10) is applied to a transistor's biasing circuit working in weak inversion (subthreshold) mode. Taking the transconductance $g_{\text{m}} = I/V_{\text{T}} = Iq/kT$ from (7.1.80) and assuming that the transistor operates in the saturation region ($\frac{\langle Q_{\text{c}} \rangle}{Q_{\text{c}}(0)} = \frac{1}{2}$), we obtain

$$\frac{d\langle i_n^2 \rangle}{df} = 4kT\frac{Iq}{kT}\frac{1}{2} = 2qI \ . \tag{7.8.11}$$

This is just the shot noise of the bias current and corresponds to the expectation for the case of very low current when the average number of carriers in the channel is below one, so that the probability for crossing the channel for each carrier is not reduced by the presence of another carrier within the channel.

We therefore can assume general validity of (7.8.10) for thermal noise, both for weak and strong inversion operation.

Turning now to low-frequency noise, we recall that for strong inversion operation the noise has been parameterized as a gate noise voltage with a noise power spectrum given in (7.2.21). Multiplying this expression with the transconductance, we obtain the bias current noise power spectrum:

$$\frac{\mathrm{d}\langle i_n^2 \rangle}{\mathrm{d}f} = g_\mathrm{m}\frac{A_\mathrm{f}}{f} = g_\mathrm{m}\frac{K_\mathrm{F}}{WLC_\mathrm{ox}^2}\frac{1}{f} \ . \tag{7.8.12}$$

To see if such a parameterization is also reasonable for subthreshold operation, we must also consider the effect that a charge positioned at a fixed point in the channel has on the transistor current. For strong inversion this has already been done in Sect. 7.2.4. We will now look at weak inversion operation of a p-channel MOS transistor. We spread the charge q over the full width and a length Δy of the channel, so that the channel potential changes in this region by

$$\Delta\psi_\mathrm{c} = \frac{q}{W\Delta y C_\mathrm{ox}} \ , \tag{7.8.13}$$

as shown in Fig. 7.37.

We can apply (7.8.7) to this form of potential and find the diffusion current as

$$
\begin{aligned}
I &= -qAD_p p_0 \frac{1 - \mathrm{e}^{-\frac{\psi_\mathrm{S}-\psi_\mathrm{D}}{V_\mathrm{T}}}}{\int_{\eta=0}^{L} \mathrm{e}^{-\frac{\psi_\mathrm{S}-\psi_\mathrm{c}}{V_\mathrm{T}}}\,\mathrm{d}\eta + \int_{\eta=y_\mathrm{t}}^{y_\mathrm{t}+\Delta y}\left[\mathrm{e}^{-\frac{\psi_\mathrm{S}-\psi_\mathrm{c}-\Delta\psi_\mathrm{c}}{V_\mathrm{T}}} - \mathrm{e}^{-\frac{\psi_\mathrm{S}-\psi_\mathrm{c}}{V_\mathrm{T}}}\right]\mathrm{d}\eta} \\[2mm]
&= -qAD_p p_0 \frac{1 - \mathrm{e}^{-\frac{\psi_\mathrm{S}-\psi_\mathrm{D}}{V_\mathrm{T}}}}{\mathrm{e}^{-\frac{\psi_\mathrm{S}-\psi_\mathrm{c}}{V_\mathrm{T}}}\left[L + \frac{\Delta\psi_\mathrm{c}}{V_\mathrm{T}}\Delta y\right]} \\[2mm]
&= -qAD_p p_0 \mathrm{e}^{-\frac{\psi_\mathrm{c}-\psi_\mathrm{S}}{V_\mathrm{T}}}\frac{1 - \mathrm{e}^{-\frac{\psi_\mathrm{S}-\psi_\mathrm{D}}{V_\mathrm{T}}}}{L}\frac{1}{1 + \frac{q}{LWC_\mathrm{ox}V_\mathrm{T}}} \\[2mm]
&= I_0 \frac{1}{1 + \frac{q}{LWC_\mathrm{ox}V_\mathrm{T}}} \approx I_0\left(1 - \frac{q}{LWC_\mathrm{ox}V_\mathrm{T}}\right) \\[2mm]
I_0 &= -qAD_p p_0 \mathrm{e}^{-\frac{\psi_\mathrm{c}-\psi_\mathrm{S}}{V_\mathrm{T}}}\frac{1 - \mathrm{e}^{-\frac{\psi_\mathrm{S}-\psi_\mathrm{D}}{V_\mathrm{T}}}}{L} \ .
\end{aligned}
\tag{7.8.14}
$$

The current I_0 is calculated for the case that only a single charge carrier is present at a time in the channel. The difference $\Delta I = I - I_0$ between the current with and without an additional charge carrier (trapped) in the channel, being

Source   **Channel**   **Drain**

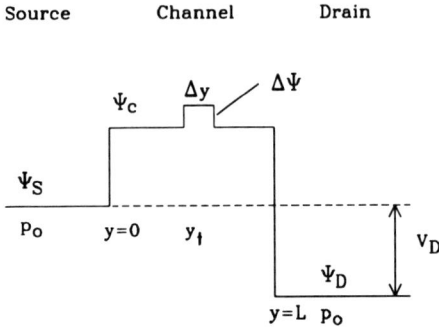

Fig. 7.37. Effect of a charge q located in the channel of a transistor working in weak inversion. The charge located at position y_t will increase the potential barrier for the other charges diffusing through the channel

also the RTS (random telegraph signal; see Sect. 7.2.4) noise signal amplitude, becomes with the transconductance in subthreshold operation $g_m = \frac{|I|}{V_T}$ (see (7.1.80))

$$\Delta I_{\text{weak}} = I - I_0 \approx -I_0 \frac{q}{LWC_{ox}V_T} = -g_m \frac{q}{LWC_{ox}} \; . \tag{7.8.15}$$

The effect of a trapped charge in the channel can also be described by a gate voltage change of

$$\Delta V_{G,\text{weak}} = -\frac{q}{LWC_{ox}} = -\frac{q}{C_G} \; .$$

The reduction of the current due to the potential change of mobile carriers present in the channel has two consequences: first, the current rises slower than the exponential rate of the the gate voltage or the channel potential ψ_c; secondly, the statistical fluctuation of the current will be damped, since the presence of one carrier in the channel will reduce the probability of other carriers crossing the channel.

We will now obtain the RTS current amplitude in strong inversion using the results of Sect. 7.2.4. The current change due to one single charge trapped in a MOSFET operating in strong inversion was given in (7.2.13) as

$$\Delta I = \frac{q}{L}v(y_t) = \frac{q}{L}\frac{I}{Q_c(y_t)W} = \frac{qI}{LWQ_c(0)}\frac{Q_c(0)}{Q_c(y_t)}$$

$$= \frac{I}{LWC_{ox}}\frac{q}{V_{G,\text{eff}}}\frac{Q_c(0)}{Q_c(y_t)} \; . \tag{7.8.16}$$

Here we have used the relations between current and velocity $I = WqQ_c v$ and between channel surface charge density Q_c and gate voltage – taken at the source end of the channel $Q_s(0) = C_{ox}V_{G,\text{eff}}$.

At saturation one has $V_D = V_{G,\text{eff}}$ and obtains from (7.1.75) and (7.1.76) the effective gate voltage and transconductance expressed as a function of the drain current:

$$V_{G,\text{eff}} = V_{D,\text{sat}} = -\sqrt{\frac{2L}{W\mu_p C_{\text{ox}}}(-I_{D,\text{sat}})}$$

$$g_{m,\text{sat}} = \sqrt{2\frac{W}{L}\mu_p C_{\text{ox}}(-I_{D,\text{sat}})} \ .$$

Thus for strong inversion and operation in saturation the current change due to a single trapped charge at position y_t in the channel becomes

$$\Delta I_{\text{strong}} = \frac{I_{D,\text{sat}}}{LWC_{\text{ox}}}\frac{q}{V_{G,\text{eff}}}\frac{Q_c(0)}{Q_c(y_t)} = \sqrt{-I_{D,\text{sat}}}\frac{q}{L}\sqrt{\frac{\mu_p}{2WLC_{\text{ox}}}}\frac{Q_c(0)}{Q_c(y_t)}$$

$$= g_{m,\text{sat}}\frac{q}{LWC_{\text{ox}}}\frac{Q_c(0)}{2Q_c(y_t)} \ . \tag{7.8.17}$$

This shows that, for strong inversion, the current step is proportional to the square root of the current, while for weak inversion (see (7.8.15)) it rises linearly with I, just as is the case for transconductance.

A difference is also seen in the position dependence. While for weak inversion the RTS amplitude is independent of the position, this is not the case for strong inversion operation. Near the center region of the channel, where the channel surface charge density $Q_c(y_t) \approx \frac{1}{2}Q_c(0)$, both formulas give identical results.

It is therefore reasonable to assume that the low-frequency $(1/f)$ noise parameterization can be extended to the subthreshold operational mode and the parameterization of (7.8.12) can be applied here.

Noise with Punch-Through Biasing

Noise in punch-through biasing at low bias current can be considered in a similar way to weak inversion FET biasing, the difference being the replacement of the single-step potential barrier by a variable-height barrier, as shown in Figs. 7.10 and 7.35 respectively.

Again, we must look at the change in potential and as a consequence also in the current, when a charge q is put somewhere close to the channel. Here, we encounter the difficulty that the potential change will not be restricted to a small region around the position of this charge, but will extend over a larger region. While for the FET case the potential change could be simply estimated from a one-dimensional calculation, a more sophisticated analysis is necessary for the punch-through case. We will leave aside the question of the method of calculating the potential change $\Delta\psi(y, y_t)$ for a charge located at a specific position y_t in the vicinity of the channel. Again we assume that the channel is of constant cross-section A, as was done in Sect. 7.8.1.

Calling I_0 and $\psi_0(y)$ the current and potential without the charge q being close to the channel, and calling $I \equiv I_0 + \Delta I$ and $\psi(y) \equiv \psi_0(y) + \Delta\psi(y, y_t)$ the same quantities with charge q being present at position y_t, we find from (7.8.7):

$$I = -qAD_p p_0 \frac{1 - \exp(-\frac{\psi_S - \psi_D}{V_T})}{\int_{y=0}^{L} \exp(-\frac{\psi_S - \psi_0(y) - \Delta\psi(y,y_t)}{V_T}) \, dy}$$

$$\approx -qAD_p p_0 \frac{1 - \exp(-\frac{\psi_S - \psi_D}{V_T})}{\int_{y=0}^{L} \exp(-\frac{\psi_S - \psi_0(y)}{V_T}) \left[1 + \frac{\Delta\psi(y,y_t)}{V_T}\right] \, dy}$$

$$= I_0 \frac{1}{1 + \int_{y=0}^{L} \frac{\Delta\psi(y,y_t)}{V_T} \exp(-\frac{\psi_S - \psi_0(y)}{V_T}) \, dy / \int_{y=0}^{L} \exp(-\frac{\psi_S - \psi_0(y)}{V_T}) \, dy} \quad .$$

Here we have taken the approximation $e^{\frac{\Delta\psi}{V_T}} \approx 1 + \frac{\Delta\psi}{V_T}$. The current change thus becomes:

$$\Delta I(y_t) \approx -I_0 \frac{\int_{y=0}^{L} \frac{\Delta\psi(y,y_t)}{V_T} \exp(\frac{\psi_0(y) - \psi_S}{V_T}) \, dy}{\int_{y=0}^{L} \exp(\frac{\psi_0(y) - \psi_S}{V_T}) \, dy} \quad .$$

The current change is seen to be proportional to the current, as was the case for a weak-inversion FET. We can also see that the potential change $\Delta\psi$ is weighted with an exponential function of the potential barrier, so that only the potential change in the vicinity of the barrier peak (where the channel potential remains within a few times the thermal voltage V_T from the peak value) is relevant for the current change.

Turning now to thermal noise, we have the equivalent situation as was the case for weak-inversion FET biasing: at very low current, when on average less than one charge carrier is present in the peak region, each charge carrier passes the barrier by diffusion independently from other carriers. One therefore will have the shot noise from the bias current. At higher current, the charge carriers already in the barrier region will raise the barrier height and therefore reduce the probability for further carriers to pass the barrier. An upward fluctuation of charge carriers in the channel will diminish the probability of further carriers passing the barrier. This negative feedback will thus reduce the current fluctuations below the value expected from simple shot noise.

The situation is qualitatively similar to that of MOSFET biasing of a gate length resembling the peak region of the punch-through device. An important difference with respect to weak-inversion MOSFET biasing is the separation of the bias current path from the oxide surface. Noise will therefore be mainly due to trapping in the bulk region, while for enhancement FET biasing traps in the oxide reached by tunneling will be dominant. Futhermore, trapping outside the peak region will still change the potential peak and therefore the bias current.

7.8.3 Noise Correlations

Noise correlation in strip detectors was already touched upon in Sect. 6.2.3. It is due to the capacitive network representing the detector, as shown in Fig. 7.33, which results in a coupling between the inputs of neighboring amplifiers. For the example shown in Fig. 7.33, in which capacitive charge-division readout is used, noise correlation exists for both serial and parallel noise.

Using theoretical arguments, noise correlation in strip detectors due to serial amplifier noise has been predicted (Lutz 1991) and the general situation with an arbitrary number of charge-division strips and an arbitrary signal–cluster size has been dealt with. Here for simplicity we will assume that there is only one charge-division strip between readout strips and restrict the treatment to two-strip signal clusters only. Each readout strip is connected to a charge-sensitive amplifier.

Serial Noise

As discussed in Sect. 7.3.2, series noise is due to noise sources within the amplifier – usually dominated by the input transistor – whose effect is represented by a single noise voltage source in the amplifier input (Fig. 7.17). This noise voltage is really present at the input, as the voltage of the high-gain (idealized) amplifier remains zero at the input. Thus an output voltage corresponding to a noise charge $Q_{n,p} = -U_n(C_{D,tot} + C_{in} + C_f)$ is seen at the proper amplifier, while a charge $U_n C_{D,tot}$ is fed into the capacitive network of the detector. $C_{D,tot}$ is the total capacitance of the strip. Part of the charge fed into the detector is seen in the neighboring readout channels. Thus serial noise simultaneously feeds noise charge of opposite sign into the own and into the neighboring readout channels.

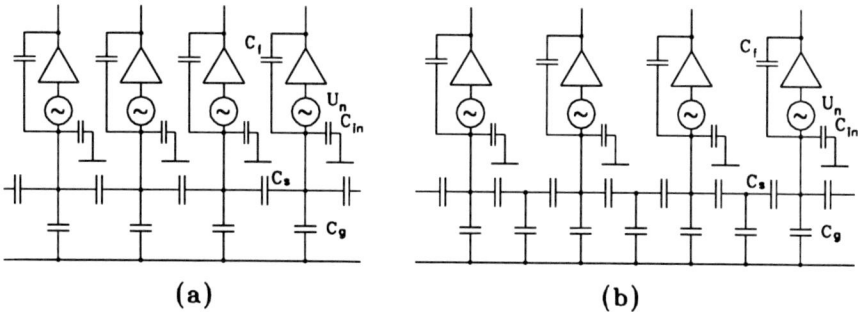

Fig. 7.38a,b. Serial noise: circuit diagram for a strip detector using charge-sensitive amplifiers (CSA) for readout of every strip (a) and using capacitive charge division for reading out every second strip only (b). The noise of the amplifier is represented by a single noise voltage source U_n at the input. C_{in} and C_f are the amplifier input and feedback capacitances. The capacitive network of the detector is composed of strip–ground (C_g) and strip–strip (C_s) capacitances. Capacitances between non-neighboring strips have been ignored. Normally one has $C_s \gg C_g$

We will first consider the simpler case of having each strip read out by a separate amplifier (Fig. 7.38a). As each amplifier input can be considered to be virtual ground, the total strip capacitance is

$$C_{D,tot} = C_g + 2C_s$$

and the noise charges at the proper input Q_p and those fed to the neighbors Q_s are

$$Q_p = -U_n(C_{D,\,tot} + C_{in} + C_f) = -U_n(C_g + 2C_s + C_{in} + C_f)$$
$$Q_s = U_n C_s \ .$$

One thus has

$$Q_p = -2Q_s - U_n(C_g + C_{in} + C_f) = -Q_s(2 + F) \qquad \text{with}$$
$$F = \frac{C_g + C_{in} + C_f}{C_s} \ .$$

Introducing the factor $F = -\frac{Q_p}{Q_s} - 2$ allows for easy generalization to more complicated capacitive networks, such as that shown in Fig. 7.38b.

Serial noise correlation adds to the proper noise of a strip contributions from neighboring strips, which have to be added in quadrature, as the noise sources from different amplifiers are independent from each other:

$$\begin{aligned}
Q_{n,i} &= Q_{s,i-1} + Q_{p,i} + Q_{s,i+1} \\
&= U_{n,i-1} C_s - U_{n,i} C_s(2 + F) + U_{n,i+1} C_s \\
&= C_s[U_{n,i-1} - (2 + F)U_{n,i} + U_{n,i+1}] \ ,
\end{aligned}$$

and so

$$\begin{aligned}
\langle Q_{n,i}^2 \rangle &= C_s^2 [\langle U_{n,i-1}^2 \rangle + (2 + F)^2 \langle U_{n,i}^2 \rangle + \langle U_{n,i+1}^2 \rangle] \\
&= C_s^2 \langle U_n^2 \rangle [1 + (2 + F)^2 + 1] \ .
\end{aligned}$$

The last expression assumes equal noise properties of all amplifiers.

Very frequently, especially with charge-division readout (Fig. 7.38b), the signal charge will be split between two or more neighboring readout channels. In this case one speaks of a "signal cluster". The signal is summed over all the readout channels in the cluster and therefore one also has to consider the noise in the cluster sum. Considering the two-strip cluster of channels i and $i + 1$, one finds the cluster noise charge $Q_{n,c2,i}$ as

$$\begin{aligned}
Q_{n,c2,i} &= Q_{s,i-1} + Q_{p,i} + Q_{s,i} + Q_{s,i+1} + Q_{p,i+1} + Q_{s,i+2} \\
&= U_{n,i-1} C_s + U_{n,i} C_s[-(2 + F) + 1] \\
&\quad + U_{n,i+1} C_s[1 - (2 + F)] + U_{n,i+2} C_s \\
&= C_s[U_{n,i-1} - (1 + F)U_{n,i} - (1 + F)U_{n,i+1} + U_{n,i+2}] \ ,
\end{aligned}$$

and

$$\begin{aligned}
\langle Q_{n,c2,i}^2 \rangle &= C_s^2 \{\langle U_{n,i-1}^2 \rangle + (1 + F)^2 [\langle U_{n,i}^2 \rangle + \langle U_{n,i+1}^2 \rangle] + \langle U_{n,i+1}^2 \rangle\} \\
&= C_s^2 \langle U_n^2 \rangle [1 + 2(1 + F)^2 + 1] \ .
\end{aligned}$$

Had this calculation been performed with the wrong assumption about absence of noise correlation, one would have obtained for single-strip noise $C_s^2 \langle U_n^2 \rangle (2 + F)^2$, a too low value, and for two-strip cluster noise $C_s^2 \langle U_n^2 \rangle 2(2 + F)^2$, a too high value.

Noise correlation also plays a role in position resolution when charge division is applied and the position is determined from the signal center of gravity of the cluster. For a two-strip cluster, the negative correlation in the noise of neighboring strips leads to a worse position resolution than would be anticipated without taking correlation into account.

Parallel Noise

It is interesting to note that, in charge-division readout, parallel noise in neighboring channels is also correlated. This is due to the leakage current and biasing noise in the charge-division strips. The noise correlation is positive, however, as the noise signal generated in the charge-division strip is split and transferred into the two neighboring readout channels. Considering the special case that there is only one charge-division strip between two readout strips and that readout strips and charge-division strips produce exactly the same noise $\frac{d\langle i_{n,\,\text{par}}^2\rangle}{df}$ from leakage current and biasing, the noise current in a particular readout strip $\frac{d\langle i_{n,\,\text{rs}}^2\rangle}{df}$ will be obtained by quadratically adding the contributions from the two charge-division strips to the strip proper:

$$\frac{d\langle i_{n,\,\text{rs}}^2\rangle}{df} = \frac{d\langle i_{n,\,\text{par}}^2\rangle}{df}[(1/2)^2 + 1 + (1/2)^2] = 1.5\frac{d\langle i_{n,\,\text{par}}^2\rangle}{df} .$$

For a two-strip cluster, the two side charge-interpolation strips enter with half amplitude while the intermediate charge-interpolation strip enters fully. One thus has

$$\frac{d\langle i_{n,\,\text{clust}}^2\rangle}{df} = \frac{d\langle i_{n,\,\text{par}}^2\rangle}{df}[(1/2)^2 + 1 + 1 + 1 + (1/2)^2] = 3.5\frac{d\langle i_{n,\,\text{par}}^2\rangle}{df} .$$

Both values are slightly below (a factor 1.5/2 and 3.5/4) the values that would be obtained by connecting the interpolation strips with their neighboring readout strips.

7.9 Summary

The treatment of electronics has essentially been restricted to subjects relevant for the detector-front end electronics system. An understanding of this system requires a knowledge of the physics of both detectors and electronics input devices. Such an understanding is necessary for the optimization of a system's performance, in particular with respect to noise performance considering simultaneously speed and power consumption.

The principal function of the device electronics is the amplification of the detector signals, and the devices almost exclusively used for this purpose are transistors. Section 7.1 explains the working principles and provides quantitative models for the different types of transistors. These models are based on an understanding of the physics of the working mechanisms and the introduction of approximation procedures in order to arrive at fairly simple analytic expressions. One thus is able to estimate the change of transistor properties by variation of transistor parameters such as geometry, doping density and transistor current. These models – with some further sophistications to describe properties such as the finite output conductance – are used in circuit simulation programs.

The bipolar transistor (Sect. 7.1.1) consists of the sequence of one forward-biased (emitter–base) and one reverse-biased (base–collector) diode. The common base is so thin that the majority carriers of the emitter diffuse through the base into the base–collector space-charge region, where they are accelerated towards the collector. For a good transistor only a small fraction (typically 1/100) of the current is lost in the base by recombination. This lost current appears as base leakage current. The current–voltage characteristics of emitter current to base emitter voltage is very similar to that of a forward-biased diode. The transconductance in very good approximation is given by the ratio of a transistor's collector current to its thermal voltage (see (7.1.11)).

While in bipolar transistors both types of charge carriers are important for the transistor action, in unipolar (field effect) transistors only one type of carrier participates in the conduction process. The device works in first approximation like a resistor, with the resistance steered by the potential of the gate.

In junction field effect transistors (Sect. 7.1.2) the gates (both top and back gates) are insulated from the conducting channel by the space-charge regions of the reverse-biased diodes. The channel conductance is varied by changing the width of the depletion layers and/or the gate–channel reverse-bias voltages.

The insulation of the gate in MOS transistors is accomplished by a genuine insulator. The gate therefore can take an arbitrary voltage with respect to the channel and it is possible to build enhancement- and depletion-type MOS transistors. Depletion-type transistors have a channel connecting source and drain with the same type of doping as source and drain. They are conducting when the voltage between gate and source is zero, as are junction field effect transistors. In enhancement-type MOSFETs the conducting channel is an inversion layer that is formed by application of a suitable voltage on the gate. At zero gate–source voltage, the enhancement FET is nonconducting. The substrate in the MOSFET takes the role of the back gate in a JFET; however, quantitatively its effects can often be ignored compared with that of the MOS gate.

In the derivation of the transistor characteristics, the assumption of the presence of an electrically neutral channel for depletion-type unipolar transistors and of strong inversion in enhancement-type MOS transistors has been made. In such a situation, far above threshold, charge transport in the channel is dominated by drift and the voltage drop along the channel can be found by Ohm's law. Rather frequently, transistors are operated in the threshold or subthreshold region, and this is often done in order to improve on the ratio of noise and power consumption. Then carrier diffusion along the channel once more plays an important role. In the case where diffusion becomes dominant (far below threshold), the transconductance equals that of a bipolar transistor (see (7.1.80) and (7.1.11)), given by the ratio of transistor current and thermal voltage. The transconduction in the subthreshold region is proportional to the transistor current, while it increases only with the square root of the current above the threshold region.

Considerable attention has been given to noise. General characteristics of noise in Sect. 7.2 is followed by a specific discussion of noise sources in transistors (Sect. 7.2.4). Noise in the measurement of charge is treated in Sects. 7.3.2 and 7.3.3. One type of noise is unavoidable, namely the series white noise or thermal noise, which can be derived from thermodynamic considerations. For a resistor it can be represented by a serial noise voltage source with a spectral power density $\frac{d\overline{U_n^2}}{df} = 4kTR$ (see (7.2.1)). In the transistor one divides the channel into a number of short resistor pieces and obtains, with inclusion of the interplay of the channel with the gate for the noise voltage source to be put in the front of the transistor gate, $\frac{d\langle V_n^2 \rangle}{df} = 4kT\frac{2}{3}\frac{1}{g_m}$ (see (7.2.12)). As the serial noise in a charge-sensitive amplifier is proportional to the serial noise voltage at the input of the amplifier and the total capacitive load at the input (see (7.3.3)), one recognizes the need for a high transconductance of the amplifier input transistor.

The physical origin of low-frequency serial noise is rather complicated. The basic sources are imperfections in the semiconductor, in the insulator and in the semiconductor–insulator transition region. Charge carriers can be trapped at the defect sites and released with some random time delay. While being trapped they change the electric field and therefore the channel conductance. They thus modulate the transistor current. Individual traps inside the silicon have defined characteristic times (which depend not only on the defect characteristics but also on the carrier densities in the vicinity). The noise spectral density is of Lorentzian form (see (7.2.19)).

Whereas a changing of the charge state of traps in the semiconductor is due to direct charge-carrier capture and emission, the change of the charge state of traps in the oxide requires tunneling. Characteristic times for tunneling are strongly dependent on the thickness of the barrier. A large number of traps is present in the oxide and the noise power spectrum is an average of many Lorentzian spectra, with different characteristic times resulting in an approximate $1/f$ spectrum.

Common electronics circuits were presented in Sect. 7.4. Due to space, power and cost requirements most of the electronics used for semiconductor detector readout is of the integrated circuit type. A short description of some common integrated circuit technologies was given in Sect. 7.5. Examples of circuits for for the readout of strip detectors and pixel detectors were shown in Sects. 7.6 and 7.7.

The last section, Sect. 7.8, was devoted to noise in the strip detector front-end electronics system. After completing the the description of biasing circuits, including resistors, FETs and punch-through structures, their noise properties were investigated. Both white and low-frequency noise sources were considered. An additional topic concerned the correlation of noise in neighboring readout channels, which is due to capacitive coupling between the amplifier inputs through the capacitive network representing the detector.

8 The Integration of Detectors and Their Electronics

As electronics and detectors are in many cases both built on silicon using similar technology, it seems natural to try to integrate them on a single wafer. This, however, brings about some problems, which will be described in this chapter, together with solutions in Sect. 8.2. In the first section of the chapter we will consider the more conservative approach of connecting separately processed detectors with their auxiliary electronics.

8.1 Hybrid Systems of Detectors and Their Electronics

A variety of techniques for connecting detectors and their electronics has been used, and further development is in progress. Not all of them will be described here and emphasis will be given to strip and pixel detectors, however, the same techniques are applicable to other types of detectors such as drift detectors and CCDs.

As demands and operating conditions for detectors vary significantly, it is difficult to provide general guidlines on the design of hybrid systems and technology. Detectors may, for example, work in air or a vacuum, and radiation may be completely absorbed in the detector or high-energy particles may penetrate the detector so that attention has to be given to minimize the amount of material along the path of the particle in order to reduce scattering. Therefore only a few examples will be given, so as to allow the reader to draw his or her own conclusions for a specific application.

General considerations cover the need to:

- avoid excessive capacitance in the connection between detector and electronics in order to limit serial amplifier noise – this is especially important for very low capacitance detectors such as drift detectors;
- shield the detector and electronics input connection from capacitively coupled pickup noise; and
- cool the electronics and sometimes also the detectors.

8.1.1 Strip Detectors

In the early days of development, strip detectors were read out with discrete or hybrid electronics, each readout channel having its own separate amplifier. Therefore the fine readout pitch of typically $100 \,\mu m$ had to be fanned out to a pitch suitable for connecting to the electronics, typically $1 \,mm$.

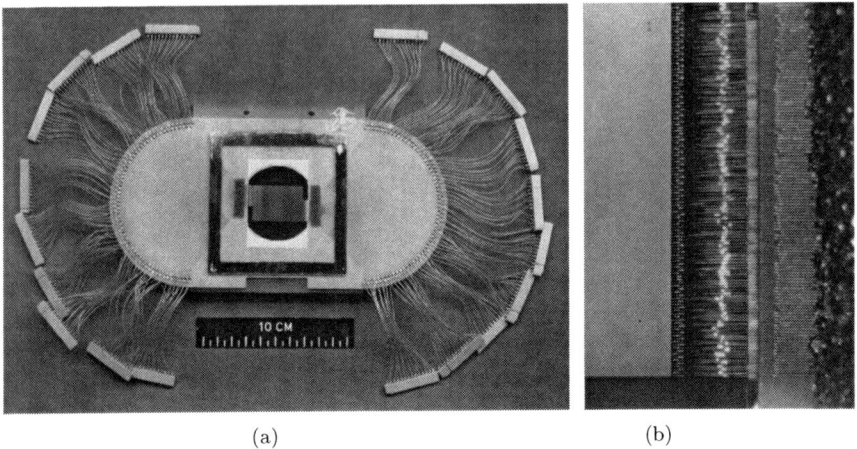

(a)   (b)

Fig. 8.1a,b. Photo of the first type of strip detectors used in a high-energy physics experiment. The left part of the figure (a) shows the assembly up to the connectors for the electronics. The right part (b) is a closeup of the wire bonding region. The detector size was $24 \times 36 \, mm^2$, strip pitch was $20 \, \mu m$, and readout pitch was $120 \, \mu m$ when read out from one side only and $60 \, \mu m$ in the center region, which was read out from both sides. A printed board fanout widened the pitch to a width where soldering of wires was possible, leading towards a hybrid electronics system. Connection between detector and fanout was made by ultrasonic wedge bonding. A total area of roughly one square meter of electronics was needed for the $24 \times 36 \, mm^2$ strip detectors. The assembly was glued to a precision quartz block, which in the experiment was located on an optical bench

An example of such an arrangement is shown in Fig. 8.1. It is the first strip detector used in a high-energy physics experiment (Belau et al. 1983b). The (single-sided) $24 \times 36 \, mm^2$ detector produced on a two-inch wafer was glued to a fiberglass copper fanout, the ends of the lines being gold-plated. Aluminum-covered detector strips with bond pads of $50 \, \mu m$ width were connected to the fanout lines by ultrasonic wedge bonding with $25 \, \mu m$-diameter aluminum wires. This method of connection is still the most common. Furthermore, thin copper wires, providing the connection to the amplifiers, were soldered to the outer ends of the fanout lines, thus providing mechanical decoupling between detectors and electronics.

With the introduction of specially developed integrated electronics for strip detector readout, the electronics pitch could be matched to the detector readout pitch and arrangements were then possible to avoid the fanout and drastically reduce space requirements. The detector strips were directly bonded to the electronics inputs. The electronics chips typically contained 64 or 128 electronics channels providing amplification, filtering, parallel storage and sequential readout. The readout chips required power lines and various control signals to fulfill their functions. Therefore a ceramic substrate carrying a whole set of electronics, containing several readout chips, output drivers and sometimes other electronics generating the logic signals for operating the readout chips, was attached to a detector to form a complete detector–electronics hybrid.

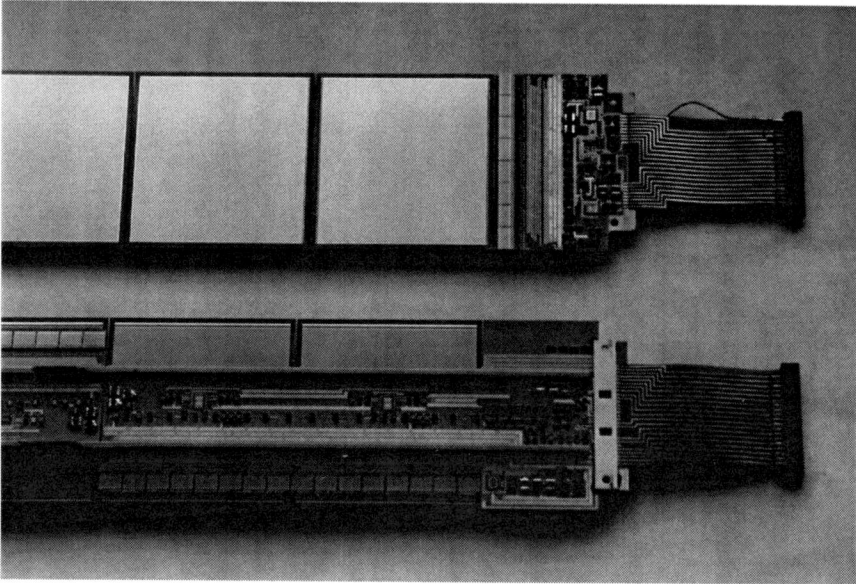

Fig. 8.2. Photo of the first double-sided strip-detector module used in a high-energy physics experiment. The capacitively coupled $6 \times 6\,\text{cm}^2$ double-sided detectors, with orthogonal strip directions on the two surfaces, are read out with the CAMEX64 readout chip. The matching readout pitch of $100\,\mu\text{m}$ of detector and electronics allowed direct bonding from detector to the front-end electronics. The strip pitch is $25\,\mu\text{m}$ on the junction p side and $50\,\mu\text{m}$ on the ohmic n side. On the p side the electronics is located at the end outside the active area of the detector, and the readout strips of the two detectors are connected to each other. On the n side, the electronics hybrid is glued onto the active areas of the two detectors

An example of such an arrangement is shown in Fig. 8.2. Double-sided detectors with orthogonal strip directions on opposing surfaces have been used. On one detector side, two detectors are daisy-chained and connected with ultrasonic wire bonding to the electronics located at the end. On the other side, the hybrid is located right on the detector and glued to it, and the detector strips are again directly connected to the electronics. The front and backside hybrids are glued back to back and a metal shield is introduced between backside hybrid and detector so as to prevent pickup in the strips of large amplitude (digital) signals from the electronics.

8.1.2 Pixel Detectors

Standard pixel detectors consist of two-dimensional diode arrays and electronics, which are usually built on a separate substrate. For each pixel an electronics channel provides amplification and some other functions such as data storage and sparse readout (suppression of readout of pixels without signals).

There is a variety of different concepts for the readout, which have been partially described in Chap. 7 on readout electronics. The geometry provided

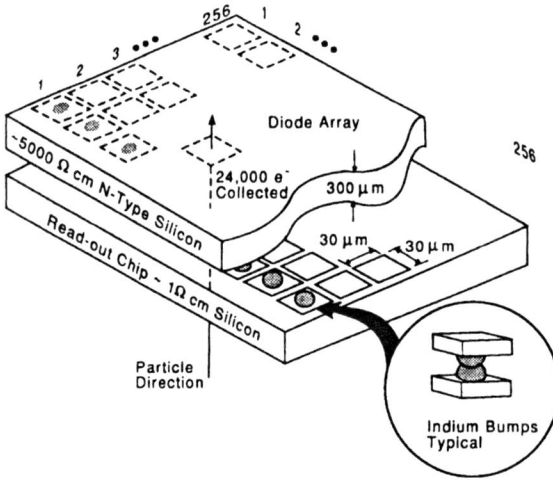

Fig. 8.3. Schematics of a flip-chip pixel detector, showing the attachment of the detector and readout electronics. (After Shapiro et al. 1989, Fig. 1)

for the electronics channel matches the diode pixel, so that electronics and detector can be assembled face to face after having one of the devices "flipped" to the other surface. Each diode is connected to the electronics input pad by a conducting "bump". In the earliest versions this bump consisted of indium, which was deposited and structured on the detector; nowadays, solder and glueing techniques also exist. This type of "flip-chip" assembling is in a state of rapid development and has reached a reasonably high level of reliability. The schematics of a hybrid pixel detector are shown in Fig. 8.3. Photos of

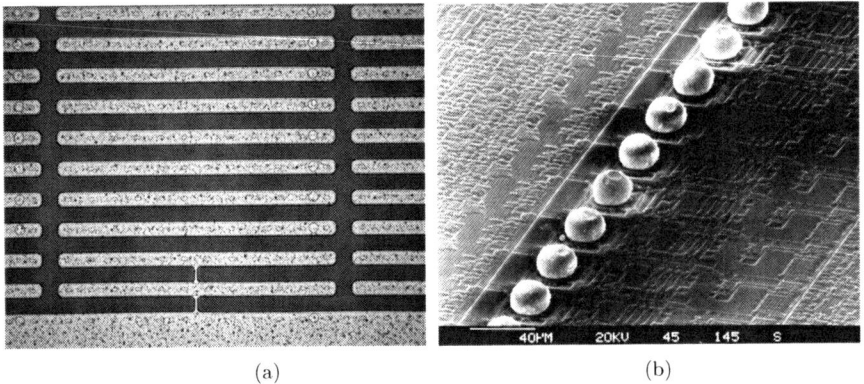

(a)   (b)

Fig. 8.4a,b. Photo of the first pixel detector used in a high-energy physics experiment. As the interesting structures of both detector and electronics are not visible after assembly, a close-up view of the detector (a) and of electronics with bumps for bonding already applied (b) are shown separately

the detector and its electronics (with bond bumps already deposited) used for the first hybrid pixel detector applied in a high-energy physics experiment are shown in Fig. 8.4.

8.2 Detector-Technology-Compatible Electronics

The importance of minimizing the capacitive load on the readout electronics in view of the noise performance of the detector–amplifier system has been mentioned before. For a drift detector with point anode or a p–n CCD, the detector capacitance is in the range 10–100 fF while the connection from detector to the electronics represents a capacitance in the picoFarad range. A large improvement can therefore be expected from the integration of detector and its electronics on the same substrate, with simultaneous optimization of the electronics for low-capacitive load.

The integration of a detector and its electronics is not, however, straightforward. The difficulties lie in the different types of substrate used and in the incompatibility of standard technological processing of detectors and electronics. Detectors are built on low-doped (high-resistivity) silicon with large minority carrier (generation) lifetime (assuring low reverse-bias currents). Standard electronics is built on approximately two orders of magnitude more strongly doped substrates with several orders of magnitude shorter minority carrier lifetime. As minority carrier lifetime is not an important parameter for electronics, technological processes have not been optimized for conserving or even improving this property, which is endangered by (for example) high-temperature processing.

Additional difficulties arise from double-sided processing, which is needed for some types of detectors, and the need for electrical insulation of electronics from the generally fully depleted bulk of the detector. The latter problem can – in principle – be solved by locating the electronics in a region with doping type opposite to the detector substrate, thus insulating it from the detector by a reverse-biassed junction.

The monolithic pixel detector shown in Fig. 8.5 is an interesting example in which such a solution has been tried. The n-type well within the p substrate contains the PMOS readout electronics and provides simultaneously a field-shaping electrode for the collection of the signal electrons. Very small area ($\approx 1\,\text{mm}^2$) prototype pixel detectors of this construction have been tested in a particle beam and shown to work (Parker et al. 1994).

This approach leads, however, to a rather complicated technology because the electronics processing has to be added to the already delicate detector processing. This is expected to lead to severe yield problems.

In an alternative approach, electronics elements are produced with the same technological steps that are already used for detector production. As such devices did not exist beforehand, they had to be developed from scratch, inventing in the process new device structures. The critical element in such an approach is the first amplifying element, namely the transistor, which is required to have very good noise performance.

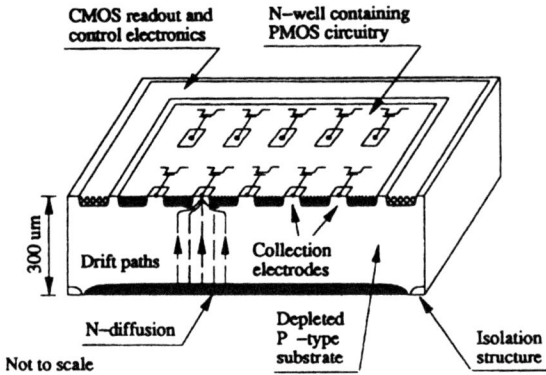

Fig. 8.5. Monolithic pixel detector prototype. (After Parker et al. 1994, Fig. 2)

An example of such a device is the Single-Sided Junction Field Effect Transistor (SSJFET), built on fully depleted and highly-resistive silicon (Radeka et al. 1989; Pinotti et al. 1993). Another example is the DEPFET, to be described in Chap. 9, and this has in addition detector properties but may also be used as an amplifier only.

The cross-section of a cylindrically symmetric SSJFET built on n-type silicon is shown in Fig. 8.6. The device is surrounded by a p-doped region (labeled Guard) and also the backside of the wafer is covered by a large-area diode. These two p regions, together with an n-doped contact on the surface (labeled Anode), are used to deplete the wafer starting from both wafer surfaces using the principle of sidewards depletion which in Chap. 6 has led to drift detectors (Sect. 6.5) and fully depleted p–n CCDs (Sect. 6.6.6). They may also be part of the detector simultaneously. The proper SSJFET consists of concentrically built

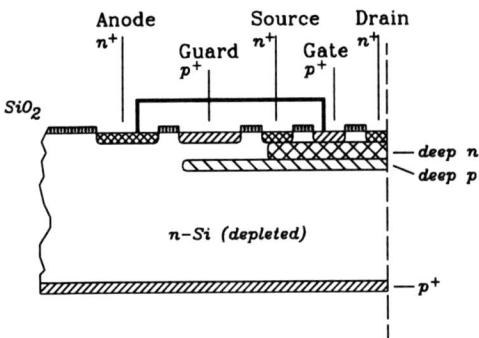

Fig. 8.6. Cross-section of a cylindrically symmetric SSJFET built on n-type silicon. A deep p implantation has been added to limit the transistor channel on the backside. The same technological steps are used as in detector fabrication. The volume below the device is fully depleted

n^+-doped source and drain connected by a deep n-implanted channel which is steered by a flat p^+ gate. The channel is limited on the top by the space-charge region starting from the top gate and on the back by the space-charge region extending all the way from the rear diode through the n^--bulk to the n-doped channel. The complete structure can be produced using the same technological steps already needed for the production of p–n CCDs (see Sect. 6.6.6).

Transistors originally built this way, i.e. without the deep p layer of Fig. 8.6 showed rather good properties, with the exception of not turning off completely even when using large negative gate voltages. The reason for this behavior was the soft boundary of the channel towards the bulk. This problem could be solved completely and transistor properties improved simultaneously by the introduction of a buried layer of p doping which limited the channel from the back and which in normal operating conditions is fully depleted.

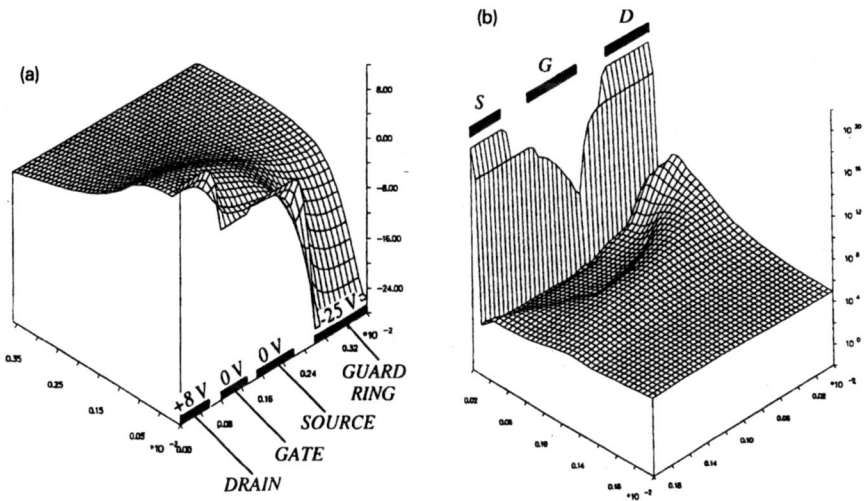

Fig. 8.7a,b. Simulation of an SSJFET built on n-type silicon with backside channel limitation: potential distribution (a) and electron concentration (b). Results are displayed up to a depth of 35 µm into the substrate. Operating conditions are chosen in the saturation region

Device simulations with the TOSCA program (Gajewski et al. 1992) demonstrate the functioning of the device. In Fig. 8.7 potential and electron concentration are shown for the device working in the saturation region. The results of the cylindrically symmetric simulation, with the drain located in the center, are displayed up to a depth of 35 µm. In the electron concentration one notices the n channel close to the surface connecting source and drain, pinched off near the drain. Also simulated was the transistor characteristics, which resemble closely the experimental results displayed in Fig. 8.8.

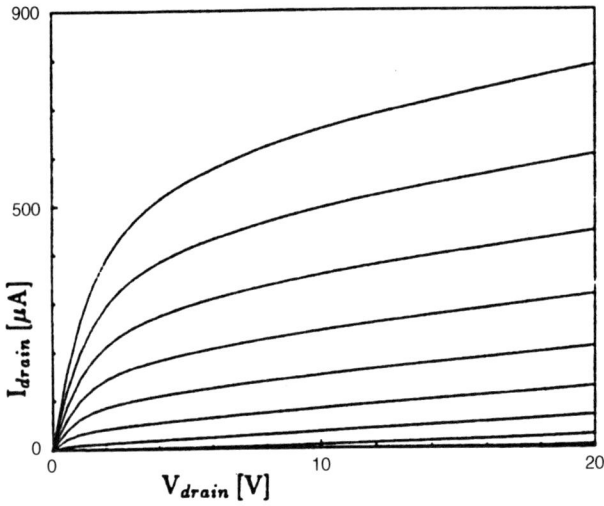

Fig. 8.8. Experimental results of an SSJFET built on n-type silicon with backside channel limitation: current–voltage characteristics

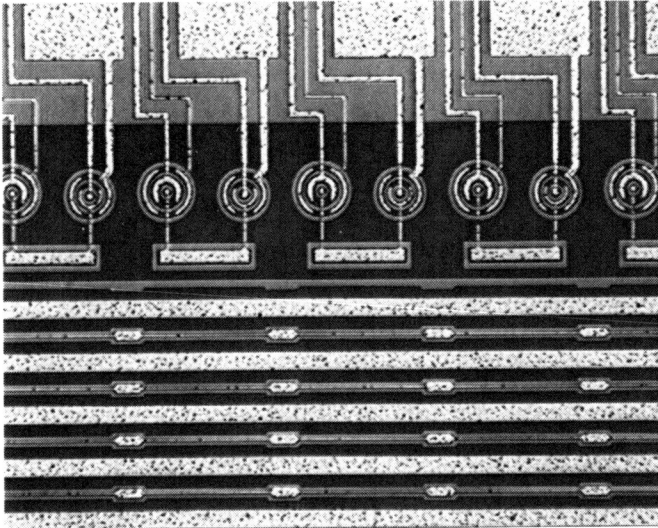

Fig. 8.9. Photo of the readout region of a fully depleted p–n CCD consisting of a source follower and a reset switch in each readout channel. The last part of the proper p–n CCD region is seen on the bottom of the photo. Charge is transferred to the rectangular anodes at the end of the channels. Only input and reset transistors have been monolithically integrated in the detector. The cylindrical transistors are connected to their respective anodes and to bond pads

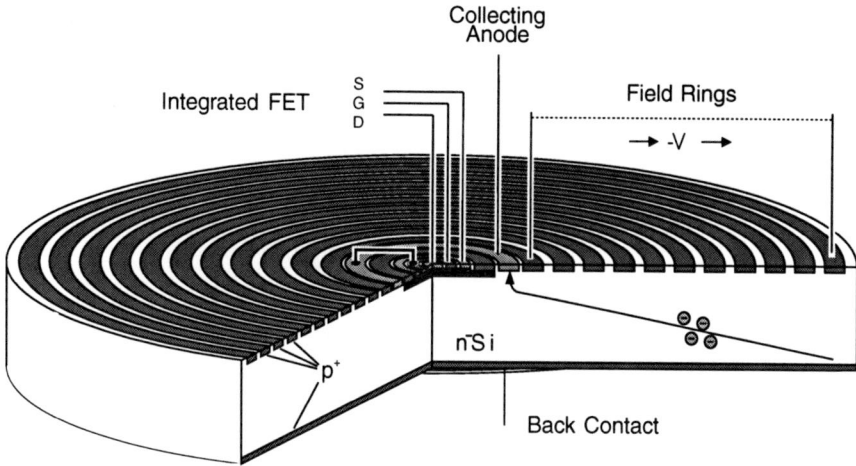

Fig. 8.10. Perspective view of a single-sided structured cylindrical drift chamber with integrated SSJFET transistor, field-shaping electrodes and biasing. The integrated biasing structure and the method used for draining the surface-generated leakage current are not shown

Other circuit elements such as resistors can also be built, as has been demonstrated by the development of a complete charge-sensitive amplifier (Rehak et al. 1990). Implanted resistor structures were used that were insulated from the surrounding semiconductor region by reverse-biassed diodes. Using these resistors and other circuit elements, all of which can be built without significantly increasing the complexity of the technology, restricts the freedom in circuit design. Therefore up to now detector technology-compatible electronics has been used for first-stage amplification only.

The on-chip readout electronics of a p–n CCD built in this technology is shown in Fig. 8.9. Only two transistors per readout channel have been monolithically integrated in the detector. Current source and rest of the front-end electronics are located on a separate chip built in CMOS-JFET electronics.

The quality to be achieved in spectroscopic applications by integrating a detector with its electronics is demonstrated on the device shown in Fig. 8.10. A single-sided junction field effect transistor (SSJFET) has been integrated into the center of a single-sided structured cylindrically symmetric drift chamber. This drift chamber in addition has a built-in structure for biasing the field-shaping electrodes and a mechanism for leading the electrons generated in the SiO_2–Si surface region towards a draining electrode separated from the signal anode. The appreciable surface-generated current thus does not contribute to noise.

In the example shown in Fig. 8.11, the device is enclosed in a hermetically sealed housing having a thin beryllium radiation entrance window and a Peltier element for cooling the device in order to reduce the thermally generated leakage current. The excellent spectroscopic performance achieved at high

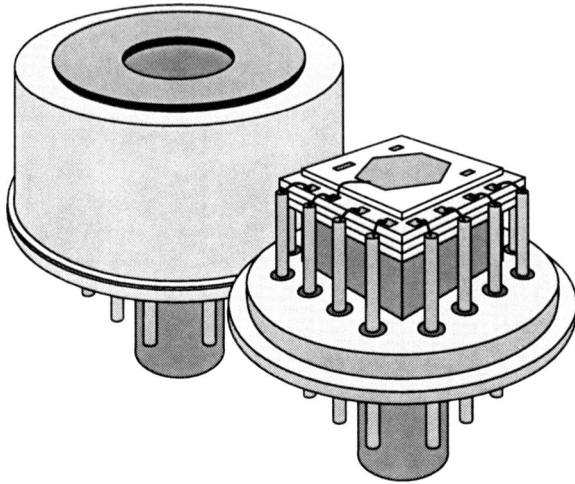

Fig. 8.11. Perspective view of a spectrometric module consisting of a single-sided structured cylindrical silicon drift chamber with integrated SSJFET transistor, cooled by a Peltier element. Modules of this type are used for X-ray fluorescence measurements

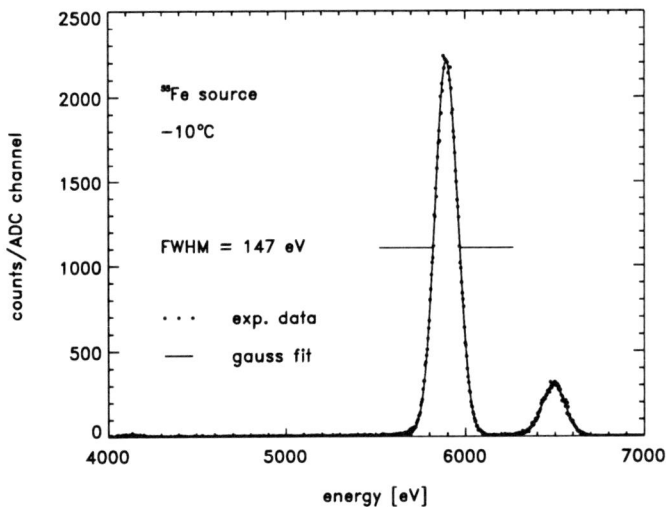

Fig. 8.12. Iron 55 spectrum of a $7\,\text{mm}^2$ single-sided structured cylindrical drift chamber with integrated SSJFET transistor, taken at $-10°\text{C}$ with shaping time constant of $\tau = 0.5\,\mu\text{s}$. The width FWHM $= 147\,\text{eV}$ corresponds to an electronic noise of 8 electrons

rate near room temperature is shown in Fig. 8.12. The width of the spectral lines is due to the statistical fluctuation in the pair-creation process by X-rays (see Sect. 4.4.1) and the electronics noise (see Sects. 7.2 and 7.3).

9 Detectors with Intrinsic Amplification

Contrary to gas detectors, semiconductor detectors usually provide only the primary ionization as signal charge. This mode of operation is possible because of the low energy needed for producing a signal electron (3.6 eV in silicon compared with ≈ 30 eV in gases) and the availability of very low noise electronics. The measurement of the primary ionization avoids the dependence on gain variation of the detector and therefore leads to stable operation in spectroscopic measurements.

High speed and/or very low noise requirements nevertheless may make intrinsic amplification of the detectors desirable. A rather old and well known device is the avalanche diode, with several different operating modes. A more recent device also possessing intrinsic amplification properties is the DEPFET (depleted field effect transistor) structure. Both structures and some further developments will be described in this chapter.

9.1 Avalanche Diode

An avalanche diode possesses a region with a field of sufficient strength so as to cause multiplication of signal charges (electrons and/or holes) that traverse the region (see Sect. 2.6.5).

An example of such a device is shown in Fig. 9.1. The base material is lowly doped p-type silicon. The junction, consisting of a thin highly doped n-type layer on top of a moderately doped p layer, may also be used as an entrance window for the radiation, especially when the bulk material is only partially depleted.

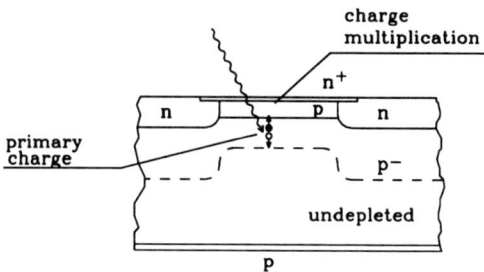

Fig. 9.1. Avalanche diode built on p-type silicon with high field region right below the top surface

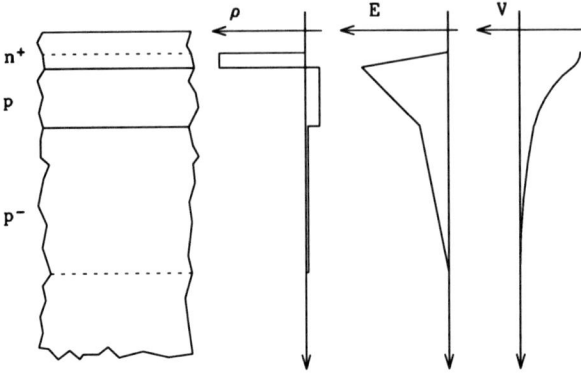

Fig. 9.2. Amplification region of the avalanche diode shown in Fig. 9.1. Also shown are charge density, electric field and potential

An enlarged view of the center top region of Fig. 9.1, in which multiplication takes place, is shown in Fig. 9.2. Also shown are charge density, electric field and potential for the idealized assumption of uniform doping in the n^+, p and p^- regions. The middle p region is fully depleted and the space-charge region extends partially into the thin n^+ top region and the low doped p^- bulk. The maximum of the electric field occurs at the n^+–p diode junction.

Electrons produced below the n^+–p junction (and holes produced above the junction) will have to pass the high field region of the junction when driven by the electric field towards the collecting electrode on top (on bottom) of the device. If the electric field is strong enough to accelerate electrons (or holes) between collisions with the lattice imperfections, so that the kinetic energy is sufficient to create another electron–hole pair, then the primary ionization charge is amplified.

One important aspect to be considered in designing or operating avalanche diodes is the different behavior of electrons and holes with respect to charge multiplication. This is seen in Figs. 3.8 and 9.3, where the ionization rate as a function of the electric field strength is given for several semiconductor materials. In silicon, the onset of charge amplification for holes occurs at higher electric fields than is the case for electrons. The opposite situation appears in germanium, while in GaAs the difference between electrons and holes is comparatively small.

Therefore several working regimes exist that vary depending on the strength of the high electric field region. In the case of silicon one finds:

- at low electric field, no secondary electron–hole pairs are generated. The device has the characteristics of a simple diode;
- at higher electric field, only electrons are able to generate secondary electron–hole pairs. The amplified signal will be proportional to the primary ionization signal, with some statistical fluctuation from the multiplication process added to the fluctuation in the primary ionization process; and

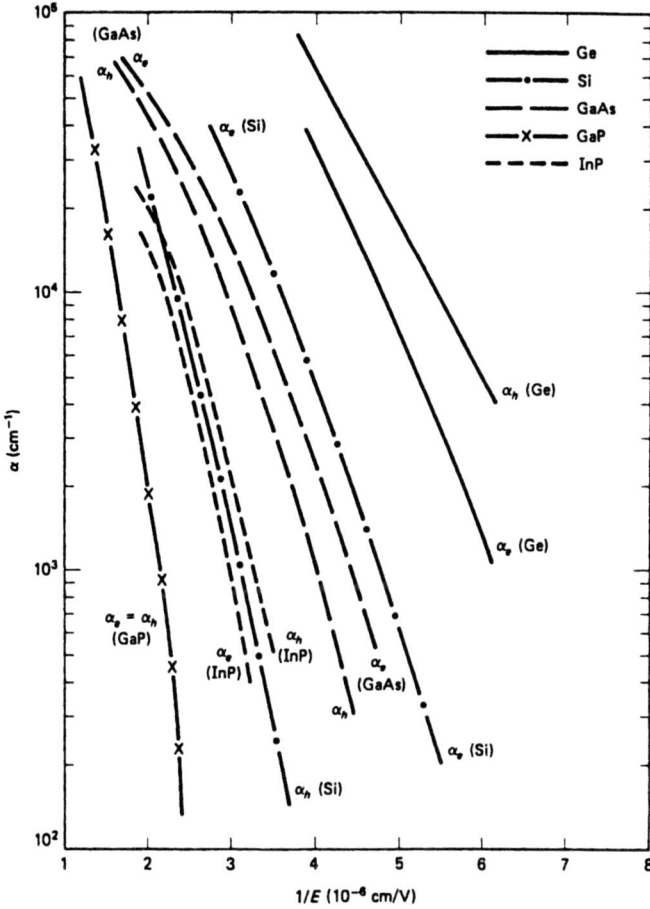

Fig. 9.3. Ionization rates at 300 K versus reciprocal electric field for Ge, Si, GaAs, GaP and InP. (After Wang 1989, p. 479 Fig. 10.4; reference to original literature given there)

- at even higher field, holes will also start to generate secondary electron–hole pairs. Secondary electrons generated by holes will again pass through (part of) the amplification region, thereby possibly generating other (tertiary) electron–hole pairs. This avalanche process will continue until it is either stopped by a statistical fluctuation in the multiplication process or a sufficiently large drop of the externally supplied voltage. This drop may be due to the increased current or an external enforcement by, for example, a feedback circuit. It should also be mentioned that the generation of a large number of movable charge carriers in the multiplication region will reduce the electric field strength and therefore also decrease charge multiplication in later stages of the avalanche generation. A highly nonlinear response of the output signal is expected in this operational mode, which is well suited for single photon detection.

As the charge multiplication probability is a strong function of the electric field strength, high uniformity over the active area is required and high field regions at the edge of the device have to be avoided by proper design of the device. A possible way is shown in Fig. 9.1: the less strongly doped n region at the rim leads to a space-charge region extending deeper into the bulk and simultaneously to a reduction of the maximum field.

If the structure of Fig. 9.1 is to be operated in proportional mode (with only electrons multiplying), primary charge has to be produced below the high field multiplication region.

In choosing the width of the depleted region, one has to consider several partially conflicting requirements. Based on noise considerations, this region should be large in order to reduce the capacitive load to the amplifier. The same is true for the detection of deeply penetrating radiation such as x-rays or highly energetic charged particles. One may even extend the depleted region all the way to the bottom surface. Then the backside p-doped surface can also be used as a radiation entrance window. This can be an advantage for low penetrating radiation such as optical photons, since such an entrance window can be made rather thin. The disadvantage of a large depleted region is the large volume available for thermal generation of electron–hole pairs, the electrons being capable of initializing the avalanche process and, depending on the application, a simultaneous sensitivity to deeply penetrating radiation.

The electric field configuration in the avalanche region is shown in an idealized way in Fig. 9.2, assuming abrupt doping changes. Such a distribution is not only unrealistic but also far from optimal for proportional operation. Then one would like to stay away from breakdown as far as possible but would rather have an extended amplification region with a lower hole-to-electron multiplication ratio, as is the case for lower fields. Such a situation can be reached by suitable doping in the avalanche region (Farell et al. 1994).

The properties of avalanche photodiodes as proportional particle detectors is discussed in Tapan et al. 1997.

A variety of avalanche detector structures has been developed, most for applications in optical data transmission. Many of them are based on compound semiconductors such as GaAs. They often make use of the possibility of varying the gap width. It is then possible to make the (large gap) avalanche region transparent for photons with energy below the band gap and convert them in a deeper lying region with a narrower gap.

Production of avalanche diodes often requires rather sharp doping changes over short distances, sometimes fairly deep inside the detector. This is difficult to achieve with the techniques of diffusion and ion implantation. Similar to bipolar transistor production, it is often necessary to apply epitaxial growth techniques. As the charge multiplication coefficient is a very strong function of the electric field, small inhomogeneities in doping and crystal defects in the avalanche region will have severe effects on device performance: one may have avalanche breakdown in some regions, while proportional amplification has still been reached somewhere else. It is therefore extremely difficult to produce large-area devices (more than a few mm^2) with homogeneous properties.

9.2 Depleted Field Effect Transistor Structure

This structure, which has detector and amplification properties simultaneously, has been proposed by J. Kemmer and G. Lutz (1987) and experimentally confirmed (Kemmer et al. 1990). It is based on the combination of the sidewards depletion principle (as used in a semiconductor drift chamber) and the field effect transistor principle. The sidewards depletion principle has already been presented in Sect. 6.5, and the field effect transistors in Sects. 7.1.2 to 7.1.4. They are again sketched in Fig. 9.4, together with a MOS version of the DEPFET device.

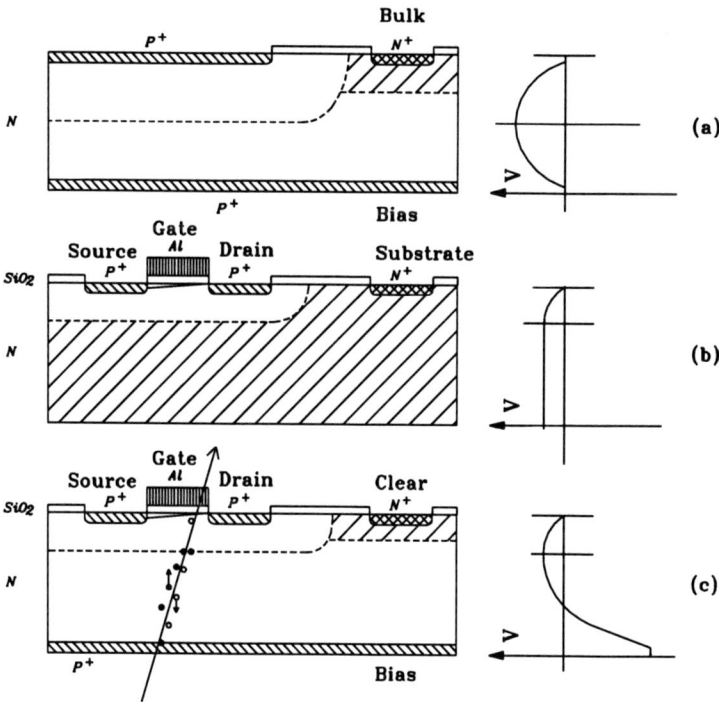

Fig. 9.4a–c. Concept of DEPFET detector–amplification structure (c) developed by combining the sideways depletion method (a) with the field effect transistor structure (b)

The sidewards depletion is obtained when diodes are located on both sides of a wafer and a substrate contact, located somewhere on the side, is polarized in the reverse bias direction with respect to the large-area diode junctions (Fig. 9.4a). A potential minimum for majority carriers (electrons in n-type silicon) forms between the two diode junctions. Its position can be moved towards the top surface by applying differing potentials to the top and bottom diode contacts.

A standard MOS enhancement-type transistor built on top of an undepleted bulk is shown in Fig. 9.4b. Two diodes (with respect to the semiconductor bulk) are connected by the transistor channel formed by biasing of the MOS gate in such a way as to create an inversion layer at the oxide–semiconductor interface. The conductivity of the channel can be steered not only by the gate voltage but also by the bulk potential. This latter possibility, which in electronics is in most circumstances an undesired effect, will be essential for the DEPFET structure shown in Fig. 9.4c.

There are several ways to explain the functioning of this structure. Starting from the center transistor structure, one may consider the changes occurring from the addition of a reversely biased diode on the back of the device. As long as the space-charge regions below the transistor and on the backside are separated from each other, no change in the transistor function is expected (although this is only true in first-order approximations when, for example, the noise characteristics are not considered). This also applies for the case where the lower space-charge region just touches the space-charge region below the transistor and a potential valley is formed below the transistor, as the transistor characteristics are only dependent on the potential in the valley and not on the state of depletion or nondepletion below this boundary.

Having now a potential valley for majority carriers (electrons in Fig. 9.4) below the transistor, we may consider what happens when electron–hole pairs are created in the depleted bulk below the transistor by ionizing radiation. Minority carriers (holes) will move towards the surface while the majority carriers will assemble in the potential valley below the transistor channel. They will induce charges of roughly the same amount in the channel, thereby increasing the channel conductance and the transistor current. The DEPFET structure thus has detector and amplifier properties simultaneously.

9.2.1 Depleted p-Channel MOS Field Effect Transistor (DEPMOSFET)

It is straightforward to produce a practical detector–amplifier device using the DEPFET principle if some precautions are taken. They concern the sideways limits of channel and internal gate, the prevention of carrier injection from source and drain through the bulk towards the rear bias contact, and the introduction of a clearing mechanism to remove the signal charge from the internal gate.

In the example shown in Fig. 9.5 the first of the problems has been avoided altogether by using a closed (cylindrically symmetric) structure. Charge injection through the bulk to the backside is no problem as long as the potential valley is far enough from the frontside so that the potential barrier is of the order of 1 V. This requirement, however, may result in a deterioration of the transistor properties. If the potential valley is at a distance of the order of the channel length or larger away from the surface, a significant fraction of the signal charge will be induced in source and drain of the transistor. The effect of

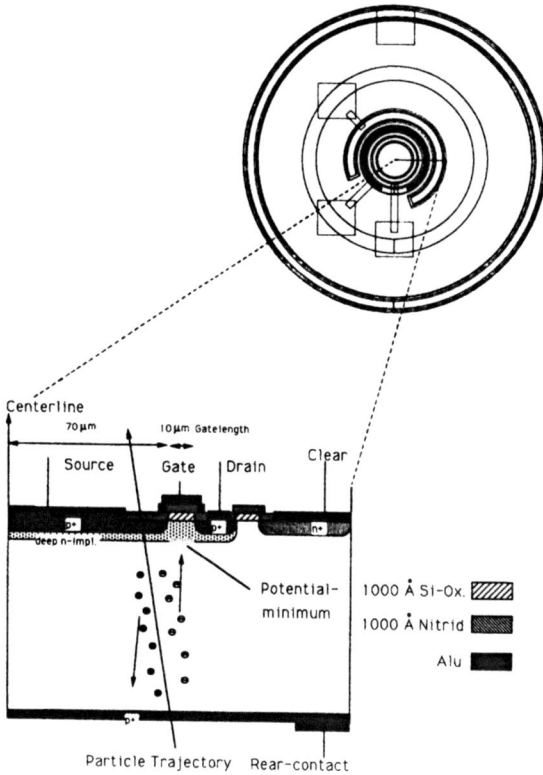

Fig. 9.5. Cylindrically symmetric PMOS-type DEPFET with increased substrate doping close to the surface and integrated clear structure. (After Kemmer et al. 1990, Fig.2)

the signal charge on the transistor current will be thereby reduced. Increasing the bulk doping near the top surface allows the moving of the internal gate closer to the channel.

Example 9.1

Problem: *Find the closest distance of the potential valley to the top surface, in order to assure a one-volt potential barrier against emission of majority carriers from the source towards the rear contact. The bulk is assumed to be uniformly doped with doping density of $N_D = 10^{12}$ cm^{-3}. Which surface doping density is required to be able to put the valley at 5 µm from the channel while keeping the same potential barrier?*

Solution: *The situation can be approximated in a one-dimensional fashion and reduces to the problem of relating depletion depth and applied voltage of a very assymmetrically doped junction diode. This situation has been dealt with in Sect. 3.1.2. One obtains from (3.1.5) with $N_A \gg N_D$:*

$$d = \sqrt{\frac{2\epsilon\epsilon_0}{qN_D} V_b} \ .$$

Here the potential difference across the space-charge region, $V_{bi} - V$, has been replaced by the potential difference between source and potential valley V_b. Putting in numerical values, we derive with $V_b = 1\,V$ and $N_D = 10^{12}\,cm^{-3}$:

$$d = \sqrt{\frac{2 \times 11.9 \times 8.85 \times 10^{-14}\,F/cm}{1.6 \times 10^{-19}\,As \times 10^{12}\,cm^{-3}}\,1\,V} = 36.3\,\mu m \ .$$

In order to arrive at a shorter distance, the doping density has to be increased by the square of the ratio of distances. The necessary doping density N_D for $5\,\mu m$ distance is

$$N_D = \left(\frac{36.3}{5}\right)^2 10^{12}\,cm^{-3} = 5.3 \times 10^{13}\,cm^{-3} \ .$$

The increase in bulk-type doping near the surface not only solves the problem of charge injection towards the back side but also confines the signal charge to a region below the gate, as it is hindered from spreading to the region below the source. The reason for this lies in the partial compensation of the surface doping by source and drain implantation, resulting in a potential maximum inside the bulk right below the gate in which signal electrons assemble. This region will be called "internal gate". The situation is shown in Fig. 9.6, where simulated potential, electron and hole densities are presented for a device in its

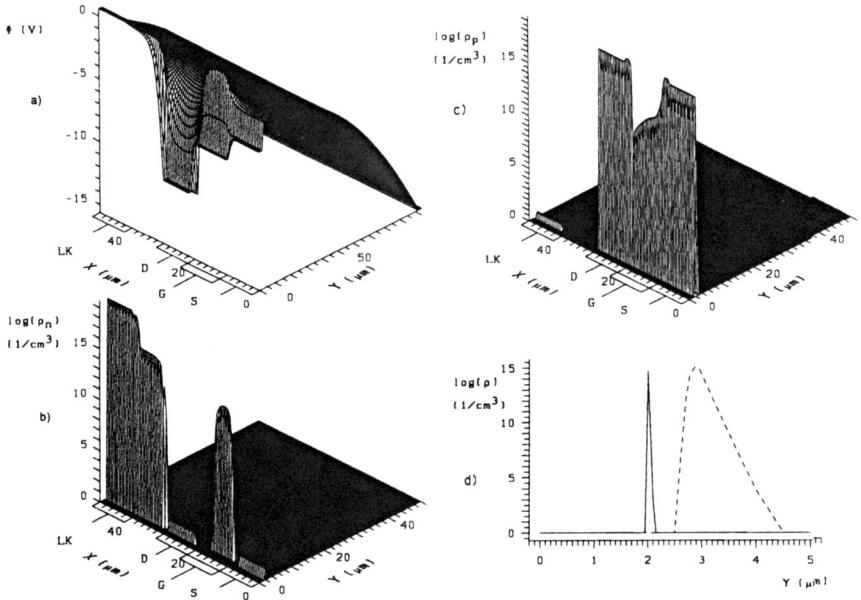

Fig. 9.6a–d. Simulated potential (a), electron and hole densities (b and c) of a p-channel DEPMOS structure in normal operating condition. Also shown are electron (*dashed line*) and hole (*continuous line*) densities in a cut through the gate (d). (After Kemmer et al. 1990, Fig. 3)

normal operating condition. Also shown is a cut-through the gate perpendicular to the surface, showing hole and electron densities. The signal electrons are well separated from the holes forming the channel inversion layer.

The remaining task of removing the signal charge from the internal gate in the device of Fig. 9.5 is accomplished by a punch-through mechanism. The n^+ doped clear electrode is pulsed to a fairly large positive voltage with respect to the source, so that the potential barrier for electrons between internal gate and clear electrode disappears temporarily and the electrons flow from the internal gate towards the clear electrode.

9.2.2 Electrical Properties and Device Schematics

As is evident from the functional description of the DEPFET at the beginning of this chapter, the electrical characteristics will be rather similar to that of standard FETs. The difference is in the space-charge region below the channel, which in the case of the standard FET (enhancement-type MOSFET in the example considered so far) extends up to the undepleted bulk, while for the DEPMOSFET the region between channel and potential valley is of relevance. While the undepleted bulk is at constant potential, the potential valley varies as long as it is not filled with signal electrons. For a completely empty internal gate, the variation of the valley potential will follow closely the variation of the channel potential, so that for this case the influence of the bulk on the channel surface-charge density, given by the integral over the space-charge density between channel and valley, is position-independent.

The simplified characteristics of a standard MOSFET in which the variation of the depletion depth below the channel has been ignored (see (7.1.72) to (7.1.76) for p-type and (7.1.67) to (7.1.71) for n-type enhancement-type MOS devices) is therefore a better starting point for development of a description of the electrical properties of DEPMOS devices than the more elaborate models developed for standard field effect transistors.

Recall that in deriving the MOS transistor characteristics in Sect. 7.1.3 we have started from the situation of source, drain and bulk contacts connected and the gate held at the flat-band voltage with respect to this common contact. The potential of the (nondepleted and uniformly doped) bulk in this condition was defined as zero. Applying an external voltage between source–drain and bulk contact resulted in a bulk potential change of exactly the same value. The bulk–source contact voltage equalled the bulk potential. The situation becomes even more complicated when the bulk doping changes as a function of position, as will be the case for the DEPMOSFET with the enhanced (deep n) bulk doping near the surface. This device can also be operated in the standard undepleted mode. Equation (7.1.72), giving the current voltage characteristics, is still valid, but the explicit expression (7.1.73) for the effective gate voltage and the threshold voltage is imprecise.

We will come back to the questions of threshold voltage and on the relationship between V_{valley} and backside bias voltage V_{back} in the case of nonuniform bulk doping after considering the effect of a charge collected in the internal gate on the transistor current.

The easiest way to come to a quantitative answer on the signal is by consideration of charges that are induced in the neighboring conductors, such as channel, source, drain and backside contact. The induced charge, being equal in magnitude and opposite in sign to the signal charge, divides in the ratio of the respective capacitances. For the internal gate being close to the channel, most of the charge (a fraction f) will be induced in the channel. With the very rough approximation that this induced charge is uniformly spread over the channel, the effect is equivalent to an external gate voltage change by an amount of $f Q_{\rm sig}/C_{\rm G}$ with $C_{\rm G} = W L\, C_{\rm ox}$, the gate–channel capacitance. With this approximation the current–voltage characteristics (see (7.1.72)) become

$$I_{\rm D} = -\frac{W}{L}\mu_p C_{\rm ox}\left\{\left(\frac{f\,Q_{\rm sig}}{C_G} + V_{\rm G,\,eff} - \frac{V_{\rm D}}{2}\right)V_{\rm D}\right\}\ , \tag{9.2.1}$$

$$V_{\rm G,\,eff} = V_{\rm G} - V_{\rm T}\ . \tag{9.2.2}$$

The threshold voltage $V_{\rm T}$ is the gate voltage necessary for the onset of strong inversion in the case of an empty internal gate. Its value and its dependence on the backside bias voltage will be derived below for the condition of nonuniform bulk doping. The saturation voltage follows from the condition

$$\frac{\partial I_{\rm D}}{\partial V_{\rm D}} = 0\ ,$$

leading to

$$V_{\rm D,\,sat} = \frac{f\,Q_{\rm sig}}{C_G} + V_{\rm G,\,eff}\ , \tag{9.2.3}$$

$$I_{\rm D,\,sat} = I_{\rm D}(V_{\rm D,\,sat}) = -\frac{W}{L}\mu_p C_{\rm ox}\frac{V_{\rm D,\,sat}^{\,2}}{2}\ , \tag{9.2.4}$$

$$g_{\rm m,\,sat} = \frac{\partial I_{\rm D,\,sat}}{\partial V_{\rm G,\,eff}} = -\frac{W}{L}\mu_p C_{\rm ox}V_{\rm D,\,sat} = \sqrt{\frac{2W\mu_p C_{\rm ox}}{L}}\sqrt{-I_{\rm D,\,sat}}\ , \tag{9.2.5}$$

$$g_{\rm q,\,sat} = \frac{\partial I_{\rm D,\,sat}}{\partial Q_{\rm sig}} = -\frac{W}{L}\mu_p f\frac{C_{\rm ox}}{C_G}V_{\rm D,\,sat} = f\sqrt{\frac{2\mu_p}{WL^3 C_{\rm ox}}}\sqrt{-I_{\rm D,\,sat}}\ . \tag{9.2.6}$$

Here, in addition to the transconductance of the external gate $g_{\rm m} = \frac{\partial I_{\rm D}}{\partial V_{\rm G}}$, we have introduced the charge "steilheit" of the internal gate $g_{\rm q} = \frac{\partial I_{\rm D}}{\partial Q_{\rm sig}}$.

Returning to the threshold voltage, we define potential zero as the bulk voltage at the position of the oxide–semiconductor interface at flat-band condition with source, drain and clear contact connected. At flat-band condition, we have zero field inside the semiconductor at the semiconductor–insulator boundary. At the onset of strong inversion, the potential at the semiconductor–oxide interface will have changed from zero to $\psi_{\rm s} = -2\varPsi_{\rm B}$, two times the original distance between intrinsic level and Fermi level. Calling $d_{\rm s}$ the distance between the oxide–semiconductor interface and undepleted region in the internal gate (or the electron potential valley for the fully depleted internal gate), the electric field normal to the surface $\mathcal{E}_s$ at the interface and the potential in the internal gate ψ_{valley} can be calculated from the boundary condition of zero field in the valley:

$$\mathcal{E}_s = \int_0^{d_s} \frac{qN_D(x)}{\epsilon\epsilon_0}\,dx \tag{9.2.7}$$

$$\psi_{valley} - \psi_s = \int_0^{d_s} \frac{qN_D(x)}{\epsilon\epsilon_0}\,x\,dx \quad. \tag{9.2.8}$$

As the charge density of the channel is approximated as zero at the onset of strong inversion, one has for the change of the electric field with respect to the flat-band condition near the Si–SiO$_2$ boundary:

$$\epsilon\,\mathcal{E}_s = \epsilon_{ox}\Delta\mathcal{E}_{ox} = \epsilon_{ox}\frac{V_T - \Psi_s - V_{FB}}{d_{ox}} \quad,$$

$$V_T = V_{FB} + \psi_s - \frac{\epsilon}{\epsilon_{ox}}d_{ox}\mathcal{E}_s = V_{FB} - 2\Psi_B - \frac{\epsilon}{\epsilon_{ox}}d_{ox}\int_0^{d_s}\frac{qN_D(x)}{\epsilon\epsilon_0}\,dx \quad. \tag{9.2.9}$$

The backside potential required for keeping the internal gate at a distance d_s from the top surface is found by integration between valley at $x = d_s$ and back surface at $x = d_w$:

$$\psi_{valley} - \psi_{back} = \int_{d_s}^{d_w} \frac{qN_D(x)}{\epsilon\epsilon_0}(x - d_s)\,dx \quad. \tag{9.2.10}$$

Example 9.2
Problem: A p-channel DEPMOS device is built on a n-type silicon substrate with original doping density $N_{D0} = 10^{12}\,cm^{-3}$ and thickness of $d_w = 300\,\mu m$. An additional surface bulk doping has been applied with an integrated dose of $Q_{n,deep} = 10^{11}\,cm^{-2}$ in the silicon and an exponential depth dependence with average depth of $\lambda = 1\,\mu m$. The SiO$_2$ thickness is 0.1 μm, the flat-band voltage $-2\,V$. Find the gate threshold voltage V_T, the potential of the internal gate and the potential at the backside diode when the internal gate is located at $d_s = 2\,\mu m$ depth.

Check if the assumed conditions are reasonable; evaluate at which depth the internal gate can be positioned in order to arrive at reasonable operating voltages.
Solution: For $\lambda \ll d_w$ the bulk doping is described as $N_D(x) = N_{D0} + \frac{Q_{n,deep}}{\lambda}e^{-\frac{x}{\lambda}}$, so that the electric field at the silicon boundary and the potential difference between valley and the Si–SiO$_2$ interface according to (9.2.7) and (9.2.8) are given as

$$\mathcal{E}_s = \frac{q}{\epsilon\epsilon_0} \int_0^{d_s} \left[N_{D0} + \frac{Q_{n,\text{deep}}}{\lambda} e^{-\frac{x}{\lambda}} \right] dx$$

$$= \frac{q}{\epsilon\epsilon_0} \left[N_{D0}d_s + Q_{n,\text{deep}} \left(1 - e^{-\frac{d_s}{\lambda}} \right) \right]$$

$$= \frac{1.6 \times 10^{-19}\,\text{As}}{11.9 \cdot 8.854 \times 10^{-14}\,\text{F/cm}}$$

$$\times \left[10^{12}\,\text{cm}^{-3} \cdot 2 \times 10^{-4}\,\text{cm} + 10^{11}\,\text{cm}^{-2} \cdot \left(1 - e^{-2} \right) \right]$$

$$= 1.518 \times 10^{-7}\,\text{Vcm} \cdot \left[2 \times 10^{8}\,\text{cm}^{-2} + 0.865 \times 10^{11}\,\text{cm}^{-2} \right]$$

$$= 13.1 \times 10^{3}\,\text{V/cm}\ \ ;$$

$$\Psi_{\text{valley}} - \Psi_s = \frac{q}{\epsilon\epsilon_0} \int_0^{d_s} \left[N_{D0}x + \frac{Q_{n,\text{deep}}}{\lambda} x e^{-\frac{x}{\lambda}} \right] dx$$

$$= \frac{q}{\epsilon\epsilon_0} \left[N_{D0} \frac{d_s^2}{2} + Q_{n,\text{deep}} \left(\lambda \left(1 - e^{-\frac{d_s}{\lambda}} \right) - d_s e^{-\frac{d_s}{\lambda}} \right) \right]$$

$$= \frac{1.6 \times 10^{-19}\,\text{As}}{11.9 \cdot 8.854 \times 10^{-14}\,\text{F/cm}} \cdot \left[10^{12}\,\text{cm}^{-3} \cdot \frac{(2 \times 10^{-4}\,\text{cm})^2}{2} \right.$$

$$\left. + 10^{11}\,\text{cm}^{-2} \left(10^{-4}\,\text{cm} \left(1 - e^{-2} \right) - 2 \times 10^{-4}\,\text{cm} \cdot e^{-2} \right) \right]$$

$$= 1.518 \times 10^{-7}\,\text{Vcm} \left[2 \times 10^{4}\,\text{cm}^{-1} + 10^{7}\,\text{cm}^{-1}(1 - 0.135 - 0.2706) \right]$$

$$= 0.902\,\text{V}\ \ .$$

The bulk doping directly at the $\text{Si} - \text{SiO}_2$ interface is $N_{D0} + Q_{n,\text{deep}}/\lambda = 10^{12}\,\text{cm}^{-3} + 10^{11}\,\text{cm}^{-2}/10^{-4}\,\text{cm} = 10^{15}\,\text{cm}^{-3}$, so that Ψ_B according to (3.3.1) is $\Psi_B = \frac{kT}{q} \ln \frac{N_D}{n_i} = 0.0259\,\text{V} \cdot \ln \left(\frac{10^{15}}{1.45 \times 10^{10}} \right) = 0.289\,\text{V}$. The threshold voltage according to (9.2.9) is then

$$V_T = V_{FB} - 2\Psi_B - \frac{\epsilon}{\epsilon_{ox}} \mathcal{E}_s d_{ox}$$

$$= -2\,\text{V} - 2 \cdot 0.289\,\text{V} + \frac{11.9}{3.84} 13.1 \times 10^{3}\,\text{V/cm} \times 10^{-5}\,\text{cm}$$

$$= (-2 - 0.578 - 0.406)\,\text{V} = -2.98\,\text{V}\ \ .$$

The backside potential is obtained from (9.2.10) as

$$\Psi_{\text{valley}} - \Psi_{\text{back}} = \frac{q}{\epsilon\epsilon_0} \int_{d_s}^{d_w} \left[N_{D0} + \frac{Q_{n,\text{deep}}}{\lambda} e^{-\frac{x}{\lambda}} \right] (d_w - x)\, dx$$

$$= \frac{q}{\epsilon\epsilon_0} \left[N_{D0} \frac{(d_w - d_s)^2}{2} + Q_{n,\text{deep}} \left((d_w - d_s + \lambda) e^{-\frac{d_s}{\lambda}} - \lambda e^{-\frac{d_w}{\lambda}} \right) \right]$$

$$= \frac{1.6 \times 10^{-19}\,\text{As}}{11.9 \cdot 8.854 \times 10^{-14}\,\text{F/cm}} \left[10^{12}\,\text{cm}^{-3} \frac{(298 \times 10^{-4}\,\text{cm})^2}{2} \right.$$

$$\left. + 10^{11}\,\text{cm}^{-2} \left(299 \times 10^{-4}\,\text{cm}\,e^{-2} - 10^{-4}\,\text{cm}\,e^{-300} \right) \right]$$

$$= 1.518 \times 10^{-7} \, \text{Vcm} \left[4.44 \times 10^8 \, \text{cm}^{-1} + 4.047 \times 10^8 \, \text{cm} \right]$$
$$= (67.4 + 61.4) \, \text{V} = 128.8 \, \text{V} \; .$$

One notices that with the chosen conditions the electron potential valley is just deep enough $(0.9 \, \text{V})$ to ensure separation of the signal electrons in the internal gate from the channel and that the backside bias voltage to be applied is almost double the full depletion voltage. Shifting the internal gate closer to the surface would require an increase of the surface bulk doping $Q_{n,\text{deep}}$ combined with a steeper depth dependence.

A much simpler estimate can be used to find these conditions. The backside voltage can be estimated by the sum of full depletion voltage for the homogeneously doped wafer $V_{\text{FD}} = \frac{q}{\epsilon \epsilon_0} N_{\text{D}0} \frac{d_{\text{w}}^2}{2}$ plus the voltage needed for depleting the tail of the additional bulk doping below the potential valley $Q_{n,\text{deep}} e^{-d_s/\lambda}$, which is approximately $V = \frac{q}{\epsilon \epsilon_0} Q_{n,\text{deep}} \, e^{-d_s/\lambda} d_{\text{w}}$. It should also be mentioned that the emission of holes from the source towards the rear contact has to be prevented. For the surface bulk implantation extending over the source region, the potential barrier for holes is lower than the electron potential valley depth, due to partial doping compensation by the source doping.

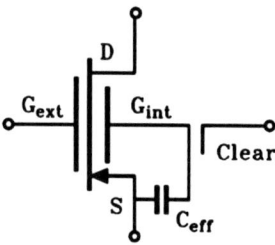

Fig. 9.7. Device schematic for the p-channel enhancement DEPMOSFET structure

Having understood the working principles of the DEPFETs, it is clear that we will use the device symbols of standard FETs, adding the additional features, the steering of the transistor current by the charge assembled in the internal gate and the clearing mechanism for the internal gate. One obtains for the p-channel enhancement DEPMOSFET the schematic shown in Fig. 9.7.

9.2.3 Other Types of DEPFET Structures

As the DEPFET differs from the standard FET essentially by depletion of the substrate from the backside, all types of FETs described in Sect. 7.1.5 have their analog as DEPFET. The device structures shown in Fig. 7.11 can be

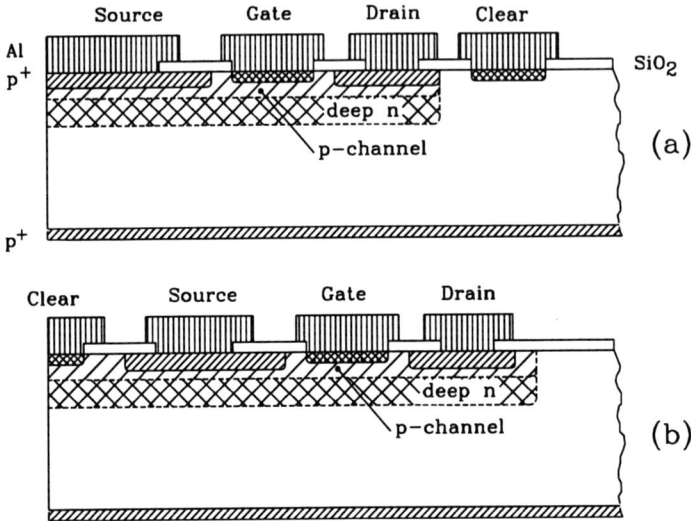

Fig. 9.8a,b. Device structure of p-channel junction-type DEPJFETs with lateral (a) and vertical (b) punch-through clear structures. The vertical n–p–n^+ punch-through clearing structure uses the same doping structures already present in the transistor part. It can also be used in continuous operation mode

adapted to the DEPFETs by replacing the backside ohmic substrate contact with a reverse-biased diode and adding a clear contact.

The clearing method employed in the examples of Fig. 9.4 and Fig. 9.8a is based on a lateral punch-through mechanism between the internal gate and the clear contact. A voltage pulse has to be applied to the clear contact in order to temporarily remove the potential barrier between the internal gate and the clear contact. It is also possible to use a vertical punch-through mechanism. This is easily accomplished in a junction-type DEPFET, as shown in Fig. 9.8b. The deep n doping providing the internal gate extends over a small n^+ clear electrode. The two n regions are electrically separated by a p region with the same doping concentrations as the channel. Pulsing the clear electrode to positive voltages will draw the electrons out of the internal gate. The function is then similar to the lateral punch-through clear structure. An alternative way of using the structure is the application of a constant positive potential to the clear electrode, with adjustment in such a way that with a partially filled internal gate the average current flowing into the internal gate is taken away through the punch-through structure. The punch-through structure has then a similar function to the feedback resistor in a charge-sensitive amplifier.

The operational principles, including noise properties of punch-through structures, have been discussed in the context of strip-detector biasing in Sect. 7.8. Essentially the same approach can be applied for the DEPFET continuous clear vertical punch-through structure.

An important property of the vertical clear structure is the electrical separation of the clear electrode from the charge collection volume. This avoids the potential danger of signal charges flowing directly to the clear electrode instead of the internal gate.

Other significant variations are in the topology of such devices. The use of open (e.g. rectangular) geometries allows the design of smaller structures, with their advantages in position resolution and decrease in capacitance. Precautions have to be made, however, to confine the lateral extension of the internal gate, so as to keep the signal charge below the gate. In addition, the appearance of high-conducting parasitic channels between source and drain at the edge of the transistor structure have to be avoided.

9.2.4 DEPFET Properties and Applications

Compared with a standard combination of separated detector and electronics amplifier, the DEPFET structure has several unique properties:

- the detector is simultaneously the amplifier. From a noise point of view, detector and amplifier input capacitance do not have to be added for estimating the series noise as is the case in standard detector–amplifier combinations. For separated detector and amplifier, the optimum series noise performance is obtained by equalizing detector and "cold" amplifier input capacitance. The optimum capacitive input load in standard configurations is therefore twice the detector capacitance, while for the DEPFET this doubling of series noise does not apply;
- parasitic capacitances from the connection between detector and amplifier input are omitted;
- the DEPFET gate capacitance can be reduced to very small values, and this simultaneously increases amplification and reduces serial noise;
- use in pulsed clear mode will completely empty the internal gate, thus avoiding reset noise due to statistical variation of the residual charge; and
- the signal charge being confined in the potential well of the internal gate is not destroyed by any reading out of the device, and multiple readout of the signal charge is therefore possible.

Many applications of DEPFET devices are possible. DEPFETs can be used as amplifiers of fully depleted devices such as drift chambers or CCDs. In these cases the charge generated somewhere in the device will be moved laterally into the internal gate of the DEPFET. The most interesting application is as an element of a pixel detector. An area of the silicon wafer will be covered with DEPFET devices and they will be electronically connected in such a way that individual transistors can be turned on separately, thus measuring the signal charge collected in the internal gate by comparison of the transistor current with the value obtained with empty gate.

DEPFET type pixel detectors of this and other concepts will be discussed in the next section.

9.3 DEPFET Pixel Detectors

In contrast to strip detectors, pixel detectors measure unambiguously two co-ordiates even when several hits are simultaneously present in the detector. We have already encountered several devices with this property: matrix drift devices (in Sect. 6.5.2), charge coupled devices (CCDs, in Sect. 6.6) and hybrid pixel detectors (in Sect. 8.1.2).

All of these devices have advantages and drawbacks. Drift devices need an external time signal in order to derive the position from the drift time. In CCDs the signal charge is collected and stored in local potential minima and then shifted one by one to one or several output nodes. Their drawbacks are the slow readout speed, signal losses during transfer, and the continuous sensitivity during transfer, leading to incorrect assignment of position for data generated during the readout cycle. Hybrid pixel detectors do not have these disadvantages; however, a full electronics channel is required for each individual pixel. As the size of the readout electronics in the "flip-chip" technique has to be matched to the pixel size, only moderately small pixel sizes are possible. In addition, the capacitive load to the (low power) input amplifier cannot be made too small (a significant fraction of $1\,\mathrm{pF}$) due to size requirements of the bump-bond technique.

The DEPFET concept, due to its intrinsic amplification and the possibility of repeated nondestructive readout, offers the possibility of circumventing some of these problems. In the following possible applications of the DEPFET principle, four types of pixel detector are proposed. Two of them (described in Sects. 9.3.1 and 9.3.3) are under development at present; the others, requiring more elaborate technology for their realization, exist only as concepts.

9.3.1 DEPFET Pixel Detector with Random Access Readout

In the most straightforward application of DEPFETS for pixel detectors, a large number of DEPFETs are placed in a two-dimensional matrix array. The devices can be of the MOS type, as shown in Fig. 9.5, or the junction type as shown in Fig. 9.8. They are electrically connected in such a way that charge can be collected simultaneously in all internal gates of the device, while during readout one transistor is turned on at a time.

An example of how this can be done is shown in Fig. 9.9 for the case of DEPMOSFETs with pulsed clear structures. Sources are connected in rows, drains in columns, and gates in a diagonal direction. Turning on an individual DEPFET requires the application of proper voltages to source gate and drain. It is thus possible to select a specific DEPFET or a group of DEPFETs (e.g. a line of DEPFETs) for readout. One can start with a coarse scan of the device in order to find out the regions of interest, and these can then be investigated in a follow-up precise scan of the interesting regions. This is possible since the signal charge is not destroyed by the readout procedure.

A certain problem of the device is due to the fact that the signal charge is obtained from two measurements of the transistor current taken at different

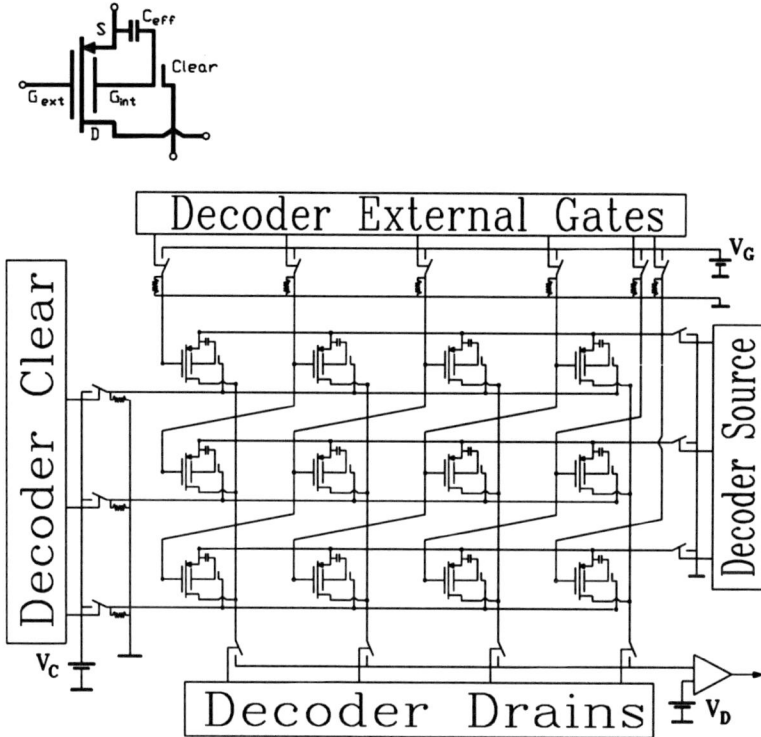

Fig. 9.9. Principle of a DEPMOSFET pixel detector with pulsed clearing

times, corresponding to the situations of cleared (empty) and filled internal gate. Part of the charge obtained this way will be due to the detector leakage current integrated in the time difference between the two current measurements. Although it is possible to correct for this effect, the statistical fluctuation of this leakage current charge remains.

In the device presented in Fig. 9.9, the clearing procedure can also be performed row by row. For pure image applications, an operating mode in which the device is read out row by row, each readout followed immediately by the clearing step is also possible. In such a case the signal can be taken as the difference in current levels before and after the clearing operation.

9.3.2 DEPFET Pixel Detector for Continuous Operation

The development of this concept has been initiated in view of the high-rate applications in which repeated pulsed clearing is not possible.

The method of continuous clearing shown in Fig. 9.8b is applied and the DEPFETs are permanently active. In order to limit power consumption they have to work at very low current. However, as the individual transistors will show individual differences in the gate threshold voltage, one cannot simply

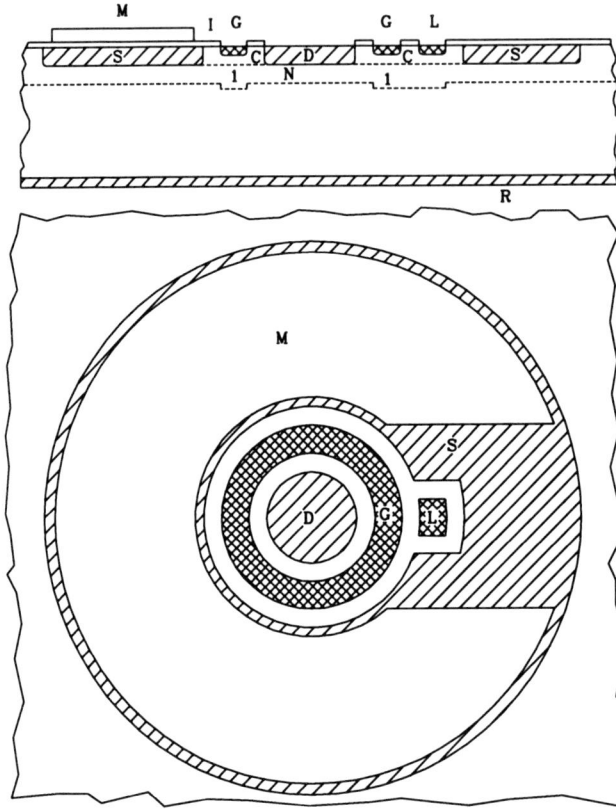

Fig. 9.10. Basic cell of a DEPFET pixel detector with vertical punch-through clearing and threshold compensation suitable for continuously sensitive operation. The signal charge is stored in the internal gate (1) within the deep-n region (N) having an increased doping concentration below the external gate (G). While the drain is connected directly to the outside, the source is coupled through the built-in capacitor formed by the source implant (S), insulator (oxide, I) and metal electrode (M). Clearing is performed by vertical punch-through from the internal gate (1) through the channel implant (C) to the clear electrode (L), which is held at constant potential and thus defines the potential of the partially filled internal gate

operate all DEPFETs with common source and gate voltages but instead has to find a way to compensate for the individual device variations.

In the device shown in Fig. 9.10 this is accomplished by coupling the source through an integrated capacitor C_s to a fixed external potential. On the assumption that the transistor was originally in a conducting stage, this capacitor will be charged until the transistor becomes nonconductive. All of the transistors will then be just at threshold. If a charge Q_{sig} is deposited in the internal gate of an individual DEPFET, its channel will become conducting until a charge $Q = Q_{sig}\frac{C_s}{C_G}$ has been transported through the transistor channel. Measuring this charge with a charge-sensitive amplifier, which can be connected either to the capacitance C_s or the drain of the DEPFET, one obtains a charge amplification $a = \frac{C_s}{C_G}$.

This method of operation leads in some applications to a signal rise time that is slow and in addition depends on the signal size. The signal rise time can be shortened by providing a defined standing current in each transistor. This can be done with the help of a resistor in the MΩ range, for example.

As the signal provided by the DEPFET pixel is already amplified, it is possible to split it into several parts without seriously affecting noise properties. The split signals can be connected in such a way as to uniquely identify the pixels from which they originated. Splitting is possible by, for instance, dividing the coupling capacitor in each pixel into three parts and connecting them with other pixels corresponding to three readout directions. The device can be read out in the same manner as three strip detectors.

A further modification of this concept leads to the introduction of a drift field into each pixel, driving the signal charge towards the internal gate and thereby speeding up charge collection. Also possible is the connecting of the external gate with the internal gate: this avoids the need of providing an exernal gate voltage and simultaneously increases the charge amplification of the DEPFET.

9.3.3 Hybrid DEPFET Pixel Detector

Compared with "flip-chip" pixel detectors such as described in Sects. 8.1.2 and 7.7, the continuously sensitive device described in the previous section has the disadvantage of not being a "true" pixel detector but rather the functional equivalent of three strip detectors occupying exactly the same space. Only the combination of the information of the three readout directions unambiguously determines a point in space related to a detected event.

Restricting the electronics technology to the one used already in detector fabrication makes the implementation difficult of more sophisticated functions, requiring as they do a large number of electronic elements in a single pixel. It is therefore worthwhile to evaluate DEPFET pixels with hybrid ("flip chip") readout. Such a combination simplifies requirements for both detectors and their electronics and in addition improves performance of the system. The detector itself provides an amplified signal, therefore easing the requirements on the electronics input amplifier. As the gate capacitance of the DEPFET is much smaller than the parasitic capacitances resulting from bump bonding, lower noise and/or faster risetime can be achieved. Simplification of the DEPFET is achieved through the possibility of providing the current source by the flip chip bonded electronics.

9.3.4 DEPFET Pixel Detector
with Three-Dimensional Analog Memory

An interesting concept (Fig. 9.11) – whose technical realization may, however, pose severe challenges – is a three-dimensional device that was proposed some time ago (Kemmer et al. 1988). It combines the DEPFET pixel readout structure with a three-dimensional memory located on top of a fully sensitive depleted detector volume.

Signal charges produced in the sensitive volume are collected in local potential minima of the lowest grid plane. By varying the potential of these grids, it is possible to move them upwards in a vertical CCD-like fashion towards the DEPFET readout device on the top surface.

With such a device it is e.g possible to take several images in rapid sequence and read them out slowly afterwards.

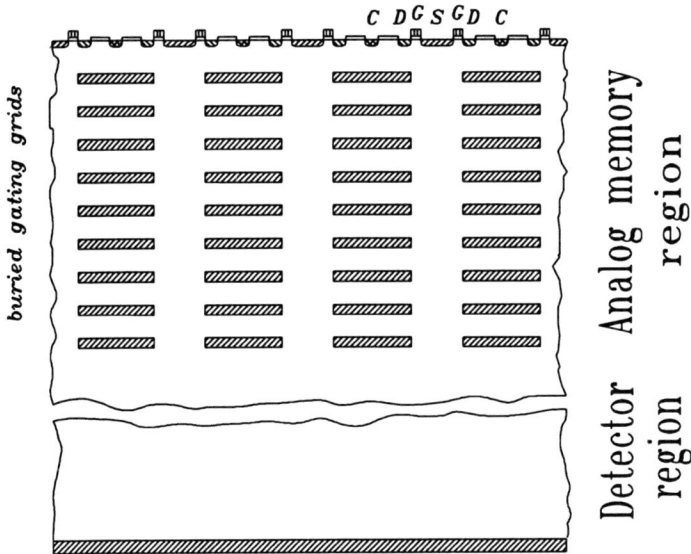

Fig. 9.11. Concept of a DEPFET pixel detector with a three-dimensional analog memory

10 Detector Technology

Today's semiconductor detector technology is to a large extent based on planar technology, which was originally developed in the field of microelectronics. It has been adapted and introduced to detector grade silicon processing by J. Kemmer (1980). Although these detectors were by today's standards fairly simple in design, the requirement of maintaining the high-quality semiconductor properties during processing did not allow a straightforward application of the technology. It was the general belief at that time that planar technology with its high-temperature processing steps would destroy the quality of the silicon material. Since then, detectors have become much more sophisticated and approach the complexity found in microelectronics. Therefore many of the methods used in microelectronics have been introduced into detector processing and diagnostics.

Although structure sizes are still fairly coarse in comparison with microelectronics, the requirements on technology in some aspects are more stringent. This concerns in particular the processing of ultrapure silicon in a way that does not cause its properties to deteriorate by, for example, the introduction of impurities or the creation of defects, the need of processing both wafer sides without damaging the opposite surface, and the production of wafer-sized defect-free detectors. These requirements demand that microelectronics technology and processing cannot be applied in a straightforward manner, and that special equipment and procedures have to be applied.

It is beyond the scope of this book to present in enough detail this technology in order to be helpful in detector processing. Instead, it is the intention to indicate to nonspecialists the possibilities and complexities involved in detector fabrication. General information on semiconductor processing can be found, amongst other references, in Sze 1983. It is intended to publish a book (by a different author) that specifically deals with detector processing.

10.1 Production of Detector Substrates

Detector substrates, usually referred to as wafers, are normally bought directly from manufacturers who also supply this material to the semiconductor and microelectronics industry. Detector substrates in general are single crystals with as perfect as possible a crystal structure and as few as possible foreign atoms, with the exception of dopants that are added intentionally in order to change the extrinsic material properties (see also Chaps. 2 and 4).

Growth techniques of semiconductors in general and of silicon and germanium in particular have been developed and perfected in the framework of (micro) electronics. Single-element semiconductors (Si and Ge) are grown from the melt of high-purity polycrystalline material with the help of a seed crystal of defined orientation, which then determines the orientation of the full-grown ingot.

For microelectronics application the melt is usually kept in a crucible (made of SiO_2 for Si growth) and the seed crystal is pulled from the top. This Czochralski crystal-growing method is not suited to detector-grade material as too many impurities – originating mostly from the crucible – are left in the crystal. This reduces the minority carrier lifetime to an unacceptably low level and makes production of high-resistivity material impossible.

Detector-grade silicon instead is produced using the float-zone technique. A high-purity polysilicon rod, vertically suspended in vacuum or an inert gas, is melted in a narrow horizontal zone using high-frequency induction. This zone is slowly moved from the seeding crystal on the bottom upwards, and the material at the bottom of the molten zone solidifies into a single crystal. During this process many of the foreign elements, having a small solubility in silicon, are pushed upwards towards the top of the rod. Repeating this zone-melting procedure several times therefore results in very pure single crystals.

Single-crystal ingots are subsequently cut into thin wafers and their surface(s) lapped and polished. In these steps, care has to be taken in order to avoid the introduction of crystal defects by the mechanical forces applied to the surface. These single- or double-sided polished wafers are the starting material for detector processing.

10.2 Processing Sequence in Planar Technology

The processing sequence for a simple p–n junction diode has been indicated in Chap. 5. It shows the characteristics of planar technology: growing or depositing uniform thin layers of material that are subsequently structured by photolithographic methods. In addition, doping on the semiconductor surface can be structured, making use of the structured layers on top of the semiconductor surface.

The following processing steps are widely used in various sequences:

- photolithographic structuring;
- chemical etching (wet and dry);
- doping;
- growing of an insulating layer by chemical reaction with the semiconductor surface (oxidation);
- deposition of insulating or conducting layers produced by chemical reaction between gases surrounding the detector (SiO_2, Si_3N_4);
- deposition of conductors (metals) by evaporation or sputtering;
- thermal treatments; and
- passivation.

Very important and sometimes difficult are, in addition, various cleaning and removal (e.g. photoresist) steps.

Designing a process sequence for a specific detector configuration is not straightforward as many of the processing steps change the results obtained in previous steps, as will be described in the following.

10.2.1 Photolithographic Structuring

Photolithography is used to transfer the structure from a mask (usually chrome on glass) onto the detector. For this purpose the semiconductor wafer is covered with a thin (typically $1\,\mu m$) layer of photosensitive resin. This is accomplished by dropping a small amount of liquid resin on a fast-spinning wafer. After drying, the resin is illuminated through the mask, which is either pressed against the wafer (contact illumination) or held at a close ($\approx 10\text{--}20\,\mu m$) distance (proximity illumination). Contact illumination allows high resolution and small feature size; its disadvantage is the likelihood of defect introduction by (for example) particles or resin getting stuck to the mask, thus producing faults in consecutive illuminations. Therefore proximity illumination is the standard approach in detector processing. The very expensive method of projective illumination, in which an optical system transfers the structures from the mask onto the wafer, is so far uncommon in detector production.

After illumination the resist is developed. For commonly used positive photoresist, the illuminated resist dissolves so that the wafer surface is unprotected in these regions and therefore is exposed to the subsequent processing steps of etching or doping.

Masks are also produced by photolithography, illuminating the photoresist on top of a chrome-covered glass through a collimator, which is moved step by step over the glass to produce the desired topology. Direct writing into the photoresist using a focussed electron beam results in even higher quality masks and allows the production of finer structures.

A special feature of many detectors is the need for double-sided structuring. In this case, front and backside photolithography have to be aligned with respect to each other. Two methods are commonly used for this purpose: infrared light to look through the wafer, and the electronic storage of mask images against which the other side can be aligned once the direct view is blocked by the wafer put in front of the mask. Front–back alignment of typically $5\text{--}10\,\mu m$ can be obtained quite easily.

10.2.2 Chemical Etching

As shown in the previous section, the wafer surface can be selectively protected by photoresist that is resistant against most liquid chemical etchants. The etchant is applied to the wafer as a whole and will act on all regions not protected by the resist. The etchant may be liquid or gaseous (ion etching). An important property of the etchant is the selective activity to a specific material such that the etch process stops at the underlaying layer.

Wet etching is most commonly used in detector processing, while dry ion etching is rather rare. The reasons for this difference in usage, besides lower investment costs, are the usually modest requirements on structure size and the avoidance of oxide radiation damage that is commonly connected with dry ion etching. The isotropic etching properties of wet methods lead to some underetching below the resist (typically comparable to the photoresist thickness of 1–2 μm). Underetching has to be antipicipated in the mask design.

Silicon nitride (Si_3N_4) wet etching is somewhat problematic since standard photoresists are not sufficiently resistant against the etchant used at elevated temperature. One therefore uses structured silicon oxide on top of silicon nitride as mask.

10.2.3 Doping

Usual dopants for silicon are boron (p-type), phosphorus and arsenic (n-type). One starts with a uniformly doped wafer and the aim is to change the doping in selected regions of the surface. This can be done either by diffusion, or by implantation, or by a combination of both methods.

With diffusion the donor atoms move from the open semiconductor surface into the semiconductor volume, which is held at elevated temperature during the process. Either the dopant concentration is controlled at the surface during the process, or a defined amount of dopant is deposited at the surface at the beginning of the procedure.

Implantation is done by bombarding the wafer with a (defocussed) ion beam of defined energy. Ions will be stopped in the semiconductor in the regions that have been opened in the preceding photolithographic and etching steps. In the other regions the ions will be stopped in the insulating layer(s) above the semiconductor, unless their energy is high enough to penetrate into the semiconductor. The structured photoresist may also be used as a stopping layer for the implantation.

The deposition of the donors in the crystal is not sufficient to obtain the desired electrical performance of the doped regions: the dopant atoms may, for instance, not be in a regular lattice position – a necessary condition for being electrically active in the desired fashion. On the contrary, at high dose of implantation the semiconductor lattice may even be destroyed. A heat treatment of the crystal is necessary in order to bring the dopant atoms into regular lattice positions and to regrow the lattice.

With implantation one can produce buried layers of doping by choosing high implantation energies. One may also implant through an insulating layer as, for example, with SiO_2, thus simplifying the production process. Implanting through thin insulators (SiO_2) is also used with the purpose of shaping the doping profile, for instance for placing the maximum concentration at the semiconductor surface. Diffusion introduces a much smaller amount of defects than implantation but it is less flexible than implantation and often requires the use of toxic gases.

A disadvantage of implantation is the fairly small depth (typically less than $1\,\mu m$) that can be reached with standard energies. One therefore may need a "drive-in" diffusion following implantation. One can also use as the base material a semiconductor wafer with an epitaxially grown layer of different doping. Epitaxial growing means growing of the crystal in the correct lattice structure on top of a single-crystal wafer; it is most commonly performed in a gaseous athmosphere.

10.2.4 Oxidation

In almost all detectors and other silicon semiconductor devices, a large fraction of the outer surface is covered with SiO_2, which provides a rather good passivation of the silicon surface such that most of the open bonds of silicon are saturated by the oxide in the $Si–SiO_2$ transition region, and only a small amount (approximately $10^{11}/cm^2$) of dangling bonds are left.

Oxidation is performed by heating silicon in "dry" oxygen at roughly 1000°C or in a wet (H_2O) atmosphere. In both cases oxygen diffuses through the layer of oxide already grown and reacts with silicon at the $Si–SiO_2$ boundary. The process slows down with increasing thickness of the oxide layer and is strongly temperature-dependent. Wet oxide grows much faster and at lower temperature than dry oxide.

Oxidation is usually the highest temperature process and is performed at the beginning of detector fabrication. Due to the high temperature used in the process, many of the impurities and crystal defects become mobile. Great care therefore has to be taken in order to prevent the introduction of impurities that may spoil the properties of the semiconductor material. A surface cleaning procedure therefore usually precedes the oxidation process.

One often also makes use of that high mobility during the oxidation process by combining it with impurity-gettering processes. A high concentration of defects is artificially introduced at an usually unused part of the surface of a crystal. Impurities reaching these regions get trapped and thus are removed from the sensitive detector volume. It is also possible to remove these gettering zones at a later stage of the production process.

10.2.5 Deposition from the Gas Phase

Silicon dioxide cannot only be grown from silicon as described in the previous section but may also be deposited from the gas phase by a chemical reaction between two gases, resulting in the production of SiO_2 that will cover the detector (and also the vessel in which the reaction occurs).

Deposition from the gas phase is often performed at low pressure and a variety of different materials can be produced, such as polysilicon and silicon nitride. It opens up a large variety of possibilities of sophisticated detector structures.

Deposition from the gas phase usually results in amorphous or polycrystalline layers. It is also possible to grow monocrystalline layers on the surface of a wafer, thereby extending the crystal. In this growing procedure dopants can be added, so that a monocrystal with varying doping can be produced. This method of epitaxial growth during device fabrication is rather common in bipolar electronics technology but not in detector production. However, wafers with unstructured epitaxial layers – as supplied by a silicon producer – have been used for CCD detectors, for example.

10.2.6 Metal Deposition

Metal layer(s) are used for low-resistivity interconnection and for supplying connection possibilities (bond pads) to the outside. The most common metal used is aluminum. It can be deposited either by evaporation or by sputtering. Both processes have to be performed in a vacuum. Evaporation is usually done by heating in a tungsten boat, sputtering by bombarding an aluminum target with either ions (e.g. Ar), atoms or electrons, using various methods of acceleration.

Aluminum has the ability to dissolve a small fraction of silicon. If this fraction of a few percent is not provided during the deposition procedure, it will be taken out of the silicon substrate during the thermal treatment (sintering) that is required for producing a good electrical and mechanical contact between the silicon and the aluminum.

Aluminum spikes can grow into the silicon during this process, leading sometimes to electrical shortening and malfunctions of the devices. Spiking is also strongly dependent on crystal orientation: it is much less likely on $\langle 1, 1, 1 \rangle$ orientation than on $\langle 1, 0, 0 \rangle$ surfaces. Various methods have been invented for suppression of spiking, such as the introduction of a thin layer of barrier metal or conducting polysilicon between the silicon and the aluminum.

Aluminum has the advantage of excellent electrical conductivity, but it has the drawback of a low melting temperature (660°C). This creates problems in situations in which more than one metallization layer is needed. In this case one has to either introduce between the two metalization layers an insulating layer formed at low temperature or one has to use a temperature-resistant first metal layer such as tungsten; both methods have their problems.

10.2.7 Thermal Treatments

As many of the previously described technical processing steps occur at elevated temperatures, thermal treatments are already implicitly performed – sometimes also leading to undesired results such as a broadening of doping profiles.

Nevertheless, additional thermal treatments are usually necessary for:

- *activating the dopants after implantation.* Implantation forces foreign atoms into the crystal lattice. These atoms will in general not be located on a regular lattice location but at an interstitial position, at the same time distorting or destroying the crystal lattice in its vicinity. Damage and/or destruction of the lattice will also be done by silicon recoil atoms produced in collisions with the dopant atoms or ions. A thermal treatment in controlled condition is needed to place the dopant atoms into regular lattice sites and to repair most of the damage that has occurred during the implantation process;

- *treatment of the oxide–silicon interface.* As mentioned previously (Sects. 4.3.2 and 10.1.4), due to the incomplete chemical matching on the Si–SiO$_2$ interface, dangling bonds are left at the interface, causing positive charges and corresponding flat-band voltage shifts. These dangling bonds can partially be bound with hydrogen. Heating the device in a part-hydrogen atmosphere (hydrogen annealing) enables hydrogen to diffuse through the oxide towards the interface, thus leading to a saturation of the dangling bonds and a reduction of the flat-band voltage;

- *sintering.* In order to ensure a good electrical contact between deposited conductors (e.g. aluminum) and silicon, a thermal treatment is needed after structuring the conductor. With aluminum this is done by heating the device to a temperature somewhat below the metal's melting point, i.e. to about 400°C. This step is rather delicate due to the problem of spiking which is also dependent on other parameters, as described previously (Sect. 10.1.6);

- *stress relaxation.* Due to differences in thermal expansion coefficients and the formation of layers at elevated temperatures, considerable mechanical stress can occur at room temperature. This stress can sometimes be reduced by thermal treatment, for example in slow cooling after the formation process.

10.2.8 Passivation

Detectors are in many cases used without protection against mechanical or chemical damage due to the environment in which they have to operate. Often protection may even be undesirable when, for instance, extremely thin entrance windows are required for the measurement of low penetrating radiation. In many cases, however, a protection against chemical poisoning and environmental changes such as humidity will be needed. Also needed may be mechanical protection of the soft aluminum structures during the sometimes elaborate mechanical assembly procedures.

Passivation is performed by covering the wafer (with the exception of the electrical connections to the outside) with a robust electrically insulating layer. This layer has to be formed at rather low temperature in order not to damage the detector properties in general and the aluminum in particular. The passivation layer can be structured using photolithographic methods.

The following materials can be used for passivation:

- LTO, silicon oxide deposited from the gas phase at low temperature and low gas pressure;
- silicon nitride (Si$_3$N$_4$);
- phosphorglass;
- polyimide, an organic resin that can be handled in a similar fashion to photoresist.

10.3 Technology Simulation

The various processing steps discussed are not independent of each other and therefore cannot be optimized separately. In designing the process for a specific device the interplay between the processing steps – mainly due to thermal treatments – has to be taken into account.

One- and two-dimensional simulation programs are used to predict those quantities that are relevant for the electrical characteristics of the devices such as dopant concentrations and activation fractions. They are obtained from a simulation of the sequence of processing steps (such as thermal oxidation), and they include the redistribution of dopants in the surface region, implantation either directly into silicon or through an (oxide) scattering layer, and redistribution as well as activation of dopants in the following thermal processing steps.

Many of the mechanisms involved in the redistribution of impurities are rather complicated and only partially understood. They are therefore often parameterized from systematic measurements. These measurements have been done carefully and systematically but sometimes not in a domain that is relevant for detectors but rather for microelectronics. One therefore has to be careful in applying generally available software packages for detector purposes.

11 Device Stability and Radiation Hardness

Reliable operation of semiconductor detectors requires insensitivity to environmental conditions such as humidity, temperature and working in atmospheric or vacuum conditions. The radiation to be measured may change the detector material and thus also the properties of the detector.

For improperly designed detectors property changes in many cases are of catastrophic nature leading (amongst other things) to electrical breakdown. Such a breakdown often occurs quite suddenly: a seemingly perfect detector that draws very low reverse-bias current at high voltage for a reasonably long time (hours or days) may within minutes show many orders of magnitude higher current even if it is being operated at a small voltage. These problems are mainly connected with effects involving the surface of the detector, and protective measures in the design will be proposed in this chapter.

The exposure of the detector to intense nuclear radiation poses even more severe problems. In this situation, which has become very important in recent years, not only the surface is affected but the basic material bulk properties may change drastically. In the design of detectors these effects have to be anticipated in order to ensure continuously reliable operation of the detectors.

11.1 Electrical Breakdown and Protection

Electrical breakdown occurs when the electrical field is high enough to initiate avalanche multiplication, i.e. where movable charge carriers (electrons and holes) are accelerated strongly enough between collisions with lattice atoms to cause the production of electron–hole pairs. This avalanche process (see Sects. 2.6.5 and 3.1.5) is rather complicated to describe in detail; typically at room temperature in silicon a field of $30\,\mathrm{V}/\mu\mathrm{m}$ becomes dangerous. Obviously the avalanche condition will depend both on the temperature (scattering occurs due the thermal displacement of the atoms from the ideal lattice) and on crystal defects (for instance caused by doping or lattice distortion near the surface). Also the spatial extent of the high field region plays a role as the electrons or holes have to gain enough energy between collisions with the lattice. The mean free path is in the range of $0.1\,\mu\mathrm{m}$ in silicon, although it can be considerably shorter very close to the surface (interface) region than deep inside the bulk due to the higher density of crystal defects. Another effect which in addition complicates a quantitative treatment is the change of the electric field due to the avalanche generation of movable charge carriers, which to some extent limits the breakdown current.

The electric field strength maxima, and therefore the breakdown probability, will depend strongly on the detailed geometric layout and on the operating conditions of the detector. It will also depend on inhomogeneities in the material such as doping density variations and crystal defects. Defaults in the production process, such as lithographic errors, in addition can lead to strong electric fields. Very strong effects are due to radiation-induced changes in the material properties (Sect. 11.2).

In designing a detector one should therefore aim at the lowest possible electric fields compatible with the functional requirements. Within limits, the situation can be improved by technological measures such as optimizing doping profiles. However, even more important is the topological layout.

We will illustrate the effects to be considered on a silicon strip detector with capacitively coupled readout, considering the proper detector region (Sect. 11.1.1) and the detector rim (Sect. 11.1.2). It is then straightforward to apply similar considerations to other type of detectors.

11.1.1 Breakdown Protection in Diode Strip Detectors

A cross-section through a simple capacitively coupled diode strip detector built on n-type silicon is shown in Fig. 11.1. The readout capacitances are obtained by separating the aluminum from the p^+ implants of the strips by means of the insulating oxide. The region from the center of one strip to the next is used in the simulation, applying boundary conditions corresponding to an infinite mirrored repetition of this structure. The electric field maximum, located close to the edge of the strip, will depend strongly on the gap width W_{gap}, the oxide charge N_{ox} and the boundary condition applied on the top of the bare oxide.

While we are able to go a long way with essentially one-dimensional approximations, this is not possible for the situation considered here. A true two-dimensional numerical simulation has to be applied. Then one encounters the question of the boundary condition to be applied on the uncovered oxide

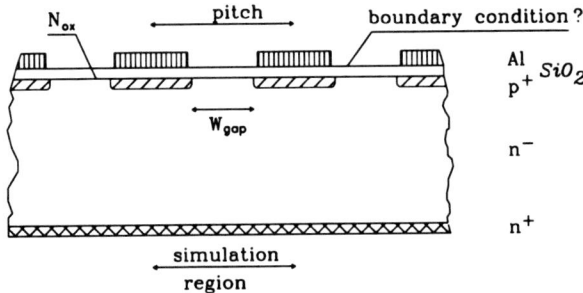

Fig. 11.1. Cross-section through a capacitively coupled strip detector extending from the center of one strip to the next

in the gap region. Frequently the Neumann boundary condition, the assumption of zero electric field component normal to the surface, is applied, but this boundary condition does not represent a realistic situation. It is a rough approximation for the assumption that no electric charge is present at the outer surface of the oxide and that the oxide has a very high dielectric constant ($\epsilon_{ox} \gg 1$). The assumption of zero charge on the outer oxide surface has been shown invalid for normal operating conditions (Longoni et al. 1990); only immediately after application of the bias voltage to an initially unbiased detector does it give a reasonable description.

Keeping the detector biased, the small surface conductivity on the outer oxide layer will eventually lead to a situation in which the surface assumes the same potential as the neighboring metal strips. This process can take seconds to days, depending on the surface conditions, which are themselves strongly dependent on the environment in forms of humidity, surface contaminations, etc.

The assumption of defined electric potential at the outer oxide surface, henceforth called the "gate boundary condition", therefore is the correct condition to be applied to a detector in a "steady state" condition. It results in strongly different field conditions compared with the Neumann boundary condition. It also leads to the conclusion that changing the relative voltage between strip implant and strip metal stongly influences the electrical field distribution inside the semiconductor. The gate boundary condition therefore has been applied in all simulation examples discussed below.

Turning now to the question of topology, we focus on the qualitative difference between narrow- and wide-gap configurations in a p^+–n diode strip detector (Fig. 11.2). Consideration of the effect of the positive oxide charge is essential for understanding the resulting electric field configuration. We will assume for the moment that for capacitively coupled strip detectors implanted strip and metal are both at ground potential and that the bias voltage is applied to the backside n^+ contact. The metal potential will also spread to the uncovered outer oxide surface. From an electrical point of view the situation is that of a MOS structure (see Chap. 3, Sect. 3.3 and in particular Sect. 3.3.2). An electron-accumulation layer compensates for the oxide charge.

Applying a positive reverse-bias on the backside makes the space-charge region grow from the strips, both in depth and laterally. The increase of the voltage across the oxide in the gap region is accompanied by a decrease in the strength of the electron-accumulation layer. The reduction of the electron layer is slowed down drastically when the space-charge regions of two neighboring strips join in the depth of the bulk, thereby forming a barrier for the electrons caught near the surface. Such is the case for the narrow-gap situation (Fig. 11.2a). One then has an undepleted region near the surface in the gap region, separated from the main part of the bulk. Its potential is determined by the potential saddle point at midgap, slightly inside the bulk, and changes only slowly with the backside potential.

For the wide-gap situation (Fig. 11.2b) the electron layer will disappear completely before such a situation arises and with reverse-bias voltage above

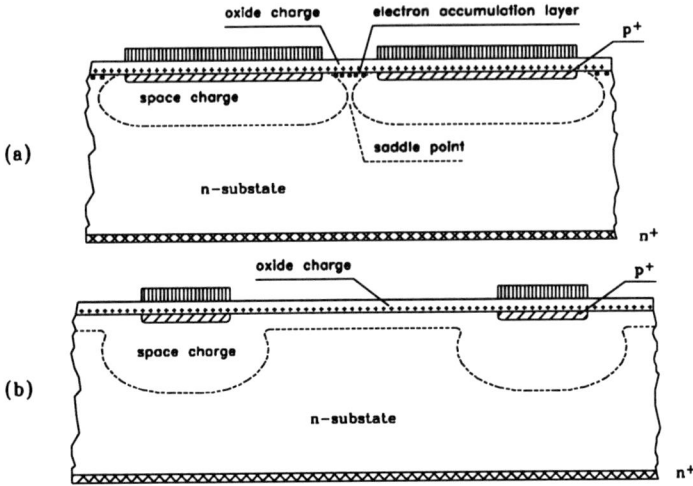

Fig. 11.2a,b. Diode strip detector: schematic illustration of the properties of small- and wide-gap strip detector topology. In the narrow gap situation (a) an undepleted region below the oxide is separated from the undepleted region in the bulk by the space-charge regions growing from neighboring strips and an electron layer forms below the oxide in the gap region. In the wide-gap situation (b) the bulk is depleted up to the oxide–semiconductor interface and no electron layer forms below the oxide

the flat-band voltage V_{FB} the space-charge region in the gap region will grow from the oxide–silicon interface downwards.

Comparing the two situations, we can see that the potential in the narrow-gap case is essentially controlled by the gap width, while in the wide-gap situation it is the flat-band voltage (added to the offset voltage of the metal strip in capacitively coupled detectors) that has the effect.

As an illustration, simulations for the two situations are shown in Fig. 11.3 for the case of high oxide charge density $N_{ox} = 2 \times 10^{12}$ /cm^2, such as is expected after considerable radiation damage (see Sect. 11.3).

As expected, we can also see the presence of electrons below the narrow-gap region (the peak of the distribution is off-scale in the display), while they are missing for the wide-gap case. Similarly, the voltage difference between gap and strip region is much smaller for the narrow-gap than the wide-gap configuration. As a consequence the electric field maxima, which in both cases are located at the edge of the strips, are considerably higher in the wide-gap situation. For the wide-gap situation they are just above the critical value of 30 V/μm. But it is necessary to remember that this example is already assuming increased oxide charge due to irradiation. For an unirradiated detector with an order of magnitude lower oxide charge, both situations are safe.

We can now discuss qualitatively the effects to be expected from changes in topology and operational conditions. Keeping the same applied voltages but increasing the fairly thin oxide (0.22 μm) to the frequently used much higher values (≥ 1 μm) will result in a large change of the gap potential in the wide-gap

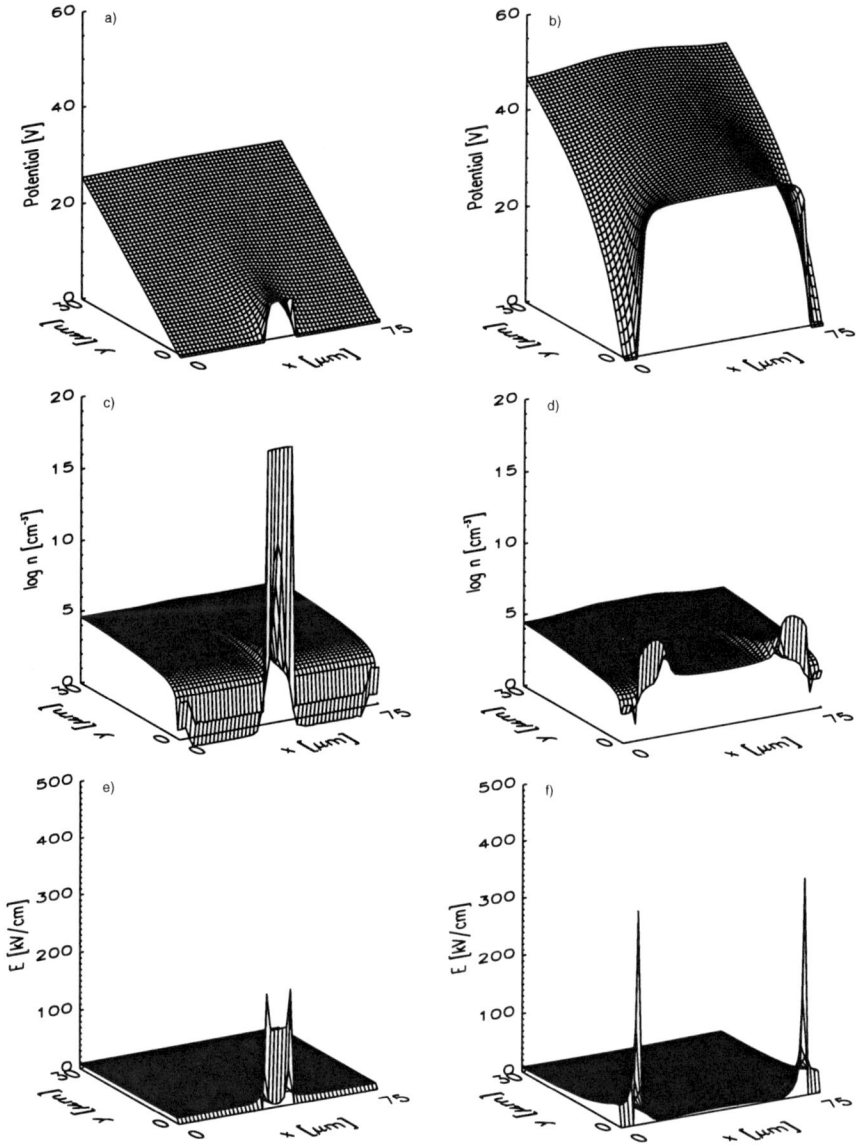

Fig. 11.3a–f. Potential, electron density and electric field of a capacitively coupled strip detector with narrow (10 μm) gap (*left column*) and wide (65 μm) gap (*right column*). A strip pitch of 75 μm, detector thickness of 300 μm, bulk doping of $N_D = 2 \times 10^{12}$ cm^{-3}, oxide charge density $N_{ox} = 3 \times 10^{12}$ cm^{-2}, oxide thickness $d_{ox} = 0.22$ μm and a reverse bias voltage of $V_{back} = 130$ V are assumed. The region from the center of one strip to the next is shown. Results are obtained with the TOSCA device simulation program. (see Gajewski et al. 1992 *re* TOSCA; diagrams above are after Richter et al. 1996, Fig. 4)

case. Assuming the same oxide charge density at the Si–SiO_2 interface, this change will be proportional to the the oxide thickness.

Changing the metal strip potential by ΔV in a positive direction has the same effect as increasing the oxide charge by $\Delta N_{ox} = \frac{\Delta V}{qC_{ox}}$, thus only slightly changing the potential for the narrow-gap case but (for the wide-gap case) increasing the gap potential by the same amount, ΔV, while the maximum electric field increases in both situations. Increasing (in the wide-gap situation) ΔV further, as one might attempt to do for a double-sided detector when wanting to operate the readout electronics of both detector sides on ground potential, the gap potential will follow until eventually the situation is reached that the surface region is pinched off by the space-charge region extending from the neighboring strips, thus reaching in effect a narrow-gap situation. Such a situation will lead to strong electric fields and is therefore prone to electrical breakthrough.

Turning now to n-type strips, either on p-type bulk or n-type bulk (as is the case in double-sided detectors), one has to consider in addition the strip isolation method needed to interrupt the electron layer formed below the oxide. Two isolation methods will be considered: the "p-blocking" and the "p-spray isolation" techniques (Fig. 11.4). With the p-spray isolation technique (Fig. 11.4b) a shallow unstructured p implant provides a doping density that, integrated over depth, exceeds somewhat the oxide surface charge density, thus preventing the buildup of an electron layer below the oxide. In the strip region it is over-compensated for by the high-dose n^+ implant. The situation is rather similar to the previously considered p^+ strips: the highest electric fields will occur at the edges of the n^+ strips. One also will have distinctly different small- and wide-gap situations.

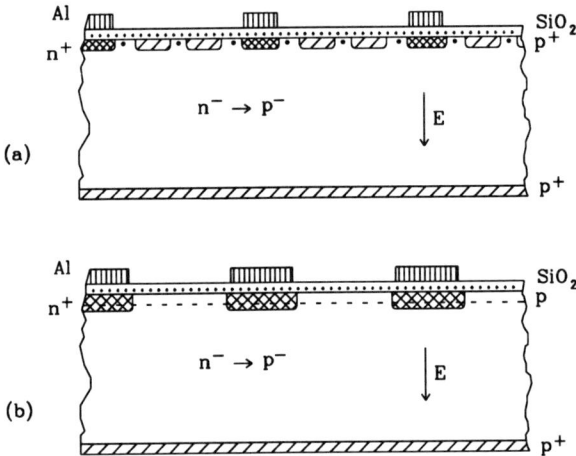

Fig. 11.4a,b. Isolation methods for n^+ doped strips: isolation by p-doped strips (p-blocking implants) (a); unstructured p-type compensation implant for the oxide charges (p-spray implant) (b)

In the p-blocking isolation technique (Fig. 11.4a), a fairly highly doped p region isolates neighboring n strips from each other. The n strips are effectively widened by the oxide-charge induced electron layer and the field maxima occur at the edge of the blocking implantation. Here the equivalent wide-gap situation is reached when the blocking implant is competely depleted. For high blocking implant doping densities, this situation will not occur; instead, the potential difference between blocking and strip implants will increase such that the "narrow-gap" field configuration is retained. The large potential difference and corresponding high electric field once more makes this topology prone to electrical breakdown.

We will come back to n-side insulation problems when discussing radiation hardness.

11.1.2 Breakdown Protection of the Detector Rim

Consider a simple large-area diode or a more complicated detector such as the one shown in Fig. 11.1. The bias voltage will be applied between a bulk contact and the diode. Without further measures, the full bias voltage will drop in a region from the edge of the diode towards the cutting edge (we assume that the detector has been cut to the required geometrical size out of a processed wafer) of the detector, which for the moment we assume to be on bulk potential.

The situation differs from the previously considered case of the strip detector in a very important way. We have assumed for the capacitively coupled detector that all aluminum strips are on the same potential and that the oxide surface eventually assumes the same potential. Now the potential of the oxide surface at the cutting edge and the diode edge differ by the bias voltage, and the surface potential distribution is unpredictable as it depends on local variations of the surface resistivity – which itself is dependent on cleanliness of the surface, humidity, etc.

We will consider two extreme cases that both lead to severe problems. In the first case (left side of Fig. 11.5), we assume that the voltage at the oxide surface drops to bulk potential close to the diode edge. Then an electron-accumulation layer will be present at the Si–SiO$_2$ interface and the bulk below the oxide will not be depleted except for a small distance next to the diode edge. The complete bias voltage will drop over this short region and the corresponding high electric field will cause avalanche breakdown of the detector. This situation is similar to the situation previously considered involving the space between strips of capacitively coupled strip detectors and the metal strips biased at backside potential.

In the second case to be considered (right side of Fig. 11.5), we assume that the oxide surface gets charged up to the diode potential almost all the way to the rim of the wafer (the cutting edge of the detector). Then an inversion layer will form below the oxide that will essentially enlarge the space-charge region all the way up to the cutting edge. As the crystal is heavily damaged close to the cutting edge, a huge generation current will be produced at this site.

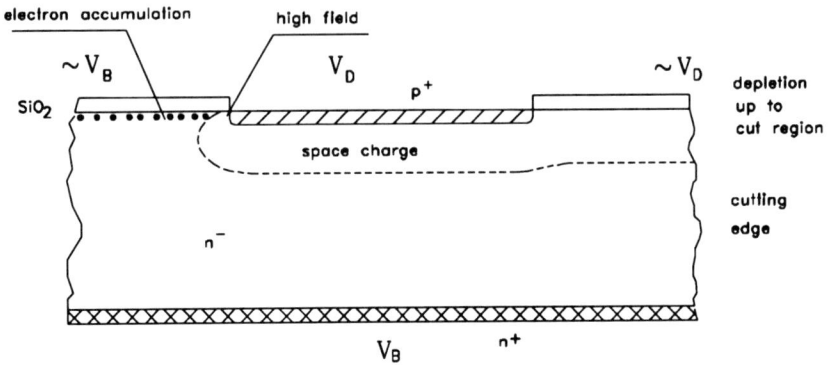

Fig. 11.5. Edge breakdown mechanisms of a large-area diode detector. The breakdown is due to the undefined outer surface condition of the silicon dioxide. On the left the situation is shown that the surface potential drops fast towards bulk potential. Due to the positive oxide charge, an electron-accumulation layer forms below the oxide and the full bias voltage $V_D - V_B$ drops over a short distance next to the edge of the diode. Avalanche breakdown is likely in this region. On the right side the surface potential is assumed to remain at diode potential almost all the way to the cutting edge. High field breakdown will occur on the left diode edge. A very large reverse-bias current is generated due to crystal defects near the cutting edge inside the space-charge region

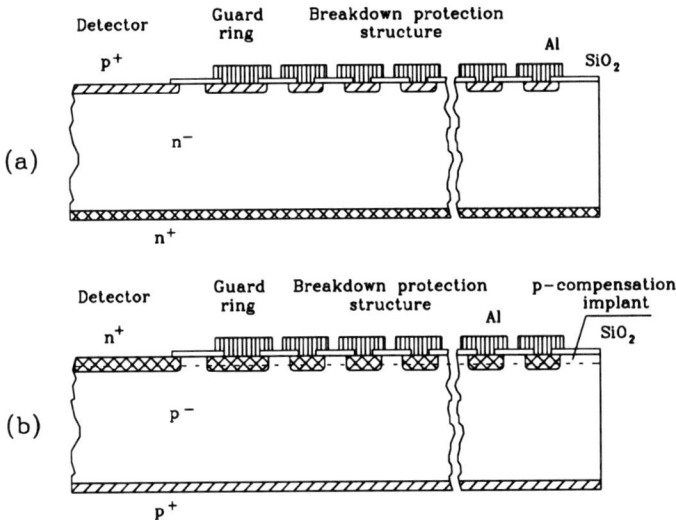

Fig. 11.6a,b. Edge breakdown protection structures for n- (a) and p-type (b) detectors. The structures can be built and operated without additional complications, compared with the unprotected detector. Strip and implant are connected so that the outer oxide surface potential is defined. The aluminum overlaps the gap between implants. In the case of n strips on p material the electron-accumulation layer has to be interrupted by p implantation, which can either be a large-area implantation (as shown) or ring implants between n-rings. The same structures can be used on opposite sides of a double-sided detector

From previous considerations it has become clear that it is important to ensure a gentle drop of the oxide surface potential from the edge of the diode towards the cutting edge of the detector, and there are many ways to achieve this. In the simplest case one may put individually biased metal rings surrounding the active area of the detector on top of the oxide. Another method would be the use of only two rings, which are connected by a highly resistive conducting sheet deposited on top of the oxide.

In general it is preferable to have a solution that does not complicate operation and/or production of the detector. A simple arrangement of guard rings fulfilling this requirement is shown in Fig. 11.6 both for p strips on n-type silicon and for n strips on p-type silicon. The detector is surrounded by several rings consisting of implantation and aluminum. The biasing of the individual rings is achieved by punch-through biasing in a very similar way as was done in strip biasing of capacitively coupled strip detectors (Sect. 6.4). An important aspect of the structure is the connection of the implant with the aluminum on top of the oxide surface. In this way the oxide surface potential is defined in the vicinity of the rings, thus ensuring proper biasing of the rings. The geometry may be chosen in such a way as to drop the voltage from ring to ring by a moderate amount of a few tens of volts, thereby ensuring that no breakdown occurs. In the version shown the aluminum is partially overlapping the gap between implants, thus ensuring the prevention of any formation of a connecting inversion layer between strips and also the prevention of high fields at the other edge of the implanted rings.

11.2 Radiation Damage in Semiconductors

Although radiation damage mechanisms are important and somewhat similar in many semiconductors, we will restrict ourselves essentially to silicon, the most important and most extensively studied semiconductor detector material. Nevertheless, much of the general material presented in this section applies also to other semiconductors.

As we are concerned with radiation detectors, it is the radiation itself, which we want to detect, that may also cause damage to the detectors. Nuclear radiation interacts with the electron cloud, but also with the nuclei in the lattice. While the interaction with the electron cloud in silicon is a transient effect that is in fact used for the detection of the radiation, the interaction with the lattice may lead to permanent material changes, which often are of detrimental nature.

The following processes are of importance for the lattice:

- displacement of lattice atoms, leading to interstitials (atoms between regular lattice sites) and vacancies (empty lattice sites);
- nuclear interactions (e.g. neutron capture and nucleus transmutation);
- secondary processes from energetic displaced lattice atoms, respectively defect clusters from cascade processes.

Most of these primary defects are not stable. Interstitials and vacancies are mobile at room temperature and will therefore partially anneal if by chance an interstitial fills the place of a vacancy. However, there are also chances for the formation of other (room temperature) stable defects. Examples are the well known A-center, a combination of a vacancy and oxygen (a certain concentration of oxygen interstitials is present in the crystal after crystal growing), the E-center, a vacancy phosphorus complex, and the divacancy (two missing silicon atoms right next to each other). These stable defects may then change the electrical properties of the semiconductor. In general they will worsen the detector properties.

In the following we will discuss the formation and electrical properties of lattice defects in more detail.

11.2.1 The Formation of Primary Lattice Defects

In order to displace a silicon atom from its lattice site, a minimum recoil energy of 15 eV is needed. This energy can be provided by elastic scattering of a high-energy charged or neutral particle. The threshold is not sharp, however, as it depends on the direction of the recoil. If the direction of the recoil points towards a neighboring atom, a much higher threshold is expected than for a direction pointing between neighboring atoms. Looking at the displacement probability as a function of the recoil energy, one uses a displacement energy E_d, the energy at which the displacement probability is roughly one half (for silicon $E_d = 25$ eV (van Lint et al. 1980)). Recoil energies below E_d will predominately lead to lattice vibrations only, while values above E_d will in addition create a vacancy–interstitial pair. If the energy of the recoil is far above E_d, the recoiling silicon atom is able to create additional vacancy–interstitial pairs.

Before looking at the primary process, which depends on the type of irradiation, we will consider the fate of the recoiling silicon atom. For recoil energies below roughly 1–2 keV, only isolated (point) defects will be created; between 2 keV and 12 keV the energy is high enough to create one defect cluster and additional point defects; and above 12 keV several clusters and additional point defects will be produced. A cluster is a dense agglomeration of point defects that appear at the end of a recoil silicon track where the atom loses its last 5–10 keV of energy and the elastic scattering cross-section increases by several orders of magnitude. A typical size for a cluster is 5 nm diameter with 100 lattice displacements.

All these numerical values come from model calculations; it is not known to the author that experimental proof for the existence of defect clusters exists.

Dependence on Type of Radiation

The probability for creation of a primary knock-on atom, a silicon atom displaced from its lattice location by the impinging radiation, as well as its energy distribution, depend on the type and energy of the radiation concerned. This for two reasons:

- the elastic cross-section for scattering on silicon atoms depends on the type of radiation. Charged particles such as protons scatter by electrostatic interaction with the (partially screened) nucleus, while neutral particles such as neutrons scatter elastically with the nucleus only; and
- the kinematics and the energy transferred to the silicon atom (in particular) is strongly dependent on the mass of the impinging radiation.

In Table 11.1, some important characteristics of the primary interactions of electrons, protons, neutrons and (for comparison) silicon knock-on atoms are given. Looking at this table, one realizes that the light electrons will produce point defects but almost no clusters.

Table 11.1. Characteristics of interaction of radiation with silicon and of primary knock-on atoms. The radiation energy is 1 MeV. T_{max} is the maximum kinematically possible recoil energy, T_{av} the mean recoil energy and E_{min} the minimum radiation energy needed for the creation of a point defect and for a defect cluster

Radiation	Electrons	Protons	Neutrons	Si^+
Interaction	Coulomb scattering	Coulomb and nuclear scattering	Elastic nuclear scattering	Coulomb scattering
T_{max} [eV]	155	133700	133900	1000000
T_{av} [eV]	46	210	50000	265
E_{min} [eV] point defect	260000	190	190	25
defect cluster	4600000	15000	15000	2000

Scaling of Radiation Damage

Although the primary interaction of radiation with silicon is strongly dependent on the type and energy of the radiation, this dependence is to a large extend smoothed out by the secondary interaction of primary knock-on silicon atoms. It is therefore customary to scale measurements on radiation damage from one type of radiation and energy to another. As the interaction of radiation with electrons produces ionization but no crystal defects, the quantity used for scaling is the non-ionizing energy loss (NIEL). The dependence of this scaling variable on energy and type of irradiation (normalized to 1 MeV neutrons) is shown in Fig. 11.7.

11.2.2 Formation and Properties of Stable Defects

The primary defects, Si interstitials and vacancies, are – at room temperature – still mobile and cannot be considered stable. Part of these defects will anneal either by an interstitial filling a vacancy or by diffusing out of the surface. They may, however, also interact with another defect and form a new type of defect complex that becomes immobile at room temperature. The other defect may have been already present in the crystal or can also have been produced by the radiation.

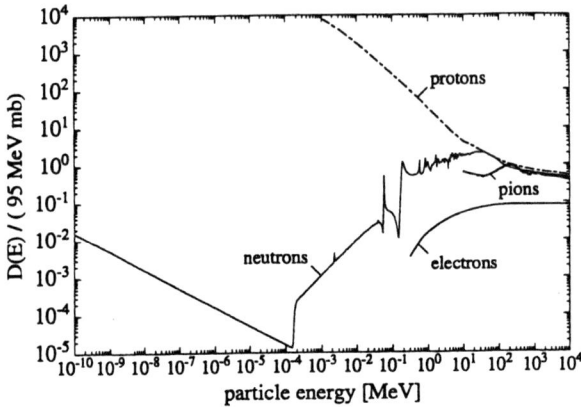

Fig. 11.7. Energy dependence of non-ionizing energy loss (NIEL) in silicon for various types of radiation. (After Vasilescu 1998 et al.)

An example of the first kind of defect is the formation of a vacancy–phosphorus complex (E-center). The standard dopant for n-type silicon is phosphorus. A vacancy right next to the phosphorus dopant (on a regular lattice site) forms a stable complex with new electrical properties. These properties will be dealt with in the next section. The phosphorus atom does not fulfill its original role of donor any more, and the process is therefore also called donor removal.

A second example is the formation of vacancy–oxygen complexes (A-center). Oxygen is always present within silicon to a certain degree as a remnant from the crystal-growing process. Oxygen as an interstitial (between regular lattice sites) is electrically inactive. The stable oxygen–vacancy complex becomes electrically active, i.e. it forms an acceptor state in the upper half of the band gap that, although electrically neutral in the space-charge region, acts as a trapping center for electrons (see Sect. 11.2.4).

An example of a stable defect complex that can be produced from radiation-generated primary defects alone is the divacancy, namely two missing silicon atoms right next to each other.

The formation of defect complexes in semiconductors is complicated and only partially understood. Defect complexes are not only produced in two-step processes, as was the case in the examples given above; processes involving several steps also occur. Important roles in these processes are played not only by the radiation-generated primary defects but also by defects already abundantly introduced during the crystal-growing process. Particular well known examples in silicon are oxygen and carbon interstitials. The formation processes are also dependent on the temperature. Defect complexes that are stable at room temperature may become mobile or may even break up into their constituents at elevated temperatures. This opens up the possibility of reducing the apparent radiation damage (annealing) and the formation of defects of a different nature.

It will be shown below that the defect complexes change the macroscopic properties of the semiconductor. A large effort has been invested to correlate the macroscopic property changes with specific microscopic defects. Several experimental methods have been invented to recognize the presence of defects and to measure their properties. Most of them are based on creating a thermal disequilibrium and observation of the transition to thermal equilibrium as a function of temperature. This subject will, however, not be considered in depth in this book.

11.2.3 Electrical Properties of Defect Complexes

Contrary to simple flat donors and acceptors, which are usually intentionally introduced into regular lattice sites in order to change the properties of the semiconductor, radiation-generated defect complexes have much more complicated electrical properties. It is the purpose of this section to review the most important properties and to discuss the consequences to be expected for detector operation. This includes, in particular, the behavior of defects in the space-charge region.

Before going into depth, it is worth mentioning that defects will have the following main consequences:

- *they act as recombination–generation centers.* They are able to capture and emit electrons and holes. In the space-charge region of a detector, alternate emission of electrons and holes leads to an increase of the reverse-bias current;
- *they act as trapping centers.* Electrons or holes are captured and re-emitted with some time delay. In the space-charge region of the detector, signal charge is trapped and may be released too late for efficient detection, thus causing a reduction in the signal; and
- *they can change the charge density in the space-charge region,* thus requiring an increased bias voltage to make the detector fully sensitive.

A list of well known defects and defect complexes is compiled in Table 11.2. One sees that some defects can exist not only in two but frequently in more than two charge states.

Defects with a Single Energy Level

We will start with a simple donor, a defect that can be positively charged or neutral depending (in thermal equilibrium) on the position of the Fermi level. The simplest example for a donor is the replacement of a silicon atom by phosphorus with one more electron in the outer shell. The additional electron may either be bound to the phosphorus atom (occupied or neutral defect state) or emitted to the conduction band (empty or positively charged defect state). The charge state of the defect is changed not only by electron emission to the conduction band and electron capture, but it may also be changed from positive to neutral by hole emission to the valence band or from neutral to positive by hole capture.

Table 11.2. Characteristics of important defects in silicon

Impurity	Charge state	Energy level	Charge in s.c. region	
phosphorus	P^0 P^+	$E_C - 0.045$	+	
boron	B^- B^0	$E_V + 0.045$	−	

Defect	Charge state	Energy level	Charge in s.c. region	Annealing temp. T_{ann} [K]
interstitial	I^- I^0 I^+	$E_C - 0.39$ $E_V + 0.4$	0	140-180 540-600 370-420
vacancy	$V^=$ V^- V^0 V^+ V^{++}	$E_C - 0.09$ $E_C - 0.4$ $E_V + 0.05$ $E_V + 0.13$	0	≈ 90 150
divacancy	$V_2^=$ V_2^- V_2^0 V_2^+	$E_C - 0.23$ $E_C - 0.39$ $E_V + 0.21$	0	≈ 570 ≈ 570 ≈ 140 ≈ 570
A-center	$(V-O)^-$ $(V-O)^0$	$E_C - 0.18$	0	≈ 600
E-center	$(V-P)^-$ $(V-P)^0$	$E_C - 0.44$	0	≈ 420
boron interstitial	B_I^- B_I^0 B_I^+	$E_C - 0.45$ $E_C - 0.12$	0	420
vacancy boron	$(V-B)^0$ $(V-B)^+$	$E_V + 0.45$	0	≈ 300
vacancy arsenic	$(V-As)^-$ $(V-As)^0$	?	?	440

The average electron occupation probability for the donor will be determined by the probability of the four charge-changing processes. Although we have *a priori* no knowledge on the individual probabilities – they should be determined experimentally – we are able to find relationships between them from thermal equilibrium considerations. In thermal equilibrium the occupation probability F is given by temperature and Fermi level as

$$F(E_{\mathrm{d}}) = \frac{1}{1 + e^{\frac{E_{\mathrm{d}} - E_{\mathrm{F}}}{kT}}} \ , \tag{11.2.1}$$

with E_{d} the defect energy level, E_{F} the Fermi level, k the Boltzmann constant and T the temperature. For reasons of simplicity we do not consider here and in the following the degeneration of charge states, an aspect taken into account correctly when generalizing to defects with several energy states[31].

In thermal equilibrium the rates of electron emission and electron capture have to be equal as there is no net flow of electrons to or from the conduction band. The same is true for hole capture and emission. Looking first at electrons, the capture rate R_n will be proportional to the number of unoccupied donors $(1 - F(E_{\mathrm{d}}))N_{\mathrm{d}}$ and to the electron concentration $n = n_{\mathrm{i}}e^{\frac{E_{\mathrm{F}} - E_{\mathrm{i}}}{kT}}$. We have

$$R_n = nN_{\mathrm{d}}(1 - F)\sigma_n\,\nu_{\mathrm{th}\,n} = \sigma_n\nu_{\mathrm{th}\,n}\,n_{\mathrm{i}}\,N_{\mathrm{d}}\,(1 - F)\,e^{\frac{E_{\mathrm{F}} - E_{\mathrm{i}}}{kT}} \ . \tag{11.2.2}$$

The electron generation rate G_n will be given by the product of density of occupied defects $N_{\mathrm{d}}F$ and the electron emission probability ϵ_n. Thus

$$G_n = N_{\mathrm{d}}\,F\,\epsilon_n \ . \tag{11.2.3}$$

Setting the two rates equal to one another, in thermal equilibrium, we have

$$\epsilon_n = \sigma_n\nu_{\mathrm{th}\,n}\frac{1 - F}{F}n_{\mathrm{i}}e^{\frac{E_{\mathrm{F}} - E_{\mathrm{i}}}{kT}} \ , \tag{11.2.4}$$

with

$$\frac{1 - F}{F} = e^{\frac{E_{\mathrm{d}} - E_{\mathrm{F}}}{kT}} \ ,$$

so that we obtain

$$\epsilon_n = \sigma_n\nu_{\mathrm{th}\,n}n_{\mathrm{i}}e^{\frac{E_{\mathrm{d}} - E_{\mathrm{i}}}{kT}} \ , \tag{11.2.4}$$

a relationship not containing the Fermi level and valid also under nonequilibrium conditions. Similarly for the hole capture rate and probability of hole emission, we have

$$R_p = pN_{\mathrm{d}}F\sigma_p\nu_{\mathrm{th}\,p} = \sigma_p\nu_{\mathrm{th}\,p}n_{\mathrm{i}}N_{\mathrm{d}}Fe^{\frac{E_{\mathrm{i}} - E_{\mathrm{F}}}{kT}} \tag{11.2.6}$$

$$\epsilon_p = \sigma_p\nu_{\mathrm{th}\,p}n_{\mathrm{i}}e^{\frac{E_{\mathrm{i}} - E_{\mathrm{d}}}{kT}} \ . \tag{11.2.7}$$

These electron and hole capture and emission probabilities can be used to find the interesting physical quantities as average charge state, current generation and trapping probability also in nonequilibrium situations such as in the space-charge region of a detector.

---

[31] Charge state degeneration is a quite common property. For example, the simple donor in its neutral state without electrons in its outer shell is non-degenerate. However, due to the two possible electron spin orientations, the singly negatively charged state is degenerate with a degeneration factor 2.

We will return to these questions after some remarks on how to read Table 11.2. The phosphorus donor has an energy level 0.045 eV below the conduction band. Assuming it to be first in the neutral state, it may become ionized (positively) by emitting an electron into the conduction band. A minimum energy of 0.045 eV is necessary for this process. The donor state can come back to neutral again by either capturing an electron from the conduction band or by emitting a hole of minimum energy $E_g - 0.045$ eV into the valence band.

Looking at Table 11.2, we recognize other donor states caused by defects – for instance vacancy–boron or interstitial–carbon. Their energy level is, however, deep in the gap, in fact below midgap.

In thermal equilibrium a donor state will be on average half the time in the charged state and half the time in the neutral state when the Fermi level E_F coincides with the donor energy level E_D. If the Fermi level is only a few times the thermal energy ($kT = 0.025$ eV at room temperature) above the donor level, it will be almost permanently neutral; if it is a few times kT below the donor level, the state will be almost permanently positively charged.

Considering now the situation in the space-charge region of a depleted detector, the Fermi level loses its significance as we are not dealing with equilibrium any more. Instead, we have to use the physically significant quantities of electron and hole capture and emission probabilities. For low reverse-bias currents we may assume negligible electron and hole densities within the space-charge region[32] and ignore electron and hole capture, so that we have to consider emission processes only.

The ratio of electron and hole emission probabilities

$$\frac{\epsilon_n}{\epsilon_p} = \frac{\sigma_n \nu_{\mathrm{th}\,n}}{\sigma_p \nu_{\mathrm{th}\,p}} e^{2\frac{E_d - E_i}{kT}} \tag{11.2.8}$$

is strongly dependent on the energy level of the defect, while the absorption cross-sections and thermal velocities of electrons and holes will be of similar magnitude. We can therefore conclude that defects with a single energy level located above the band gap center E_i have much higher emission probability for electrons than for holes. The defect will therefore be predominately in the more positive state. Analogously, defects with energy levels below E_i will be in the more negative state when located in the space-charge region.

Example 11.1

Problem: *Find the average charge state of a single energy level (E_d) defect in the space-charge region. Consider donors and acceptors above and below the intrinsic level E_i. Assume for this consideration that the product of capture cross-section and thermal velocity is the same for electrons and holes.*
Solution: *We call E_d the energy level, N_d the density, f_1 the fraction of defects being in the more negative state, $f_0 \equiv 1 - f_1$ the fraction in the more positive state. In the space-charge region the density of electrons and holes is close to*

---

[32]This is not the case in GaAs or heavily radiation-damaged silicon detectors with appreciable reverse-bias current.

zero so that capture processes can be neglected and only electron and hole emission have to be considered. The number of electrons and holes emitted per unit volume and unit time has to be equal, and we have therefore:

$$N_d f_1 \epsilon_n = N_d f_0 \epsilon_p \ ,$$

which, with (11.2.8) and the short notation

$$c_n = \nu_{\mathrm{th}\,n} \sigma_n \quad \text{and} \quad c_p = \nu_{\mathrm{th}\,p} \sigma_p \tag{11.2.9}$$

yields

$$\frac{f_1}{1 - f_1} = \frac{\epsilon_p}{\epsilon_n} = \frac{c_p}{c_n} e^{-2\frac{E_d - E_i}{kT}} \tag{11.2.10}$$

and so

$$f_1 = \frac{1}{1 + \frac{c_n}{c_p} e^{2\frac{E_d - E_i}{kT}}} \tag{11.2.11}$$

$$f_0 = 1 - f_1 = \frac{1}{1 + \frac{c_p}{c_n} e^{-2\frac{E_d - E_i}{kT}}} \ . \tag{11.2.12}$$

Measuring charges in units of elementary charge, the average charge that the defect state will assume is the charge of the more positive state q_0 minus the occupation probability of the more negative state f_1. Thus:

$$\langle q_d \rangle = q_0 - f_1 = q_1 + f_0 \ , \tag{11.2.13}$$

and f_0 will change from one to zero, f_1 from zero to one within a few multiples of kT around E_i when the defect level is moved from the lower to the upper half of the band gap. Therefore only defects with enegy levels close to E_i will be, on average, fractionally charged.

The average charge states of donors (which can only be positively charged or neutral) and acceptors are with the assumptions of equal products of cross-section and thermal velocity of electrons and holes ($c_n = c_p$):

$$\langle q_D \rangle = f_0 = \frac{1}{1 + e^{-2\frac{E_D - E_i}{kT}}}$$

$$\langle q_A \rangle = -f_1 = -\frac{1}{1 + e^{2\frac{E_A - E_i}{kT}}} \ .$$

Example 11.2
Problem: Consider a reversely biased detector at room temperature (300 K) with a high density of bulk defects, either due to radiation damage or to an imperfect production process, such that an appreciable reverse-bias current is present. In such a case, capture processes cannot be ignored and the electron and hole concentrations will be a function of position inside the active region of the detector. For a particular position the electron and hole current densities J_n and J_p as well as the electric field $\mathcal{E}$ and the single-level (E_d) defect density N_d are given. Find the average charge state of the defect at this position.

Discuss the conditions under which the presence of leakage currents influences significantly the charge state of the defects. Assume for this discussion that the product of capture cross-section and thermal velocity is the same for electrons and holes. As a numerical example use $J_p = J_n = 10\,\mu A/cm^2$, $\mathcal{E} = 1000\,V/cm$ in a silicon and a GaAs detector at room temperature.

Solution: *The electron and hole densities can be found from the current densities and the electric field as*

$$n = \frac{J_n}{q\mu_n\mathcal{E}} \qquad p = \frac{J_p}{q\mu_p\mathcal{E}} \quad ,$$

which for the numerical example leads to

$$n = 4.6 \times 10^7\,cm^{-3} \quad p = 1.3 \times 10^8\,cm^{-3} \quad n_i = 1.45 \times 10^{10}\,cm^{-3}$$

for Si; and

$$n = 1.6 \times 10^7\,cm^{-3} \quad p = 1.6 \times 10^8\,cm^{-3} \quad n_i = 1.79 \times 10^6\,cm^{-3}$$

for GaAs.

Changing from the more positive to the more negative state is accomplished by hole emission and electron capture, so that we can write with the same notation used in Example 11.1 and σ_p, σ_n (the hole and electron capture cross-section):

$$N_d f_1(\epsilon_n + p\,c_p) = N_d f_0(\epsilon_p + n\,c_n) \quad , \tag{11.2.14}$$

which, with (11.2.4) and (11.2.7), yields

$$\frac{f_1}{f_0} = \frac{f_1}{1-f_1} = \frac{\epsilon_p + n\,c_n}{\epsilon_n + p\,c_p} = \frac{c_p\,n_i e^{-\frac{E_d - E_i}{kT}} + n\,c_n}{c_n\,n_i e^{-\frac{E_d - E_i}{kT}} + p\,c_p} \quad , \tag{11.2.15}$$

and so

$$f_1 = \frac{c_p\,n_i e^{-\frac{E_d - E_i}{kT}} + n\,c_n}{c_n\,n_i e^{\frac{E_d - E_i}{kT}} + p\,c_p + c_p\,n_i e^{-\frac{E_d - E_i}{kT}} + n\,c_n} \tag{11.2.16}$$

$$f_0 = \frac{c_n\,n_i e^{\frac{E_d - E_i}{kT}} + p\,c_p}{c_n\,n_i e^{\frac{E_d - E_i}{kT}} + p\,c_p + c_p\,n_i e^{-\frac{E_d - E_i}{kT}} + n\,c_n} \quad . \tag{11.2.17}$$

The average charge is given by (11.2.13).

It is instructive to approximate these expressions for the two cases considered, Si with $n \ll n_i$, $p \ll n_i$ and GaAs with $n \gg n_i$ and $p \gg n_i$. We will restrict our discussion for energy levels above midgap and leave it to the reader to expand it for other situations.

For Si ($n \ll n_i$, $p \ll n_i$, $E_d > E_i$, $c_n \approx c_p$), we have

$$f_1 \approx \frac{1}{1 + \frac{c_n}{c_p}e^{2\frac{E_d - E_i}{kT}}}\left(1 + \frac{c_n}{c_p}\frac{n}{n_i}e^{\frac{E_d - E_i}{kT}}\right) \quad . \tag{11.2.18}$$

For defect energy levels close to midgap, the average charge state of the defect will be unchanged from that in a true space-charge region (with $J = 0$ in

(11.2.11)). For larger distances the limiting value $f_1 \to 0$ is reached with a factor 2 flatter exponential slope:

$$f_1 \approx \frac{c_p}{c_n} e^{-2\frac{E_d - E_i}{kT}} + \frac{n}{n_i} e^{-\frac{E_d - E_i}{kT}} . \tag{11.2.19}$$

For GaAs $(n \gg n_i, p \gg n_i \; E_d > E_i, c_n \approx c_p)$, we have

$$f_1 \approx \frac{n c_n}{c_n n_i e^{\frac{E_d - E_i}{kT}} + p c_p + n c_n} . \tag{11.2.20}$$

Here again we have to distinguish between two defect energy ranges. For defect energy levels close to midgap $(c_n n_i e^{\frac{E_d - E_i}{kT}} \ll p c_p + n c_n)$, the average charge state of the defect ($\langle Q_A \rangle = f_1$ for acceptors, $\langle Q_D \rangle = 1 - f_1$ for donors) will be independent of E_d and be given by

$$f_1 \approx \frac{n c_n}{p c_p + n c_n} . \tag{11.2.21}$$

For defect energy levels sufficiently above midgap $(c_n n_i e^{\frac{E_d - E_i}{kT}} \gg p c_p + n c_n)$, the average charge state of the defect ($\langle Q_A \rangle = f_1, \langle Q_D \rangle = 1 - f_1$) will reach the asymptotic value $f_1 \to 0$ with increasing E_d such that

$$f_1 \approx \frac{c_p}{c_n} e^{-2\frac{E_d - E_i}{kT}} + \frac{n}{n_i} e^{-\frac{E_d - E_i}{kT}} , \tag{11.2.22}$$

the same functional dependence as found for the Si example.

Results for the average charge of an acceptor-type defect are plotted as a function of the defect energy level E_D in Fig. 11.8 for the numerical values of this example. Note also that for a real device the ratio of electron and hole current changes as a function of position.

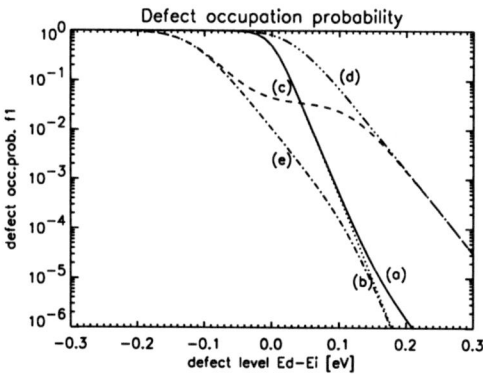

Fig. 11.8. Average defect charge state as a function of defect energy level for the conditions stated in Example 11.2. The value f_1 is the probability of finding the defect in the more negative state. This probability is shown in the semilogarithmic plot for the following situations: (a) silicon with both electron and hole currents present (continuous line), the intrinsic charge density n_i exceeding charge densities due to leakage currents; (b) GaAs in the absence of leakage currents (dotted line), the dependence being rather similar to that for Si; (c) GaAs with both electron and hole currents present (dashed line); (d) GaAs with electron current only; (e) GaAs with hole current only

Defects with Several Energy Levels
Turning now to the more complicated situation of defects that have several possible charge states, we take the divacancy (see Table 11.2) as an example. This defect has four possible charge states, ranging from double-negative to single-positive, and three energy levels, two above and one below midgap. Assuming this level to be initially in a neutral state, it may change to the positive state by capturing a hole, thereby gaining 0.21 eV of thermal energy, or by emitting an electron of energy $E_G - 0.21$ eV. Alternatively, it may change to a singly negatively charged state by capturing an electron or emitting a hole of energy $E_G - 0.39$ eV. Another hole emission of energy $E_G - 0.23$ eV or electron capture brings it to double-negative state. It may turn back to the neutral state by (for example) two electron emissions of energies 0.23 eV and 0.39 eV respectively. Each of the charge states corresponds to a particular lattice distortion and corresponding chemical binding.

The electrical characteristics of an (almost) general type defect[33] can be described by the following quantities:

- k energy levels E_l describing the energy involved in the change between charge states $l - 1$ and l;
- $k + 1$ charge states with charge $q_0 - l$ (in units of elementary charge) and degeneration factors g_l, $l = 0, 1, ...k$;
- k electron capture cross-sections $\sigma_{n,l}$ describing the change from charge state $l - 1$ to l; and
- k hole capture cross-sections $\sigma_{p,l}$ describing the change from charge state l to $l - 1$.

As was the case before for simple defects, the electron and hole emission probabilities $\epsilon_{n,l}$ and $\epsilon_{p,l}$ ($l = 1$ to k) can be inferred from the capture cross-sections by considering thermal equilibrium conditions as will be shown below.

Thermal Equilibrium Relations
Fermi statistics will give the probability for finding the defect in a particular charge state. One has for the ratio of probabilities for finding the defect in two neighboring charge states

$$\frac{P_{t,l}/g_l}{P_{t,l-1}/g_{l-1}} = e^{-\frac{E_{t,l} - E_F}{kT}} \quad . \tag{11.2.23}$$

Furthermore the sum of the probabilities for finding the defect in any charge state has to be unity:

$$\sum_{l=0}^{k} P_{t,l} = 1 \quad . \tag{11.2.24}$$

---

[33]We leave aside the possibility that a defect is able to exist in more than one configuration of the same charge but different energy. The situation that several configurations with the same charge and energy exist can be taken into account by the introduction of a degeneration factor g.

Combining the k equations (11.2.23) with (11.2.24), one is able to find the thermal equilibrium probabilities of the $k+1$ charge states.

Thermal equilibrium is kept by continuous change between neighboring charge states due to electron and hole emission and capture. Changing from charge state $l-1$ to l is accomplished by electron capture (capture cross-section $\sigma_{n,l}$) or hole emission (emission probability $\epsilon_{p,l}$). Changing in the opposite direction (l to l-1) involves a hole capture cross-section $\sigma_{p,l}$ and an electron emission probability $\epsilon_{n,l}$. With this definition the index for capture and emission constants has the range 1 to k.

As was the case for simple defects, thermal equilibrium considerations will give a relationship between emission and absorption. In order to retain constant average probabilities of the defect charge states and to have zero net flow of electrons and holes towards conduction and valence bands, the electron capture rate of charge state $l-1$ has to equal the electron emission rate of charge state l; and a similar relation holds for holes:

$$N_t \, P_{t,l-1} \, n \, c_{n,l} = N_t \, P_{t,l} \, \epsilon_{n,l} \tag{11.2.25}$$

$$N_t \, P_{t,l} \, p \, c_{p,l} = N_t \, P_{t,l-1} \, \epsilon_{p,l} \;, \tag{11.2.26}$$

from which one derives the emission probabilities with the help of (11.2.23) as

$$\epsilon_{n,l} = \frac{P_{t,l-1} n c_{n,l}}{P_{t,l}} = \frac{g_{l-1}}{g_l} \, e^{\frac{E_{t,l}-E_F}{kT}} \, n \, c_{n,l} = \frac{g_{l-1}}{g_l} \, e^{\frac{E_{t,l}-E_i}{kT}} \, n_i \, c_{n,l} \tag{11.2.27}$$

$$\epsilon_{p,l} = \frac{P_{t,l} p c_{p,l}}{P_{t,l-1}} = \frac{g_l}{g_{l-1}} \, e^{-\frac{E_{t,l}-E_F}{kT}} \, p \, c_{p,l} = \frac{g_l}{g_{l-1}} \, e^{-\frac{E_{t,l}-E_i}{kT}} \, n_i \, c_{p,l} \;. \tag{11.2.28}$$

These expressions are, with the exception of the additionally introduced degeneration factor g, identical to (11.2.4) and (11.2.7) which were derived for simple donors and acceptors.

A single defect can of course be only in one of the various charge states at a time. Finding the occupation probability for the four possible charge states in thermal equilibrium requires a more elaborate statistical treatment (Lutz 1996) than is intended in this work. We can, however, retain from the previous considerations of simple defects the following semiquantitative results:

- for a Fermi level above the highest defect energy level, the defect is predominately in the most negative charge state. The charge state changes by one unit whenever the Fermi level crosses a defect energy level;[34] and
- in the space-charge region only the energy level closest to the intrinsic energy E_i is of relevance. It will be predominantly in the more positive charge state when located above the band gap center E_i and in the more negative state when located below E_i.

---

[34]The situation is somewhat more complicated when the sequence of charge states does not follow the sequence of energy levels, as is the case, for example, for the simple vacancy. There the single positive state is suppressed in thermal equilibrium.

11.2.4 Effects of Defects on Detector Properties

The presence of radiation-induced crystal defects will change the detector properties in several ways, as should be clear already from the discussion of the electrical properties of defects in the previous section. The main changes are the increase of reverse-bias current, the change of space-charge density in the space-charge region, and the trapping of signal charges. Depending on the type of detector and application, the importance of these three effects will differ. For a diode strip detector used for position measurement, trapping is of less importance than reverse-bias current and operating voltage of the detector, while for CCDs trapping is extremely dangerous due to the signal-transfer mechanism.

In the following discussion, a qualitative explanation of the three effects will lead to a parameterization of the effects as a function of the equivalent fluence of neutrons with an energy of 1 MeV scaled according to the non-ionizing energy loss (NIEL).

In the presentation of the measurements, corrections have been applied for self-annealing (Sect. 11.2.5), the partial healing of damage that occurs during an extended irradiation period. Thus data are presented as if the full irradiation was occurring in a short time interval. For many applications, irradiation is expected to extend over periods long with respect to annealing time constants. In these circumstances a good approximation is the assumption of completed annealing, so that only the aspects of stable damage are considered.

The presentation follows closely the first extended systematic study of radiation damage of silicon detectors (Wunstorf 1992a) and concentrates on radiation damage of neutrons to n-type silicon.

Operating Voltage of Detectors

There are several radiation-damage mechanisms that lead to a change in space charge and consequently to a change in the necessary operational voltage of detectors.

The original dopants such as phosphorus or boron may be captured into new defect complexes, thereby losing their original function as flat donors or acceptors. The new defect complexes may assume a charge state within the space-charge region different from the original dopants. Phosphorus, for example, may transform into vacancy–phosphorus, thereby changing from positive to neutral space charge. In addition to complexes involving the original dopants, complexes with other impurities such as oxygen and carbon, as well as with other radiation-generated primary defects such as divacancies, can be formed. Some known defects of these types are included in Table 11.1. This table, however, is significantly incomplete, and much remains to be learned from experimental investigations.

The effective doping of an initially n-type silicon wafer is shown as a function of the irradiation fluence in Fig. 11.9. One notices that the effective doping decreases with irradiation and that the material becomes intrinsic at an irradia-

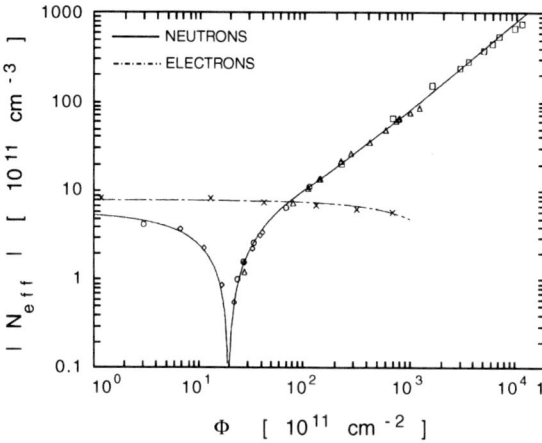

Fig. 11.9. Fluence dependence of the magnitude of the effective doping for an n-type silicon wafer irradiated with 1 MeV neutrons equivalent. The data have been corrected for self-annealing occurring already during the extended irradiation period. Also shown is the much smaller effect of irradiation with 1.8 MeV electrons also scaled to 1 MeV neutron equivalent NIEL. (After Wunstorf 1992a, Fig. 5.11)

tion fluence[35] of a few times $10^{12}\,\mathrm{n/cm^2}$. Above this value the doping becomes effectively p-type (type inversion) and eventually rises linearly with fluence (Wunstorf 1992; Pitzl 1992). This behavior matches perfectly the expectation that the following processes are responsible for the effective doping changes:

- removal of donors due to the formation of defect complexes containing donors (e.g. vacancy–phosphorus complexes[36]);
- removal of acceptors due to the formation of defect complexes containing acceptors (e.g. vacancy–boron complexes);
- creation of defect complexes assuming positive charge states in the space-charge region (effective "donors"); and
- creation of defect complexes assuming negative charge states in the space-charge region (effective "acceptors").

The following fluence dependence of the effective doping concentration is expected from an independent occurrence of these processes:

$$N_{\mathrm{eff}}(\Phi) = N_{\mathrm{D},0}\mathrm{e}^{-c_{\mathrm{D}}\Phi} - N_{A,0}\mathrm{e}^{-c_A\Phi} + b_{\mathrm{D}}\Phi - b_A\Phi \ , \qquad (11.2.29)$$

with $N_{\mathrm{D},0}$, $N_{A,0}$ donator and acceptor concentration before irradiation and c_{D}, c_A, b_{D}, b_A constants to be determined experimentally.

---

[35]The fluence at which type inversion occurs depends on the original doping, as can be seen from the parameterization in Fig. 11.10 and (11.2.30). It in addition increases when annealing occurs during irradiation (see Sect. 11.2.5).

[36]Generation of vacancy–phosphorus complexes (E-centers) is not the dominating cause for donor removal. This has been concluded from a variety of measurements including deep-level transient spectroscopy (DLTS), in which individual defects can be detected by their energy level (Matheson et al. 1996). Another yet unknown process must be responsible.

It is worth mentioning at this point that the original detector material contains acceptors and donors simultaneously and only the difference is usually known to some degree of precision. Furthermore, a variety of partially unknown defects will act as effective donors and acceptors. These defect complexes will not be completely stable and will partially self-anneal with different time constants. The vacancy–boron complex, for example, is unstable at room temperature. A simple measurement of the detector depletion voltage will only provide information on the difference of effective donors and acceptors, but not on the separate contributions.

Assuming an absence of acceptor removal and donor creation, the parameterization for effective doping simplifies to

$$N_{\text{eff}}(\Phi) = N_{\text{D},0}e^{-c\Phi} - N_{A,0} - b\Phi \ . \tag{11.2.30}$$

The linear display of the fluence dependence of the effective doping concentration in Fig. 11.10 demonstrates the excellent quality of the parameterization. The material-independent damage parameters have been determined as $c = 3.54 \times 10^{-13}\,\text{cm}^2 \pm 4.5\%$ and $b = 7.94 \times 10^{-2}\,\text{cm}^{-1} \pm 8.0\%$ (Wunstorf 1992). The original doping and doping compensation can be read off Fig. 11.10 looking at the intercept of the fitting terms with the vertical axis.

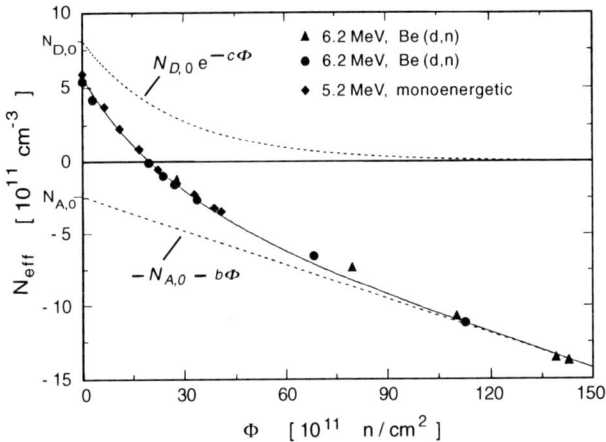

Fig. 11.10. Fluence dependence and parameterization of the effective doping according to (11.2.30) for an n-type silicon wafer irradiated with neutrons. The data have been corrected for self-annealing occurring during the extended irradiation period. (After Wunstorf 1992a, Fig. 5.12)

Reverse-Bias Current

As discussed already, in the space-charge region a crystal defect will assume predominately a single charge state unless a defect energy level is very close (within a few times kT) to the intrinsic level. Nevertheless, the defect can change for short times the charge state by emission of electrons and holes – we assume a small leakage current so that capture processes can be ignored. The alternative emission of electrons and holes is responsible for volume-generated reverse-bias currents. As the emission probability is extremely (exponentially) dependent on the position of the defect level in the band gap (see (2.6.15) and (2.6.16)), and emission of both electrons and holes are required, defects with energy levels close to the band-gap center will be most effective in generating leakage currents.

The volume-generated leakage current as a function of the 1 MeV neutron equivalent fluence is shown in Fig. 11.11 in double logarithmic form and in linear form in Fig. 11.12. As expected from a defect generation proportional to the fluence, a linear relationship between current and fluence is found:

$$\frac{\Delta I_{\text{vol}}}{V} = \alpha \Phi \ . \tag{11.2.31}$$

This linearity is seen for electrons and also for neutrons at fluences below type inversion. For fluences above inversion, a stronger rise is observed. There the volume-generated current can be parameterized as

$$\frac{\Delta I_{\text{vol}}}{V} = \alpha \Phi + \alpha^* (\Phi - \Phi_{\text{conv}}) \ . \tag{11.2.32}$$

The following numerical values have been quoted (Wunstorf 1992) for n-type silicon at 20°C:

$$\alpha = 8.0 \times 10^{-17} \, \text{Acm}^{-1} \quad \alpha^* = 9.8 \times 10^{-17} \, \text{cm}^{-1} \quad \Phi_{\text{conv}} = 4 \times 10^{12} \, \text{cm}^{-2} \ .$$

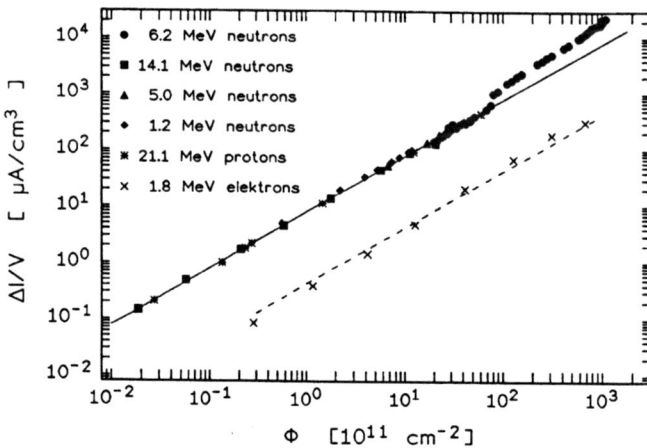

Fig. 11.11. A 1 MeV neutron equivalent fluence dependence and parameterization of the volume-generated current for an n-type silicon wafer irradiated with neutrons. The data have been corrected for self-annealing occurring during the extended irradiation period. For neutrons, straight lines with $\alpha = 8.0 \times 10^{-17} \, \text{Acm}^{-1}$ and for electrons $\alpha_e = 4.2 \times 10^{-18} \, \text{Acm}^{-1}$ have been plotted in double-logarithmic form. (After Wunstorf 1992a, Fig. 5.14)

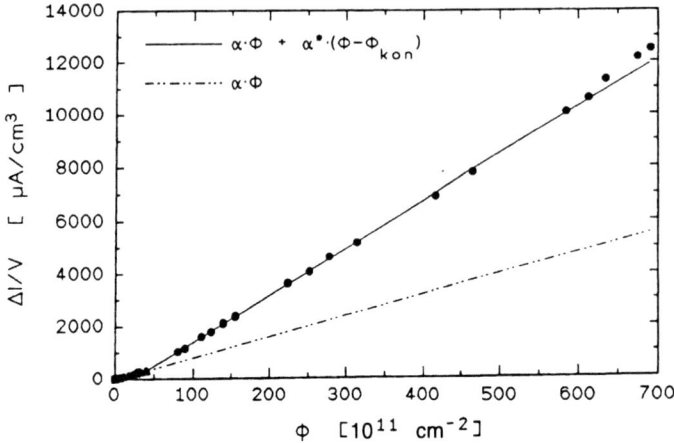

Fig. 11.12. A 1 MeV neutron equivalent fluence dependence and parameterization of the volume-generated current for an n-type silicon wafer irradiated with neutrons. The data have been corrected for self-annealing occurring during the extended irradiation period. Straight lines with different slopes before and after type inversion have been found. (After Wunstorf 1992a, Fig. 5.15)

It is interesting to note that after approximately one week of room temperature annealing the damage constants both before and after type inversion converge to the same value of $\alpha = 4 \times 10^{-17}\,\mathrm{A cm}^{-1}$ (see Fig. 11.15).

The damage constant depends not only on the type and energy of irradiation but also shows a strong trivial temperature dependence due to the variation of the intrinsic charge density n_i with temperature (see (2.3.5)). One can factor this temperature dependence by expressing the volume-generated current by the generation lifetime τ_g according to (3.1.15) and parameterizing the change of τ_g as

$$\frac{1}{\tau_g} = \frac{1}{\tau_{g,0}} + k_\tau \Phi \ , \tag{11.2.33}$$

where $\tau_{g,0}$ is the generation lifetime before irradiation. The two damage constants are related to each other as

$$\alpha = q n_i k_\tau \ . \tag{11.2.34}$$

A word of caution has to be added. After irradiation, defects are not completely stable. Part of the damage disappears with time even at room temperature. This effect – annealing – will be dealt with more fully in Sect. 11.2.6. It makes the damage constant time-dependent. For comparison of different measurements, it is therefore advisable to quote damage constants right after short irradiations and to treat annealing separately.

Trapping of Signal Charge

While in the previous two sections we have ignored electron and hole capture processes due to the small concentration of movable charge carriers in the space-charge region, it is here just the capture of moving (signal) charges we are concerned with.

Trapping is expected to be proportional to the effective trap concentration N_t and to the (thermal) velocity of the charge carriers ν_{th}, so that we write for the capture probability per unit time of electrons and holes:

$$\frac{1}{\tau_{t,n}} = \sigma_n \nu_{th,n} N_{t,n} \;\; ; \qquad \frac{1}{\tau_{t,p}} = \sigma_p \nu_{th,p} N_{t,p} \; . \qquad (11.2.35)$$

Here we have assumed that only one type of trap for each type of carrier is present and that these traps are all unoccupied. The assumption that almost all traps are unoccupied is justified in the space-charge region as long as the reverse-bias current is low and therefore the electron and hole concentrations are close to zero in the absence of signal charge. Single-level defect states above the intrinsic level – being almost entirely in the more positive charge state – will then be traps for electrons, and those below E_i will be traps for holes. In formulating the above equations, the additional assumption has been made that the thermal velocity dominates strongly over drift velocity. We leave it to the reader to consider the role of defects with several energy levels (those can be traps for electrons and holes simultaneously) and to speculate on the situations in which the other assumptions are not fulfilled.

Charge carriers that are trapped will be released again after a delay that is strongly dependent on the depth of the trap, as has to be expected by the strong energy-level dependence of the emission probabilities (see (11.2.5) and (11.2.7)), which have been obtained by applying thermal equilibrium considerations on capture and emission processes. While for the reverse-bias current only defects with energy levels close to midgap contribute significantly, all defects are capable of trapping. Very flat traps may release the charge early enough so that it still falls within the time window needed for charge collection, and in these circumstances detector performance does not deteriorate.

Example 11.3

Problem: *The charge collection time for holes in a 300 μm-thick fully depleted Si detector operated at room temperature (300 K) is $\tau_{coll,p} = 10$ ns. For which defect energy level E_d does the average time needed for re-emission of the trapped hole $\tau_{c,h}$ equal the charge collection time if the hole capture cross-section is $\sigma_p = 10^{-15}$ cm²? Discuss the dependence of this defect energy level on temperature and charge collection time.*

Solution: *The average time between hole capture and re-emission is, via (11.2.7):*

$$\tau_{c,p} = \frac{1}{\epsilon_p} = \frac{1}{\sigma_p \nu_{th,p} n_i e^{\frac{E_i - E_d}{kT}}} \; . \qquad (11.2.36)$$

The thermal velocity is obtained from the theorem of equipartition of energy[37] leading for three degrees of freedom:

$$\frac{m_p \nu_{th,p}^2}{2} = \frac{3}{2} kT \ , \qquad \nu_{th,p} = \sqrt{\frac{3kT}{m_p}} \ , \tag{11.2.37}$$

and the distance between intrinsic level and defect energy level is

$$E_i - E_d = -kT \ln[\sigma_p \nu_{th,p} n_i \tau_{c,p}] = -kT \ln \left[\sqrt{\frac{3kT}{m_p}} \sigma_p n_i \tau_{c,p} \right] \ . \tag{11.2.38}$$

Putting in the numerical values and using the hole effective mass from Table 4.1 of $m_p = 0.81 \, m_0$ (and $m_0 = 0.91095 \times 10^{-30}$ kg), one obtains

$$\nu_{th,p} = \sqrt{\frac{3kT}{m_p}} = \sqrt{\frac{3 \cdot 1.38066 \times 10^{-23} \, \text{J/K} \, 300 \, \text{K}}{0.81 \cdot 0.91095 \times 10^{-30} \, \text{kg}}} = 1.26 \times 10^5 \, \text{m/s}$$

$$E_i - E_d = -0.0259 \, \text{eV}$$
$$\times \ln[10^{-15} \, \text{cm}^2 \cdot 1.26 \times 10^7 \, \text{cm/s} \cdot 1.45 \times 10^{10} \, \text{cm}^{-3} \cdot 10^{-8} \, \text{s}]$$
$$= 0.461 \, \text{eV} \ .$$

Acceptor-type defect states with energy levels less than 0.46 eV below midgap are capable of seriously affecting the charge collection time. Using (11.2.38) to investigate temperature dependencies, one has to realize that the intrinsic carrier density n_i, which is given by $\sqrt{N_C N_V} \exp(-E_G/2kT)$ according to (2.3.5), is strongly temperature-dependent. Putting this expression into (11.2.38), one obtains

$$E_i - E_d = -kT \left(-\frac{E_G}{2kT} \right) \ln \left[\sqrt{\frac{3kT}{m_p}} \sigma_p \sqrt{N_C N_V} \tau_{c,p} \right]$$

$$= \frac{E_G}{2} \ln \left[\sqrt{\frac{3kT}{m_p}} \sigma_p \sqrt{N_C N_V} \tau_{c,p} \right] \ . \tag{11.2.39}$$

Assuming temperature independence of the capture cross section σ_p one finds a logarithmic dependence on the temperature, which is due to variation of the thermal velocity. Similarly, a factor two steeper logarithmic dependence on the charge-collection time is found.

As in the previous situations, with the present knowledge of defect properties and formation mechanisms it is not possible to describe trapping due to radiation damage in detail and we have to restrict ourselves to a global parameterization of the fluence dependence of the trapping time constant according to

$$\frac{1}{\tau_{t,p}(\Phi)} = \frac{1}{\tau_{t0,p}} + \gamma_p \Phi \ , \qquad \frac{1}{\tau_{t,n}(\Phi)} = \frac{1}{\tau_{t0,n}} + \gamma_n \Phi \ . \tag{11.2.40}$$

---

[37]We take here the root mean square value of the thermal velocity instead of the average magnitude as would be conceptually correct.

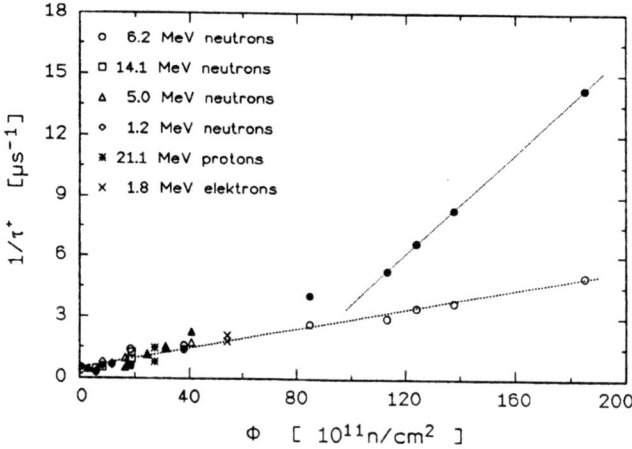

Fig. 11.13. A 1 MeV neutron equivalent fluence dependence and parameterization of the inverse trapping-time constant for holes (*open symbols*) and electrons (*solid symbols*) for an *n*-type silicon wafer irradiated with neutrons. (After Wunstorf 1992a, Fig. 5.16)

In these equations, $\tau_{t,p}$ $(\tau_{t,n})$ is the average time a hole (electron) stays trapped. $\tau_{t0,p}$ $(\tau_{t0,n})$ is the value before irradiation. This parameterization works well for hole trapping and electron trapping at moderate fluence. A value

$$\gamma_n \approx \gamma_p \approx 0.24 \times 10^{-6}\ \mathrm{cm^2 s^{-1}}$$

is found. However, as can be seen from Fig. 11.13, for electrons one observes at high fluences a stronger increase of trapping probability with fluence (Wunstorf 1992).

Properties of Nondepleted Regions

As pointed out already in Sect. 11.2.3, radiation-induced defects in general are deep-level defects. These type of defects behave differently from shallow defects in the sense that their effect in the space-charge region and in the undepleted bulk is changed. While for shallow defects the assumption that all defects are in the ionized state is always a good approximation, this is not true for deep-level localized energy states. The one-to-one correspondence between resistivity as measured from current voltage ratios of a undepleted semiconductor and the space-charge density does not hold for deep-level defects.

Imagine a semiconductor in thermal equilibrium having the density of a single deep-level state by far exceeding the density of flat donor and/or acceptor states. In order to retain charge neutrality, only part of these defects can be ionized. As the degree of ionization is determined by the Fermi level, the Fermi level has to stay close (of the oder of the thermal energy kT) to the defect energy level. It is "pinned" to the defect energy level. Thus the electron and hole concentrations will also stay "pinned" and will depend only weakly on the defect concentration, while the space-charge density is directly proportional to the defect concentration.

#366, 4-6k, 1mm thick, 0.25 cm2

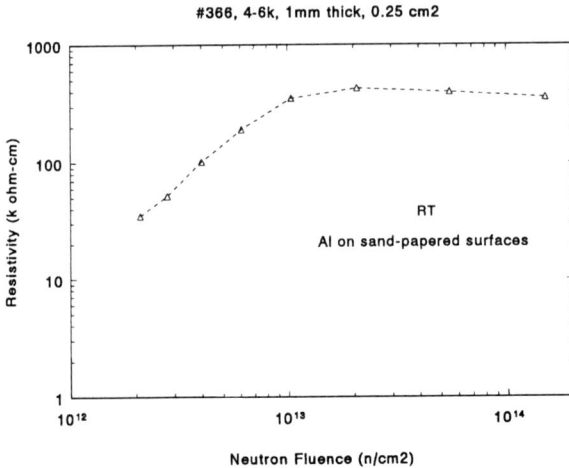

Fig. 11.14. Resistivity of originally n-type silicon irradiated with neutrons as a function of fluence. Resistivity is measured as a ratio of voltage and current in an undepleted sample of silicon. (After Li 1994, Fig. 2)

If the defect energy level is located close to midgap, one obtains simultaneously high space-charge density and high ohmic (near intrinsic) resistance in the undepleted regions of the detector. Such a situation has indeed been found experimentally (Li 1994) when measuring the resistivity as a function of irradiation, as shown in Fig. 11.14. The resistivity of n-type silicon rises with irradiation fluence to a maximum (a factor 2.5 above the intrinsic value) and then drops only slightly for higher irradiation dose. This behavior indicates that the Fermi level crosses midgap from above and stays pinned slightly below midgap.

Due to the difference in electron and hole mobilities the maximum resistivity is obtained for $E_F < E_i$ when the hole concentration exceeds the electron concentration.

For the detector behavior the high resistivity of the undepleted bulk has some surprising consequences (Lutz 1996). For high-frequency operation the non-depleted bulk is "transparent". A half-depleted diode large-area detector measuring penetrating ionizing particles will see only a quarter of the charge of a fully depleted one. This is due first of all to a halving of the charge generated in the space-charge region, and secondly to induction of half of this charge on either side of the detector by the signal charge trapped in the center plane of the detector.

11.2.5 Annealing of Radiation Damage

Observing a radiation-damaged detector after the end of the irradiation process, one notices that the observed damage to the detector (which manifests itself in an increase of leakage current, effective doping change, and trapping probability) diminishes with time. The rate of damage decrease is strongly dependent on the temperature at which the detector is kept during the waiting period.

As this observation can be naively interpreted as a partial disappearance of radiation-generated crystal defects, the effect has been called "annealing". True annealing, in which the crystal becomes perfect again, does exist – an example is the filling of a vacancy by a silicon interstitial – but in many cases defect complexes may just be transformed into other more stable defect types with changed properties.

Defects and defect complexes are in general stable only up to a characteristic temperature, the "annealing temperature", which is also included in Table 11.2. It is determined by observing the disappearance of some defect-characteristic properties with time and temperature, as for example in peaks in deep-level transient spectroscopy (DLTS). The annealing temperature is not very precisely defined as even below this temperature some annealing occurs. One assumes an exponential behavior of the annealing of the form

$$N_{\rm d}(t) = N_{\rm d}(0){\rm e}^{-\frac{t}{\tau}} \quad \text{with} \quad \tau(T) \propto {\rm e}^{\frac{E_{\rm a}}{kT}} \ , \tag{11.2.41}$$

$E_{\rm a}$ being called activation energy and also given in Table 11.2. The order of magnitude of $\tau(T_{\rm ann})$ is 10 minutes.

From the previous discussion, it seems clear that annealing is a rather complicated process involving many different and only partially understood processes between defects and defect complexes. Examples for the radiation-generated effective doping change and the volume-generated current are seen in Figs. 11.15 and 11.16.

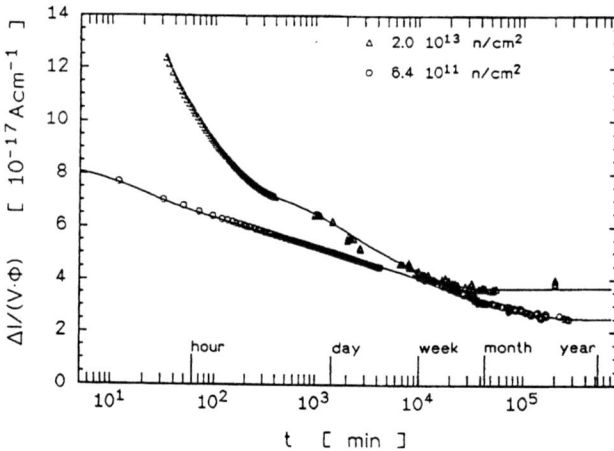

Fig. 11.15. Room temperature annealing of radiation-induced volume-generated current. Data are corrected so as to correspond to a short irradiation followed by a long-term observation at constant (20°C) temperature. Data are from a converted detector ($\Phi = 2 \times 10^{13}$ n/cm^2, upper curve) and an unconverted detector ($\Phi = 6.4 \times 10^{11}$ n/cm^2, lower curve). (After Wunstorf 1992a, Fig. 5.25)

As new defect complexes are produced, the effect of annealing may not always be beneficial for detector performance. An example of such a detrimental effect is the increase in space charge after initial annealing of intensely irradiated detectors (Fig. 11.16). This effect has been called "reverse annealing".

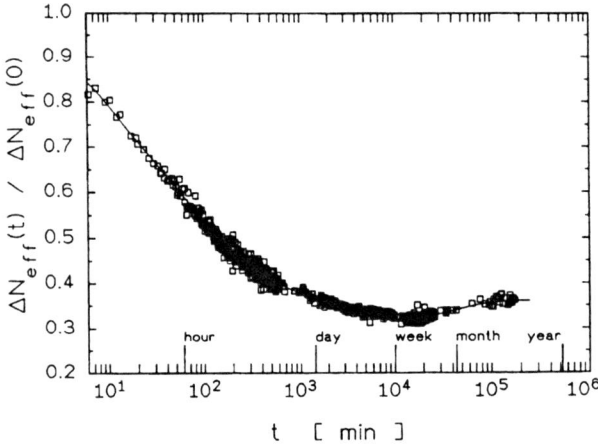

Fig. 11.16. Room temperature annealing of radiation-induced change of effective doping. Data are corrected so as to correspond to a short irradiation followed by a long-term observation at constant (20°C) temperature. Observed is a decrease on the time scale of weeks (annealing) followed by a rise (reverse annealing) on the time scale of months. (After Wunstorf 1992a, Fig. 5.18)

Parameterization of Annealing

Because in the present state of knowledge a quantitative description of the radiation damage with a physical model is not possible, one usually restricts oneself to a parameterization of the observed effects. This is done separately for the volume-generated current, the effective doping, and the trapping. The effective doping (Sect. 11.2.4) is just a simple way to describe the space-charge density in a fully depleted semiconductor region including the (partially) ionized defects. In all cases one assumes that there exist several defect types that decay with their charactristic times.

The rate of introduction of radiation-induced volume-generated current has been found to be roughly a factor two higher after type inversion than for low irradiation fluences (see (11.2.31) and (11.2.32)). The annealing behavior shown in Fig. 11.15 is also drastically different, so that after a period of roughly one week both show similar remaining damage. Parameterizing this behavior, one writes for the volume-generated current:

$$\frac{\Delta I(t)}{\Delta I(0)} = \sum_{i=1}^{n} A_i e^{-\frac{t}{\tau_i}} \quad . \tag{11.2.42}$$

Table 11.3. Relative fractions and annealing time constants for volume-generated currents of n-type silicon before and after type inversion. For p-type silicon the "after type inversion" values apply

Time constant τ_i [min]		Relative fraction A_i
before	type	inversion
$(1.78 \pm 0.17) \times 10^1$		0.156 ± 0.038
$(1.19 \pm 0.03) \times 10^2$		0.116 ± 0.003
$(1.09 \pm 0.01) \times 10^3$		0.131 ± 0.002
$(1.48 \pm 0.01) \times 10^4$		0.201 ± 0.002
$(8.92 \pm 0.59) \times 10^4$		0.093 ± 0.007
∞		0.303 ± 0.006
after type inversion	and	p-type Si
$(1.35 \pm 0.03) \times 10^1$		0.197 ± 0.010
$(8.43 \pm 0.02) \times 10^1$		0.300 ± 0.002
$(1.55 \pm 0.40) \times 10^3$		0.121 ± 0.002
$(8.74 \pm 0.48) \times 10^3$		0.139 ± 0.014
∞		0.243 ± 0.003

Numerical values for relative amplitudes and corresponding time constants are given in Table 11.3. They are obtained from measurements on n-type diode detectors. Notice that time constants change when the detector is irradiated strongly enough to become permanently inverted.

For the effective doping concentration N_{eff} one writes

$$\frac{\Delta N_{\text{eff}}(t)}{\Delta N_{\text{eff}}(0)} = \sum_{i=1}^{n} A_i e^{-\frac{t}{\tau_i}} . \tag{11.2.43}$$

Numerical values for relative amplitudes and corresponding time constants are given in Table 11.4. They are obtained from measurements on n-type diode detectors exposed to moderate irradiation ($< 4 \times 10^{12}$ n/cm^2). The corresponding experimental data includes the fitted curve shown in Fig. 11.16.

Table 11.4. Relative fractions and annealing time constants for changes in the effective doping concentration of n-type silicon

Time constant τ_i [min]	Relative fraction A_i
$(9.40 \pm 0.80) \cdot 10^0$	0.214 ± 0.030
$(6.87 \pm 0.14) \cdot 10^1$	0.262 ± 0.007
$(3.43 \pm 0.12) \cdot 10^2$	0.118 ± 0.008
$(4.00 \pm 0.04) \cdot 10^3$	0.097 ± 0.002
$(7.52 \pm 0.02) \cdot 10^4$	-0.107 ± 0.001
∞	0.415 ± 0.004

Notice the term with the longest time constant of 7.5×10^4 min with its negative amplitude. The source for this negative sign is the reverse annealing, described in the next section.

11.2.6 Reverse Annealing

When performing systematic studies of radiation damage of silicon detectors, a surprising new effect was found. Measuring the effective doping of irradiated silicon as a function of time, it was observed that the initial decrease of doping change during the first few weeks of room temperature annealing was followed by a new increase of the effective doping change, a process proceeding on the time scale of many months (Wunstorf 1992). As the effective doping change increases, this effect is usually called "reverse annealing".

A natural explanation for such behavior is the transformation of radiation-induced electrically inactive defect complexes into electrically active defects. Two different mechanisms have been proposed: the slow decay of inactive complexes of type X into electrically active (type Y) defects (first-order process); and a reaction between two kinds of inactive defects X_1 and X_2, forming an electrically active defect complex Y (a second-order process). In the first-order process the decay rate is proportional to the concentration:

$$\frac{dN_Y}{dt} = -\frac{dN_X}{dt} = \hat{k}(T)N_X \ , \tag{11.2.44}$$

$$N_Y = N_{X0}(1 - e^{-t/\tau}) \qquad \tau = 1/\hat{k}(T) \ . \tag{11.2.45}$$

For the second-order effect the interaction rate is proportional to the product of defect concentrations:

$$\frac{dN_Y}{dt} = -\frac{dN_{X_1}}{dt} = -\frac{dN_{X_2}}{dt} = \tilde{k}(T)N_{X_1}N_{X_2} \ . \tag{11.2.46}$$

Assuming nearly equal concentrations $N_{X_0}(\Phi)$ of the two types of radiation-induced primary defects (Fretwurst et al. 1994), the new type of defect being responsible for the increase in effective doping has a time dependence of the form

$$N_Y(\Phi, t, T) = N_{X_0}(\Phi) \left(1 - \frac{1}{1 + N_{X_0}(\Phi)\tilde{k}(T)t} \right) \ . \tag{11.2.47}$$

Note that both models will lead to the same saturation value $N_{Y_\infty} = N_{X_0}$. However, for the second-order process the rise time is expected to depend on the defect-concentration, thus resulting in a different form of the time dependence compared to a first order process. While earlier measurements seemed to confirm the predicted form of the time dependence of a second order process, later investigations (Feick et al. 1996) on the fluence dependence were in disagreement with this hyothesis and the interpretation as a first order process is now generally accepted.

Reverse annealing can be accelerated by raising the temperature and slowed down by cooling. Elevated temperatures are used for investigating reverse annealing in a reasonable time scale, cooling is applied to reduce or completely suppress (below approximately 0°C) reverse annealing in the operation of detectors in high radiation environments.

11.2.7 Parameterization of Radiation Damage for Low-Flux Irradiation

In some applications the duration of irradiation is very long compared with the short-duration annealing processes. In such a situation it is convenient to consider only the stable part of the damage and add to it the reverse annealing. In such a case the change of doping concentration ΔN_{eff} is given as a function of time t and fluence Φ by

$$\Delta N_{eff}(\Phi, t, T) = N_{eff0} - N_{eff}(\Phi, t, T) = N_C(\Phi) + N_Y(\Phi, t, T) \ . \qquad (11.2.48)$$

N_C represents the fluence-dependent stable part of the radiation damage, where

$$N_C(\Phi) = N_{C0}(-1 + \exp(-c\Phi)) + g_C\Phi \ . \qquad (11.2.49)$$

The first term is due to removal of original flat dopants (donor removal) and the second term is due to the radiation-generated introduction of defects with g_c the introduction constant of stable acceptor type defects. Setting the saturation value of reverse annealing proportional to the fluence Φ, one obtains for the first order process from (11.2.45)

$$N_Y(\Phi, t, T) = g_Y\Phi \left(1 - e^{-\hat{k}(T)t}\right) \ . \qquad (11.2.50)$$

11.3 Radiation Damage in the Surface Region

In Sect. 11.2.2 we considered the damage to the bulk of the detectors. This was done with the understanding that the effect of material changes is uniform over the whole volume of a detector. A very critical part of the detector is the surface region. At least part of the surface is terminated by an insulator (e.g. SiO$_2$) and already without irradiation the crystal lattice is irregular over a depth of many lattice spacings. It is near the surface at the boundaries between strongly doped and only-insulator-covered regions where the highest field strengths are present.

In silicon detectors these high field strengths are caused by the positive charges present at the oxide–semiconductor boundary. For a thorough understanding of the effect of irradiation on detectors (and electronics), the damage to the insulator and the insulator–semiconductor interface has to be known.

A further point to be discussed is the nonuniformity of the damage in the semiconductor near the surface.

11.3.1 Oxide Damage

The mechanism of oxide radiation damage is rather different from the one encountered in semiconductor bulk material. Although the principal difference between an insulator (such as SiO_2) and a semiconductor is the width of the band gap (see Sect. 2.2 and Chap. 4), there is the additional difference in material structure. While for the semiconductor one starts with an almost perfect single crystal, the oxide – and in particular the transition region between semiconductor and oxide – is highly irregular.

In contrast to that for the bulk semiconductor, irradiation damage to the crystal material structure does not play an important role as the material is already highly irregular. The additional damage to the material structure caused by the interaction of radiation with the nuclei can therefore be safely ignored. Instead, it is the supply of charge carriers created by ionizing radiation that is important. Oxide damage therefore is caused by ionizing irradiation such as photons, x-rays and charged particles, while semiconductor bulk damage, requiring damage of the crystal structure, is generated by massive particles such as neutrons, protons, and pions.

One may consider the oxide (and the oxide–silicon interface) as a region with a high density of defects whose charge state can be altered by irradiation. As is the case in semiconductors, the mechanisms for these changes are electron and hole capture processes. For deep enough traps, the emission of captured carriers into the conduction and valence bands is virtually impossible. In this connection the much larger band gap of insulators (8.8 eV for SiO_2) compared with semiconductors (1.12 eV for Si) is of primary significance.

Also important is the difference in mobility between electrons and holes in the insulator. In SiO_2, electron mobility is several orders of magnitude larger than that of holes. Compared with holes, radiation-generated electrons therefore will diffuse out of the insulator in a rather short time and the capture of holes is the dominant process that changes the oxide material's properties. Radiation damage of oxide therefore manifests itself as a buildup of positive charge due to semipermanent trapped holes, which is sensed as a shift in the flat-band voltage (see Sect. 3.3.2).

Trapping is more likely in regions with higher defect density, which is found in the semiconductor–oxide transition region. This explains the dependence of a flat-band voltage shift on the electric field direction in the oxide. For positive bias of a MOS structure (the field direction driving holes towards the Si–SiO_2 boundary), a stronger flat-band voltage shift is observed than for opposite biasing. Additionally, charges located farther from the metal have a higher influence on the flat-band voltage. The lowest flat-band shift is observed for zero field; in this condition the probability of electron–hole recombination is highest.

A saturation of the oxide charge buildup has been experimentally observed (Nicollian et Brews 1982; Di Maria et al. 1993). This is explained by a the limited number of semi-permanent traps in the oxide. As soon as essentially all of them are filled, no further increase of the oxide charge is possible.

The saturation value of the oxide charge depends on the oxide quality. A typical value for high-quality thermally grown oxide is

$$Q_{\rm ox,\,sat} \approx 3 \times 10^{12}\,{\rm cm}^{-2} \ .$$

11.3.2 Nonuniformity in Bulk Damage Near the Surface

Nonuniformity in bulk damage is naturally expected for nonuniform spatial distribution or low-penetrating irradiation. There has, however, been observed an additional effect that cannot be explained by these trivial causes.

When irradiating a silicon device with MeV-energy neutrons, the whole interior of the detector was inverted, but not the surface region. This was found by measuring the voltage–current characteristics between two neighboring original p^+–n diodes (e.g. two neighboring strips in a strip detector), which should show an ohmic behavior after conversion of the n-bulk to effectively p-type. Instead, the characteristics of two back-to-back diodes has been observed.

A natural explanation for the absence of inversion near the surface comes from consideration of the process of defect formation. The primary defects (Si–interstitials and vacancies) are mobile at room temperature. If nothing else happened to them, they would just diffuse out of the surface and the crystal would be restored to its perfect state. During their movement, however, these primary defects can interact with other primary defects and/or defects already present in the crystal before irradiation, forming in this process stable defects that alter the electrical behavior of the semiconductor.

Out-diffusion through the surface makes the primary defect density at the surface zero and reduces it near the surface. The formation of secondary stable defects therefore is suppressed near the surface (Lutz 1994). However, additional mechanisms might be needed to provide a quantitative description of the observed effect.

11.4 Radiation Damage in Detectors

Having discussed the changes in material properties induced by radiation for the semiconductor bulk material and also the insulator, we now turn to the complete detector. Here we have to consider the interplay between the radiation-induced material changes of bulk and surface:

- an effective doping change of the bulk, including type inversion of original n-type silicon;
- an increase of resistivity of undepleted bulk;
- a reduction of generation lifetime resulting in an increase of leakage current;
- the trapping of signal charge; and
- a change of oxide charge.

The basic changes, such as leakage current increase, full depletion voltage change, and signal loss, have already been discussed in Sect. 11.2.4, together with the radiation-induced semiconductor bulk material property changes.

Here, we will in addition focus on more complex effects such as those occurring at the detector surface. These very often will depend on the particular topology of detectors. A further point of interest is the consequence of near intrinsic resistivity in the undepleted regions of highly irradiated silicon.

Rather than treating this subject in a general way, we will take two well known detectors, a large-area diode and a silicon strip detector, and consider the changes expected when exposing them to irradiation.

The simple p^+–n diode detector consists of a large-area diode on a homogeneously low-doped n bulk. The backside surface layer has an unstructured n^+ layer that prevents breakdown (charge injection) when the space-charge region, growing from the top with increasing bias voltage, reaches the backside. No guard rings are shown, which might be necessary for stable operation of such a detector (Sect. 11.1.2).

Irradiation with neutrons will lead to a change of effective doping in the direction of p-type material. Simultaneously, the resistivity of the undepleted bulk will increase. The second point can be of importance if the detector is operated when partially depleted. As long as the bulk material is still effectively n-type, the minimum operating voltage for full depletion is decreasing. It starts to rise again only when type inversion is reached (at $\approx 2 \times 10^{12}$ n/cm^2 in Fig. 11.9). After type inversion, the diode junction has moved to the opposite wafer side; the space-charge region grows with increasing voltage from the n^+ backside. If the whole bulk were effectively uniformly doped, the space-charge region would extend all the way to the cutting edge of the detector, where a high density of crystal defects would produce a strong generation current and where therefore, the detector would be rather useless. Experimental observations have shown, however, that these simple diodes were functional even after type inversion. The reason for this behavior had for a long while not been understood.

A simplified explanation for the observed type inversion survival is given below. The surface region of the wafer does not invert (see Sect. 11.3.2). On the top side, the positive oxide charges, in addition, generate an electron surface layer. This conducting electron layer is connected to the backside n^+ layer through the conducting, strongly damaged cutting edge. Assuming an absence of generation current, a potential valley with depth of less than a quarter of the full depletion voltage of the wafer will form in the rim region midway between front and backside of the detector. Holes generated in the damaged region near the cut edge will move through this potential valley towards the center detector region while electrons are pulled towards the bottom, top and sideward surface regions. Here one profits from the high ohmic resistivity (low free carrier concentrations) of the undepleted bulk. The valley therefore will be filled with holes, to a larger degree near the cut edge than in the vicinity of the detector region, and the voltage drop along the bottom of the valley will provide a continuous flow of holes towards the center region. As the electron density in the space-charge region and in the (electrically neutral) valley region is close to zero, the hole density will be below the value expected in thermal equilibrium. In the undepleted valley region the excess current generation is suppressed.

Example 11.4

Problem: *Estimate the order of magnitude of the leakage current of a strongly inverted $1 \times 1\,cm^2$ large and $300\,\mu m$-thick diode with the above given hypothesis, assuming a $5\,mm$ wide cutting rim L, full depletion voltage of $100\,V$ and operation at room temperature. Discuss how the leakage current would vary with temperature.*

Solution: *Assuming the potential valley at the rim of the detector to be on average one-third filled, its resistance (see Fig. 11.14 in Sect. 11.2.4.4) is approximately*

$$R = \rho \frac{L}{A} = 0.4 \times 10^6\,\Omega cm \frac{0.5\,cm}{10^{-2}\,cm \cdot 4\,cm} = 5 \times 10^6\,\Omega \ .$$

The voltage drop along the channel is roughly $100\,V/4 = 25\,V$, so that the contribution of the rim to the leakage current is $25\,V/5 \times 10^6\,\Omega = 5\,\mu A$.

The resistivity in the undepleted region scales with the carrier concentration. As the Fermi level stays approximately pinned to the dominant defect level close to the middle of the band gap, it will scale approximately in the same way as the inverse of the intrinsic carrier concentration. Lowering of temperature by $7°C$ will increase the resistivity, and therefore also reduce the current, by a factor of two.

One can also build detectors that completely exclude this current generated at the cut edge. The simplest method is switching to p-type silicon as a base material and providing an n^+–p junction. Then the junction does not switch surfaces. Another effect, however, will cause problems: the positive oxide charge will generate an electron inversion layer that extends all the way towards the cutting edge, thus again extending the space-charge region towards the damaged cut region. Such a situation can be avoided by interrupting the electron layer with at least one ring of p-doping. If one still wants to completely suppress edge-generated currents in n-type bulk material, then it is necessary to structure both sides of the detector, so that the detector works for the diode located on either side of the wafer.

Ionizing radiation increases the oxide charge and thereby raises the electrical field strength in particular positions of the detector. For the diode shown in Fig. 11.17 the most sensitive region is next to the edge of the diode, as was already discussed in Sect. 11.1. Proper high-voltage protection structures, such as shown in Fig. 11.6, prevent electric breakdown. They have to be dimensioned in such a way as to take into account the radiation-induced saturation value of the oxide charge.

Irradiation with charged hadrons (pions or protons) leads to oxide charge increase up to the saturation value, followed by type inversion when starting with n-type silicon. Proper design of the detectors ensures that they are functioning at all conditions occurring during their lifetime.

It remains to discuss how the detectors will react to the irradiation that they are designed to measure. As long as the detector is fully depleted, the change from a nonirradiated detector is not spectacular: there is the signal loss

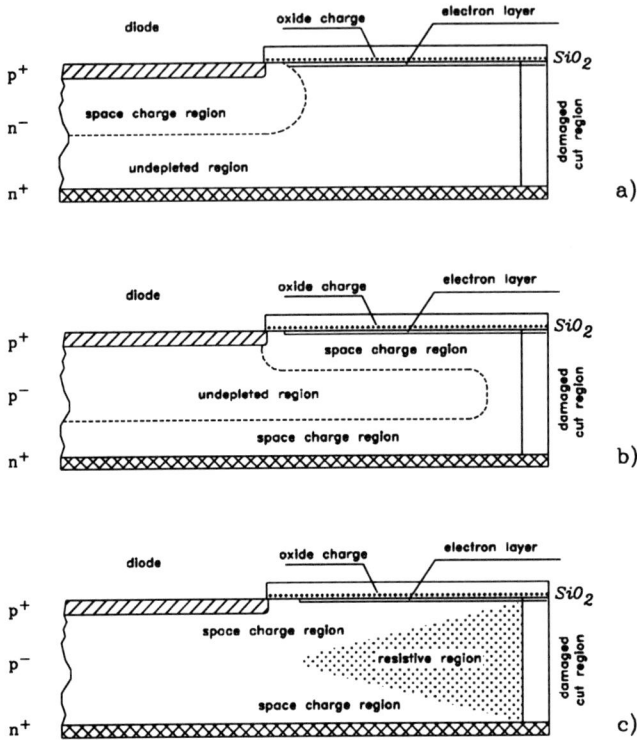

Fig. 11.17a–c. Explanation for the survival of a simple p^+–n silicon diode detector after type inversion due to strong irradiation: partially depleted detector before (a) and after (b) type inversion, ignoring carrier generation in the cut region. Fully depleted detector (c) with charge generation in the cut region. The potential valley at the rim is filled with holes, the undepleted region acts as resistor, limiting the edge leakage current

due to trapping and there is also a change in the form of the signal due to the changed electric field distribution. The situation is different for a partially depleted detector. This is due to the high resistivity of the undepleted region of the detector. For fast signals this undepleted bulk can be viewed as an insulator rather than a conductor.

Consider a homogeneously doped detector that is biased with a voltage $V_B < V_{depl}$; then a fraction $f = \sqrt{\frac{V_B}{V_{depl}}}$ will be depleted and therefore be sensitive to ionizing radiation. For a traversing charged particle, the amount of charge generated in the sensitive region will drop by the same fraction f compared with a fully depleted detector. For the inverted detector an additional effect appears: the holes will move to the edge of the space-charge region and induce a charge $f Q_{gen}$ on the p-side electrode and a charge $(1 - f) Q_{gen}$ on the n-side. It will then take a long time until the charge carriers in the undepleted region rearrange so as to bring the electric field in the undepleted region back to zero again.

The signal charge detected on a fast time scale therefore is

$$Q_{\text{sig}} = f^2 Q_{\text{sig, full}} = \frac{V_{\text{B}}}{V_{\text{depl}}} Q_{\text{sig, full}} \quad .$$

Its value rises linearly with voltage while for the nonirradiated detector the value rises with the square root.

The characteristic time for this effect can be estimated in the following way. We consider a parallelipiped (area A and thickness d) of the undepleted region and calculate the RC time constant with

$$R = \rho \frac{d}{A}$$
$$C = \frac{1}{\epsilon \epsilon_0} \frac{A}{d}$$
$$\tau = RC = \rho \epsilon \epsilon_0 \quad .$$

We find it independent of the geometry: at room temperature we obtain

$$\tau \approx 0.4 \times 10^6 \ \Omega\text{cm} \cdot 11.9 \cdot 8.85 \times 10^{-14} \ \text{F/cm} = 0.42 \ \mu\text{s} \quad .$$

Applying the same considerations to strip detectors, we have to take into account the fact that the induced charge, due to the signal charge accumulated at the edge of the space-charge region in the bulk, is distributed over several strips. The region in which significant charge will be induced will be of similar extent to the distance of the signal charge from the respective surface.

11.5 Radiation Damage in Electronics

Radiation damage in electronics has been studied extensively. In the past these studies have concentrated on failure aspects in digital electronics and were motivated by their relevance to various applications, including military ones.

Semiconductor detectors with few exeptions do not possess intrinsic amplification properties and very frequently require high-density readout. Therefore the question of radiation-induced noise degradation of integrated electronics has attracted particular attention in recent years. In the following we will give a rather short description of the principal damage mechanisms and their relevance to particular types of electronics.

As most detectors and electronics are built from silicon using similar technology, the radiation-damage mechanisms are in principle the same for electronics and detectors. There are, however, important differences due to (for example) the drastically different doping concentrations and the different function principles.

As the doping densities used in electronics is usually several orders of magnitude higher than in detectors, the radiation-induced change in doping concentration is in general not important. Important are, however, other aspects of the bulk damage and oxide damage. Their effects on devices will be discussed separately for MOS transistors, bipolar transistors and JFETs, considering first the static and then the noise properties.

As the gate oxide is a key element of MOS transistors, the device is very sensitive to its property changes. Ionizing radiation causes buildup of positive charge within the oxide and the creation of interface states in the oxide–semiconductor interface. The rate of creation is dependent on size and polarity of the electric field in the oxide and thus on the gate voltage applied during irradiation. The change in the transistor threshold voltage due to oxide charge buildup can cause catastrophic failures in circuits, in paricular in digital circuits when n-channel enhancement transistors change to depletion-type and thus cannot be turned off any more in inverters etc. The resulting continuous current flow due to the generated heat will eventually destroy the circuit. A further effect, important for devices having the conducting channel directly at the Si–SiO$_2$ interface (as is the case for most enhancement type MOS transistors), is due to the reduction of mobility near the interface, causing a reduction of the transconductance.

The static properties of bipolar and JFET transistors, being bulk devices, do not depend on oxide damage. In bipolar transistors the most prominent damage is the decrease of minority carrier lifetime, leading to an increase of recombination in the base region and therefore to a reduction of the base transport factor α_T, respectively the current gain $\beta = \frac{\partial I_C}{\partial I_B}$. JFETs, being unipolar devices are only weakly affected by bulk damage. The gate leakage current increases due to increased generation in the reverse-biased gate junctions.

Turning now to the subject of noise, we recall the three basic mechanisms described in Sect. 7.2: thermal noise, shot noise and low frequency (RTS) noise. Thermal noise can be derived from the static properties of the devices that deteriorate to some extent by irradiation, as discussed above. Shot noise does not play a significant role in MOS transistors due to the absence of gate currents through the perfect gate isolator, although it is very significant in bipolar transistors because of the reduction of current amplification and the corresponding increase of base current when keeping the emitter current constant. It also can become significant in JFETs, when the gate leakage current increases due to radiation-induced bulk damage. Low-frequency noise is important in all three devices. Recall that $1/f$ noise is explained as a superposition of RTS signals from many traps with different characteristic frequencies.

For MOS transistors the vast majority of these traps are located within the oxide layer close to the oxide–semiconductor interface and the interaction with the channel occurs by tunneling. The noise properties of MOS transistors are therefore affected by both oxide and bulk damage, although the influence of oxide damage on depletion-type MOS devices can be reduced strongly if the channel current is kept from interacting with the surface region.

In a properly designed JFET the channel is well separated from the surface; therefore only bulk traps will generate RTS noise. The frequency spectrum can deviate significantly from $1/f$ as traps with discrete energy levels and therefore characteristic emission times for trapped charge carriers are present.

Radiation-induced low-frequency noise is also observed in bipolar devices, despite the fact that these are bulk devices with very thin base regions. An explanation for this fact is the smallness of these devices, combined with the fact that each base touches the oxide.

11.6 Radiation Hardening Techniques

Radiation hardening can be done in principle by using technologies that result in materials whose properties do not change very much when exposed to radiation. Alternatively, one can design devices that tolerate the changes in material properties.

When applied to silicon detector bulk material, the first approach so far has not produced very significant results. Nevertheless, attempts to influence the creation of stable defects by controlled doping of the base material are continuing. More success has been achieved in the radiation hardness of SiO_2. Here the principal problem of oxide charge buildup can be influenced by oxide growing techniques. The main reason for this possibility is the fact that oxide defects are already there after the growth process. These defects (traps) are capable of capturing holes that are produced during irradiation with ionizing radiation. The creation of additional oxide defects due to irradiation plays a minor role. Therefore one observes a saturation of the oxide charge. Radiation hardness of oxide depends on technological parameters such as gas composition, pressure and temperature during thermal oxidation. Oxide deposited from the gas phase has significantly worse properties than thermally grown oxide. Of interest with respect to radiation hardness are also oxi-nitrides, in other words SiO_2 with an admixture of silicon nitride.

The second approach, the design of radiation-tolerant device structures, applied often together with the first approach, has led to very significant results. This is most apparent in the field of electronics, where radiation-hard devices are commercially available. In CMOS electronics the use of very thin high-quality gate oxides has strongly reduced the magnitude of radiation-induced threshold voltage shifts. Properly doped guard rings around individual transistors prevent the appearance of parasitic shorts between neighboring devices due to radiation-induced positive charge in the thick field oxide. Alternatively the SOI (Silicon On Insulator) technique provides separate small substrates for individual transistors.

No general recipe can be given for the radiation hardening of detectors. The parameterization of macroscopic material properties, as presented in Sects. 11.2 and 11.3, provide a means for anticipating the change in detector properties during the irradiation scenario. The detectors have to be designed in such a way as to remain operational during all stages of the scenario. In some cases this will require the capability of operating at very high bias voltage and/or low temperature in order to keep leakage currents (and noise) at an acceptable level.

11.7 Summary

Electrical breakdown of detectors is due to the occurrence of high electric field regions capable of inducing avalanche charge multiplication processes. The occurrence of such conditions often depends on the conditions on the outer surface of the insulator. These conditions are often poorly defined in the sense that they depend on environmental conditions and the time since application of the external potentials on the device. The time needed for arriving at a stable static condition can be very long because the conductivity on the outer surface of the insulator may be extremely low. Stable conditions are reached after a spreading of the charge on the outer surface, so that the potential at the outer surface approaches that of the neighboring conducting electrodes.

The breakdown mechanisms and ways to prevent their occurrence are described in Sect. 11.1. Special attention is given to the detector rim, the region most prone to electrical breakdown, and to measures that provide a controlled drop of potential in this region. It is advisable to avoid as far as possible undefined conditions on the outer surface of the detectors. One way of doing so is the introduction of a very low conducting layer on top of the insulator.

Radiation changes the material properties of the semiconductor and also the insulator. The primary damage in the semiconductor is the displacement of lattice atoms from their regular sites, thus creating simultaneously interstitials and vacancies. In silicon these primary defects are mobile at room temperature. They diffuse within the crystal until they either anneal by combining a vacancy with an interstitial or form a stable defect complex by combination with an impurity or other radiation-generated defect. They also disappear when reaching the surface in the process of diffusion.

The rate of generation of primary defects is dependent on the type of radiation. Scaling of the effective damage between different types of radiation and varying energies is done with the help of the non-ionizing energy loss (NIEL), which subtracts from the total absorbed energy that part which goes into the interaction with electrons.

Crystal defects can create localized energy levels within the band gap. The same defect can exist in several charge states, the energy level(s) describing the energy necessary to change the charge state by emission of an electron or a hole.

Changes to the defect charge state occur by electron and hole emission and absorption. Thermal equilibrium considerations allow the derivation of the electron (or hole) emission probability from the capture cross-section, as expressed in (11.2.5) and (11.2.7). An exponential dependence on the defect level position is found.

These relationships are then used in static nonequilibrium conditions to derive the occupation probabilities for the defect charge states and the charge generation.

The effects of semiconductor defects on detector properties (Sect. 11.2.4) are: the change in effective doping in the space-charge region, resulting in a change of the full depletion voltage; enhancement of the generation (and recombination) rate, resulting in an increase of the reverse-bias current; and the capture and delayed release (trapping) of signal charge, resulting in a loss of signal. Although in principle these properties could be derived from the microscopic changes in the crystal, the knowledge of the defect properties and production dynamics is much too limited to make this possible. Therefore one restricts oneself to a parameterization of the observed damage in effective doping, carrier lifetime and charge trapping as a function of the radiation fluence.

A similar parameterization is also done for the annealing which is due to the rearrangement of defects after the initial damage. Here the time dependence of semiconductor properties after the end of a presumed short irradiation is expressed with a superposition of several exponential decays with different decay times. This parameterization is done separately for the minority carrier lifetime (or the reverse-bias leakage current) and the effective doping observed in the space-charge region.

In the effective doping not only a decrease in the doping change with time is observed but also, on a much longer time scale, an increase is found. This increase is referred to as "reverse annealing" and is assumed to be due to an as yet unidentified and unexplained rearrangement of defects within the crystal.

Even for uniform irradiation, damage is not uniform over the semiconductor volume. For n-type silicon as the bulk material and sufficiently high fluence, the main part of the bulk inverts to p-type material while a thin surface region remains uninverted. A thorough investigation of this effect is still to be done.

Radiation damages also the insulator and the interface between semiconductor and insulator. The rate of damage is dependent on the strength and direction of the electric field.

Usually overlooked is the fact that the undepleted radiation-damaged bulk region behaves very differently from that of a normally doped region. This is due to the different effects of deep-level defects from normal shallow dopants. In strongly radiation-damaged silicon the Fermi level is pinned close to midgap so that the carrier densities are near-intrinsic and the conductivity is low. This fact is important for understanding the survival of standard diode detectors after type inversion.

Radiation hardening of detectors requires a design that takes into account the material's property changes during the irradiation. The possibilities of reducing the sensitivity of the material to irradiation is so far rather restricted.

12 Device Simulation

Throughout the book we have made quite extensive use of results from device simulations. This is due to the fact that only very few problems can be solved analytically. Those problems are usually one-dimensional and idealized assumptions (such as homogeneous doping in distinct regions of the device) are made.

Devices in general require at least two-dimensional – in some cases even three-dimensional – analysis to arrive at quantitatively correct results. At present three-dimensional analysis is only done in very exceptional circumstances due to the large demand on computer memory and power. In general, one relies on two-dimensional simulations and tries to guess the changes expected for three dimensions from several two-dimensional simulations. Two-dimensional (and a few three-dimensional) device simulation packages have been developed by a number of scientific institutes and commercial companies.

A presentation of two-dimensional methods would nevertheless exceed the scope of this book. We will instead give a one-dimensional example that involves defects as discussed in Chap. 11. In this way the methods are presented in a simpler and more transparent manner.

Numerical methods involve the replacement of differentials by finite differences between values taken on a grid. Generalization from one to two or three dimensions is relatively straightforward as long as one uses rectangular grids. However, the more modern finite-element methods use grids of different shapes (e.g. triangular) with varying grid spacing.

The presentation in this chapter will also largely be restricted to stationary situations. An example of a heavily radiation-damaged reverse-biased diode will be given. A short outline of the treatment of time-dependent (nonstationary) cases will conclude the chapter.

12.1 Mathematical Formulation

The problem to be solved involves the simultaneous finding of the (time-dependent) distributions of

- electrons;
- holes;
- potential; and
- space charge.

This is done by solving simultaneously the *continuity equation for electrons*, relating drift, diffusion and generation, the *continuity equations for holes*, and the *Poisson equation*, relating charge density and potential. The charge density depends on the distribution of all charged entities: electrons, holes, (fully ionized flat) donors and acceptors, and deep-level defects. It provides a strong coupling amongst the three differential equations.

Deep-level defects, which simultaneously are responsible for charge generation–recombination and for part of the space charge, are usually not taken into account explicitly. Their effect on generation–recombination is customarily approximated by the introduction of a carrier lifetime, while the contribution to space charge is simply ignored. We will not follow that approach here but rather give a correct treatment based on first principles.

In the following, the three coupled differential equations will be formulated, the space charge (including the partially charged defects) will be found, and the concept of Quasi-Fermi levels will be introduced.

12.1.1 Poisson and Continuity Equations

In their most general form, continuity and Poisson equations look thus:

- *continuity equation for electrons* :

$$\frac{\partial n}{\partial t} = \mu_n n \nabla \mathcal{E} + D_n \nabla^2 n + G_n - R_n \; ; \tag{12.1.1}$$

- *continuity equation for holes* :

$$\frac{\partial p}{\partial t} = -\mu_p p \nabla \mathcal{E} + D_p \nabla^2 p + G_p - R_p \; ; \tag{12.1.2}$$

- *Poisson's equation* :

$$\nabla^2 E_{\mathrm{i}}/q = \nabla^2 \Phi = \nabla \mathcal{E} = \frac{\rho}{\epsilon \epsilon_0} = \frac{q}{\epsilon \epsilon_0} Q_{\mathrm{s}} \; . \tag{12.1.3}$$

Note that the electric field $\mathcal{E}$ has been expressed by the intrinsic level $E_{\mathrm{i}} = q\phi$ and not the potential, leading to a positive sign in (12.1.3).

The excess generation rates $G_n - R_n$ and $G_p - R_p$ will be found from defect properties and concentrations in Sect. 12.1.2. The space-charge density Q_{s} (in units of elementary charge per volume) is given by

$$Q_{\mathrm{s}} = p - n + N_{\mathrm{D}} - N_{\mathrm{A}} + Q_{\mathrm{t}} \tag{12.1.4}$$

and includes a contribution from charged defects Q_{t} again to be derived in Sect. 12.1.2. All quantities, with the exception of the fully ionized flat donor and acceptor densities N_{D} and N_{A}, will be functions of space and time, to be determined by solving the coupled system of equations with the inclusion of deep-level defects, as discussed in the section following.

The situation simplifies if only stationary situations are considered, in which case one has

$$\partial n / \partial t = 0 \qquad \partial p / \partial t = 0$$

and all quantities are functions of position only.

12.1.2 Deep-Level Defects in Stationary Situations

Deep-level defects have been dealt with already in Chap. 11. Their effects on electron and hole generation–recombination and on space charge are essential for any simulation. Relations between capture and emission processes have been derived from thermal equilibrium conditions for simple defects in Sect. 11.2.3 ((11.2.5) and (11.2.7)) and for general type defects in Sect. 11.2.3 ((11.2.27) and (11.2.28)). They will be used in the treatment of nonequilibrium situations.

Considering first one single type of defect with $k + 1$ charge states and restricting ourself in addition to the stationary case, we find from the requirements of constant average charge-state occupation probability, and unity of the sum of probabilities, the occupation probability $P_{t,l}$ of the individual charge states $l = 0$ to k. In order to keep the average defect charge state constant, the sum of electron capture and hole emission rates of charge state $l - 1$ has to equal the sum of hole capture and electron emission rate of charge state l. Thus:

$$N_t P_{t,l-1}(nc_{n,l} + \epsilon_{p,l}) = N_t P_{t,l}(pc_{p,l} + \epsilon_{n,l}) \; ; \tag{12.1.5}$$

$$\alpha_l = \frac{P_{t,l}}{P_{t,l-1}} = \frac{nc_{n,l} + \epsilon_{p,l}}{pc_{p,l} + \epsilon_{n,l}} = \frac{nc_{n,l} + \left(\frac{g_l}{g_{l-1}}\right) e^{-\frac{E_{t,l} - E_i}{kT}} n_i c_{p,l}}{pc_{p,l} + \left(\frac{g_{l-1}}{g_l}\right) e^{\frac{E_{t,l} - E_i}{kT}} n_i c_{n,l}}$$

$$= \frac{c_{n,l} n + \left(\frac{g_l}{g_{l-1}}\right) c_{p,l} n_i / x_{t,l}}{c_{p,l} p + \left(\frac{g_{l-1}}{g_l}\right) c_{n,l} n_i x_{t,l}} \; , \tag{12.1.6}$$

where we have introduced the short hand notation $x = e^{\frac{E - E_i}{kT}}$.

The additional unity requirement for the sum of probabilities

$$\sum_{l=0}^{k} P_{t,l} = 1 \tag{12.1.7}$$

allows the determination of the defect charge-level occupation probabilities $P_{t,l}$ by solving the system of $k + 1$ linear equations (12.1.6) and (12.1.7). The contribution of this one type of defect to the space-charge density is found by taking the sum over defect charge levels:

$$Q_t = N_t \sum_{l=0}^{k} P_{t,l}(Q_0 - l) \tag{12.1.8}$$

with Q_0 the most positive charge state of the defect.

The excess generation rate $G - R = G_n - R_n = G_p - R_p$ is obtained by summing over the defect levels and using (11.2.27) and (12.1.6). Thus:

$$G_n - R_n = N_t \sum_{l=1}^{k} [P_{t,l}\epsilon_{n,l} - P_{t,l-1}nc_{n,l}]$$

$$= N_t \sum_{l=1}^{k} \left[P_{t,l} \frac{g_{l-1}}{g_l} c_{n,l} n_i x_{t,l} - P_{t,l-1}nc_{n,l} \right]$$

$$= N_t \sum_{l=1}^{k} P_{t,l}c_{n,l} \left[\frac{g_{l-1}}{g_l} n_i x_{t,l} - \frac{c_{p,l}p + \left(\frac{g_{l-1}}{g_l}\right) c_{n,l} n_i x_{t,l}}{c_{n,l}n + \left(\frac{g_l}{g_{l-1}}\right) c_{p,l} n_i / x_{t,l}} n \right]$$

$$= N_t \sum_{l=1}^{k} P_{t,l} \frac{c_{n,l}c_{p,l}}{c_{n,l}n + \left(\frac{g_l}{g_{l-1}}\right) c_{p,l} n_i / x_{t,l}} \left[n_i^2 - np \right] \quad ;$$

$$G_p - R_p = N_t \sum_{l=1}^{k} P_{t,l-1} \frac{c_{n,l}c_{p,l}}{c_{p,l}p + \left(\frac{g_{l-1}}{g_l}\right) c_{n,l} n_i x_{t,l}} \left[n_i^2 - np \right] \quad . \tag{12.1.9}$$

This may be written in the more familar form

$$G - R = \beta(n_i^2 - np) \quad , \tag{12.1.10}$$

with

$$\beta = N_t \sum_{l=1}^{k} P_{t,l} \frac{c_{n,l}c_{p,l}}{c_{n,l} \, n + \left(\frac{g_l}{g_{l-1}}\right) c_{p,l} \, n_i / x_{t,l}}$$

$$= N_t \sum_{l=1}^{k} P_{t,l-1} \frac{c_{n,l}c_{p,l}}{c_{p,l} \, p + \left(\frac{g_{l-1}}{g_l}\right) c_{n,l} \, n_i x_{t,l}} \quad . \tag{12.1.11}$$

Note, however, that β, the coefficient describing the excess generation rate, depends on the charge carrier concentrations n and p as the charge-state occupation probabilities $P_{t,l}$ are functions of both carrier concentrations. Therefore, in addition to Poisson's equation the continuity equations will also become nonlinear.

Extension to several types of defects is accomplished easily by finding occupation probabilities separately for each defect type (using (12.1.6) and (12.1.7)) and taking the sum over defect types in evaluating the space charge (see (12.1.8)) and excess generation rate (see (12.1.10)).

Example 12.1
Problem: *Simplify the formalism for the special case of a simple (single-level nondegenerate) defect type.*
Solution: *Only two charge states are possible. With $P_{t,0} = 1 - P_{t,1}$, from (12.1.7) equation (12.1.6) simplifies with the notation $P_t \equiv P_{t,1}$ to*

$$\frac{P_t}{1 - P_t} = \frac{nc_n + n_i c_p / x_t}{pc_p + n_i c_n x_t} \quad ; \quad P_t = \frac{nc_n + n_i c_p / x_t}{nc_n + n_i c_p / x_t + pc_p + n_i c_n x_t}$$

With $Q_0 = 1$ for donors and $Q_0 = 0$ for acceptors, the charge density becomes via (12.1.8)

$$Q_{t,\text{donors}} = N_t(1 - P_t) \qquad\qquad Q_{t,\text{acceptors}} = -N_t P_t \ .$$

The excess generation rate, via (12.1.10), simplifies to

$$G - R = \beta(n_i^2 - np) \ ,$$

with

$$\beta = N_t P_t \frac{c_n c_p}{c_n\, n + c_p\, n_i/x_t} = N_t \frac{c_n c_p}{c_n\, n + c_p\, n_i/x_t + c_p\, p + c_n\, n_i x_t} \ .$$

12.1.3 Quasi-Fermi Levels

The concept of the Quasi-Fermi level is a convenient way of describing the strongly varying charge carrier (electron and hole) concentrations by smoothly varying parameters. Besides this convenience, it has no significance in physics, even though deeper significance is sometimes wrongly attached to it. Its usefulness is found when trying to introduce in an approximate fashion the coupling of the continuity and Poisson equations into the numerical solution method.

The concept of Quasi-Fermi Levels (QFLs) will be extended in the following narrative so as to be applicable also to describe the charge state of deep-level defects. In thermal equilibrium the charge carrier densities are completely determined by the Fermi level. One has via (2.3.7):

$$n = n_i e^{\frac{E_F - E_i}{kT}} \qquad\qquad p = n_i e^{\frac{E_i - E_F}{kT}} \ . \tag{12.1.12}$$

The basic idea of the QFL is to extend the description by a Fermi level also to the nonequilibrium case. One then has to introduce separate QFLs for electrons (E_F^n) and holes (E_F^p) and to use the same functional dependence:

$$n = n_i e^{\frac{E_F^n - E_i}{kT}} \qquad\qquad p = n_i e^{\frac{E_i - E_F^p}{kT}} \ . \tag{12.1.13}$$

Recall that for electrons and holes we have approximated in Chap. 2 the Fermi distribution by Boltzmann distributions (equation (2.3.2)), which is justified for Fermi levels not too close to either band.

Extending the concept to deep-level defects, we have to use true Fermi distributions. In thermal equilibrium one has for the ratio of probabilities for finding the defect in two neighboring charge states (via (11.2.23)):

$$\frac{P_{t,l}/g_l}{P_{t,l-1}/g_{l-1}} = e^{-\frac{E_{t,l} - E_F}{kT}} = e^{-\frac{E_{t,l} - E_i + E_i - E_F}{kT}} = e^{-\frac{q\Psi_{t,l} + E_i - E_F}{kT}} \ , \tag{12.1.14}$$

with $q\Psi_{t,l}$ representing the distance of the lth defect level from the intrinsic level. Extending the concept to the nonequilibrium situation in an analogous fashion leads to the introduction of a separate QFL for each energy level of the defect:

$$\frac{P_{t,l}/g_l}{P_{t,l-1}/g_{l-1}} = e^{-\frac{E_{t,l}-E_{F,l}^t}{kT}} = e^{-\frac{E_{t,l}-E_i+E_i-E_{F,l}^t}{kT}} = e^{-\frac{q\Psi_{t,l}+q\phi-E_{F,l}^t}{kT}} \quad , \qquad (12.1.15)$$

$\phi = E_i/q$ being the electron potential. The ratio between occupation probabilities of neighboring charge states is found from consideration of the charge-changing processes in (12.1.5) and (12.1.6). and can thus be used for the determination of the defect QFLs.

The assumption of constant QFL for small changes of potential in stationary situations has proven to be an easy and efficient way to implement in an approximate fashion the changes in carrier densities and therefore space charge into the solution of the Poisson's equation. In the following, the change of space-charge density with potential will be derived by assuming constant QFL for electrons, holes and deep-level defects.

From (12.1.4) we have

$$\frac{\partial Q_s}{\partial \phi} = \frac{q}{\epsilon \epsilon_0} \left[\frac{\partial p}{\partial \phi} - \frac{\partial n}{\partial \phi} + \frac{\partial Q_t}{\partial \phi} \right] \quad . \qquad (12.1.16)$$

Having chosen $\phi = \frac{E_i}{q}$, we derive from (12.1.13) under the assumption of constant QFL E_F^n and E_F^p with $V_T = \frac{kT}{q}$:

$$\frac{\partial n}{\partial \phi} = -\frac{1}{V_T} n \qquad \frac{\partial p}{\partial \phi} = \frac{1}{V_T} p \quad . \qquad (12.1.17)$$

Similarly we find from (12.1.15)

$$P_{t,l} = \frac{g_l}{g_{l-1}} e^{-\frac{q\Psi_{t,l}+q\phi-E_{F,l}^t}{kT}} P_{t,l-1} = \alpha_l P_{t,l-1} \qquad (12.1.18)$$

by taking the derivative with respect to ϕ:

$$\frac{\partial P_{t,l}}{\partial \phi} - \alpha_l \frac{\partial P_{t,l-1}}{\partial \phi} = -\frac{q}{kT} \alpha_l P_{t,l-1} \quad ; \qquad (12.1.19)$$

and, from (12.1.7):

$$\sum_{l=0}^{k} \frac{\partial P_{t,l}}{\partial \phi} = 0 \quad . \qquad (12.1.20)$$

The set of $k+1$ linear equations (12.1.19) and (12.1.20) can be used for finding the variation of defect charge-state occupation probabilities with the intrinsic level change $\frac{\partial P_{t,l}}{\partial \phi}$. The change of the defect-type charge density is found from (12.1.8) as

$$\frac{\partial Q_t}{\partial \phi} = N_t \sum_{l=0}^{k} \frac{\partial P_{t,l}}{\partial \phi} (Q_0 - l) \quad . \qquad (12.1.21)$$

Inserting (12.1.17) into (12.1.16), the total change of space charge with electron potential becomes

$$\frac{\partial Q_s}{\partial \phi} = \frac{q}{\epsilon \epsilon_0} \left[\frac{1}{V_T} (n + p) + \frac{\partial Q_t}{\partial \phi} \right] . \qquad (12.1.22)$$

This expression will be used in the linearized Poisson equation, as will be shown later. It should be pointed out, however, that calculating the derivative of the occupation probability with respect to the potential from the assumption of constancy of defect QFLs (see (12.1.19) and (12.1.20)) does not agree completely with a direct calculation from the electron and hole concentration. Employing the latter method, one obtains

$$\frac{\partial P_{t,l}}{\partial \Phi} = \frac{\partial P_{t,l}}{\partial n} \frac{\partial n}{\partial \phi} + \frac{\partial P_{t,l}}{\partial p} \frac{\partial p}{\partial \phi} , \qquad (12.1.23)$$

with $\frac{\partial n}{\partial \phi}$ and $\frac{\partial p}{\partial \phi}$ given by (12.1.17), while the derivatives of the occupation probabilities with respect to the carrier densities can be derived from (12.1.6) and (12.1.7) by taking the derivatives; after slight rearrangement we have the following two sets of equations:

$$\frac{\partial P_{t,l}}{\partial n} - \alpha_{t,l} \frac{\partial P_{t,l-1}}{\partial n} = \frac{c_{n,l}}{c_{p,l}p + \left(\frac{g_{l-1}}{g_l}\right) c_{n,l} n_i x_{t,l}} P_{t,l-1} \qquad (12.1.24)$$

$$\sum_{l=0}^{k} \frac{\partial P_{t,l}}{\partial n} = 0$$

$$\frac{\partial P_{t,l-1}}{\partial p} - \frac{1}{\alpha_{t,l}} \frac{\partial P_{t,l}}{\partial p} = \frac{c_{p,l}}{c_{n,l}n + \left(\frac{g_l}{g_{l-1}}\right) c_{p,l} n_i / x_{t,l}} P_{t,l} . \qquad (12.1.25)$$

$$\sum_{l=0}^{k} \frac{\partial P_{t,l}}{\partial p} = 0 .$$

These two sets of $k + 1$ equations can be solved separately and the results inserted in (12.1.23).

12.2 Numerical Solution of Stationary Situations

For reasons of simplicity and transparency, this presentation will be restricted to the one-dimensional stationary case. As mentioned before, the three coupled differential equations are nonlinear. The method of linearization will be dealt with in Sect. 12.2.1, while the methods for numerical solutions, replacing differentials by finite differences on a grid, are presented in Sect. 12.2.2.

12.2.1 Linearization of the Problem

Linearization of differential equation is possible if one starts already with a solution that is close enough to the final solution so that an iterative procedure for solving the differential equation does not diverge. The method will first be demonstrated with Poisson's equation, taking into account simultaneously the coupling to the continuity equations which will alter the carrier densities and therefore the space-charge density, which is a dominant feature of Poisson's equation.

There and in the following we choose to use voltages instead of energies and introduce the following notation:

$\phi = \frac{E_i}{q}$, the electron potential (numerically equal to the intrinsic energy level, expressed in eV);

$\psi_t^j = \frac{1}{q}(E_t - E_i)$, the defect level of defect type j relative to the intrinsic level;

$\phi_n = \frac{E_F^n}{q}, \phi_p = \frac{E_F^p}{q}, \phi_t^j = \frac{E_F^{t\,j}}{q}$, the Quasi-Fermi Levels of electrons, holes and deep-level defects respectively; and

$V_T = \frac{kT}{q}$, the thermal voltage.

Poisson's Equation

With this notation Poisson's equation (12.1.3) is written as

$$\frac{\partial^2 \phi}{\partial x^2} = \frac{q}{\epsilon \epsilon_0} Q_s \tag{12.2.1}$$

with the space charge Q_s (see (12.1.4)) expressed by electron, hole, doping densities and in addition the charge densities of the deep-level defects, thus:

$$Q_s(x) = (N_D(x) - N_A(x) - n(x) + p(x) + Q_t(x)) . \tag{12.2.2}$$

Taking $h(x)$ as a small deviation of the true solution of Poisson's equation $\phi(x)$ from the approximate solution $\bar{\phi}(x)$, we have:

$$\phi(x) = \bar{\phi}(x) + h(x) \tag{12.2.3}$$

and one may linearize (12.2.1) as

$$\frac{\partial^2 \bar{\phi}(x)}{\partial x^2} + \frac{\partial^2 h(x)}{\partial x^2} \approx \frac{q}{\epsilon \epsilon_0} \left[\bar{Q}_s(x) + \frac{\partial Q_s}{\partial \Phi} h(x) + ... \right] . \tag{12.2.4}$$

Reordering (12.2.4) and ignoring higher-order terms, we obtain the linearized differential equation for the difference $h(x)$ between true $\phi(x)$ and approximate $\bar{\phi}(x)$ solutions, the space charge Q_s and its derivative with respect to the electron potential ϕ given for simple defects in (12.2.2) and (12.1.22):

$$\frac{\partial^2 h(x)}{\partial x^2} - \frac{q}{\epsilon \epsilon_0} \frac{\partial Q_s}{\partial \Phi} h(x) = -\frac{\partial^2 \bar{\phi}(x)}{\partial x^2} + \frac{q}{\epsilon \epsilon_0} \bar{Q}_s(x) \qquad (12.2.5)$$

$$\frac{\partial^2 h(x)}{\partial x^2} - \frac{q}{\epsilon \epsilon_0} \left[\frac{\bar{n}(x) + \bar{p}(x)}{V_T} + \frac{\partial Q_t(x)}{\partial \Phi} \right] h(x)$$

$$= -\frac{\partial^2 \bar{\phi}(x)}{\partial x^2} + \frac{q}{\epsilon \epsilon_0} \left[N_D(x) - N_A(x) - \bar{n}(x) + \bar{p}(x) + \bar{Q}_t(x) \right] . \quad (12.2.6)$$

The latter form has been used in order to show explicitly the difference with and without deep-level defects. Notice that electron and hole concentrations enter with the same sign on the left- and with opposite sign on the right-hand side.

Equations (12.2.5) and (12.2.6) can be used to solve Poisson's equation in an iterative way. Under the approximation of constant QFLs, electron and hole densities are rescaled according to (12.1.13) after each iteration step as

$$\phi'(x) = \phi(x) + h(x) \qquad (12.2.7)$$

$$n'(x) = n(x) e^{-\frac{h(x)}{V_T}} \qquad (12.2.8)$$

$$p'(x) = p(x) e^{+\frac{h(x)}{V_T}} . \qquad (12.2.9)$$

A similar rescaling for the defect charges would also be possible, however, the Boltzmann approximation implicit in (12.2.8) and (12.2.9) is no longer valid and the true Fermi distribution has to be used. It therefore is simpler to directly calculate after each iteration step the defect charge density Q_t and its partial derivative $\frac{\partial Q_t}{\partial \phi}$ from electron and hole densities and capture cross-sections according to the procedures described in Sects. 12.1.2 and 12.1.3.

As linearization of the Poisson equation has been reached by replacing

$$e^{\frac{h(x)}{V_T}} \approx 1 + \frac{h(x)}{V_T} ,$$

one expects convergence problems for the case that the potential correction function $h(x)$ rises above the thermal voltage $V_T = kT/q$. In such a situation it is advisable to scale down the correction function $h(x)$ by a suitable factor.

Continuity Equations

In the standard approach, the excess generation rate $G - R$ given by (12.1.10) is calculated with the assumption of fixed excess generation constant β. With this assumption the continuity equations (12.1.1) and (12.1.2) are linear. Inserting (12.1.10) into (12.1.1) and (12.1.2) and assuming $\frac{\partial n}{\partial t} = \frac{\partial p}{\partial t} = 0$ as well as equal electron and hole excess generation rates ($G_n - R_n = G_p - R_p = G - R = \beta(n_i^2 - pn)$), as is the case in stationary situations, one obtains after reordering

$$-D_n \frac{\partial^2 n}{\partial x^2} + n \left[-\mu_n \frac{\partial^2 \phi}{\partial x^2} + p \sum_{\text{defects}} \beta^j \right] = n_i^2 \sum_{\text{defects}} \beta^j \qquad (12.2.10)$$

$$-D_p \frac{\partial^2 p}{\partial x^2} + p \left[\mu_p \frac{\partial^2 \phi}{\partial x^2} + n \sum_{\text{defects}} \beta^j \right] = n_i^2 \sum_{\text{defects}} \beta^j . \qquad (12.2.11)$$

These equations are valid for general stationary situations. However, as has been pointed out already, the generation–recombination factor β is dependent on the carrier concentrations n and p so that the excess generation rate is a nonlinear function of the carrier concentations n and p in (12.1.10). Subsequently the continuity equations are nonlinear equations, which can be linearized by expanding β around the approximate carrier concentrations $\bar{n}$ and $\bar{p}$. This possibility will, however, not be followed up here.

12.2.2 The Finite Difference Method

Numerical solutions of linear differential equations are usually derived by replacing the derivatives of a function by differences taken on a grid, with either equal or unequal spacing. Taking equal spacing simplifies the mathematics somewhat; however, it will in most cases not be "economical" as it will lead in general to many more grid points. The method is illustrated in the one-dimensional case for a function $f(x)$ and grid points x_i using the shorthand notation $f(x_i) = f_i$, $\frac{\partial f(x_i)}{\partial x} \equiv f'(x_i) = f_i'$, $f''(x_i) = f_i''$ etc. The Taylor series expansion of $f(x)$ around x_i

$$f(x) = f(x_i) + f'(x_i)(x - x_i) + \frac{1}{2}f''(x_i)(x - x_i)^2 + \ldots\ldots \tag{12.2.12}$$

results in

$$f_{i-1} = f_i - f_i'(x_i - x_{i-1}) + \frac{1}{2}f_i''(x_i - x_{i-1})^2$$
$$- \frac{1}{6}f_i'''(x_i - x_{i-1})^3 + \ldots \tag{12.2.13}$$

$$f_{i+1} = f_i + f_i'(x_{i+1} - x_i) + \frac{1}{2}f_i''(x_{i+1} - x_i)^2$$
$$+ \frac{1}{6}f_i'''(x_{i+1} - x_i)^3 + \ldots \tag{12.2.14}$$

Combining these two equations, one may (ignoring higher-order terms) express first- and second-order derivatives by differences. For equal grid spacing $x_i - x_{i-1} = x_{i+1} - x_i = a$, one gets by subtracting and summing the two equations

$$f_i' = \frac{f_{i+1} - f_{i-1}}{2a} \tag{12.2.15}$$

$$f_i'' = \frac{f_{i+1} - 2f_i + f_{i-1}}{a^2} . \tag{12.2.16}$$

Note that with this method the precision is one order higher than needed, as for the first derivative the second-order terms, and for the second derivative the third-order terms, cancel exactly.

For nonuniform grid one may take first and second derivatives as

$$f_i' = \frac{(f_{i+1} - f_i)\frac{x_i - x_{i-1}}{x_{i+1} - x_i} + (f_i - f_{i-1})\frac{x_{i+1} - x_i}{x_i - x_{i-1}}}{x_{i+1} - x_{i-1}} \tag{12.2.17}$$

$$f_i'' = \frac{\frac{f_{i+1} - f_i}{x_{i+1} - x_i} - \frac{f_i - f_{i-1}}{x_i - x_{i-1}}}{\frac{x_{i+1} - x_{i-1}}{2}}, \tag{12.2.18}$$

where again the first derivative is exact to second order.

For reasons of transparency in the presentation, constant grid spacing $(x_{i+1} - x_i = a)$ will be assumed in the following.

Poisson's Equation

Replacing the differentials in the linearized Poisson equation (12.2.5) by finite differences, one obtains at grid point x_k using (12.2.16) (for nonuniform grid spacing use (12.2.18)):

$$\frac{h_{k+1} - 2h_k + h_{k-1}}{a^2} - h_k \frac{q}{\epsilon\epsilon_0} \frac{\partial Q_{s,k}}{\partial \Phi} = -\frac{\partial^2 \phi_k}{\partial x^2} + \frac{q}{\epsilon\epsilon_0} Q_{s,k} \tag{12.2.19}$$

with $Q_{s,k}$, the space charge at position x_k derived from (12.1.4) as

$$Q_{s,k} = \left[N_{Dk} - N_{Ak} - n_k + p_k + \sum_j Q_{t,k}^j \right] \tag{12.2.20}$$

$$\frac{\partial Q_{s,k}}{\partial \phi} = \frac{1}{V_T} \left[n_k + p_k + \sum_j \frac{\partial Q_{t,k}^j}{\partial \phi} \right] \tag{12.2.21}$$

$$\frac{\partial^2 \phi_k}{\partial x^2} = \frac{\phi_{k+1} - 2\phi_k + \phi_{k-1}}{a^2}. \tag{12.2.22}$$

The change of space-charge density $\frac{\partial Q^j}{\partial \phi}$ due to defect type j is obtained from (12.1.21). All quantities are fixed with the exception of the potential correction function $h(x)$ taken at the grid points $h_k = h(x_k)$.

Taking as a boundary condition fixed potentials at first and last grid points x_0 and x_n, one has $h(x_0) = h_0 = 0$ and $h(x_n) = h_n = 0$ and (12.2.19) is valid for $k = 1$ to $k = n - 1$. It can be written in the form of an $(n - 1)$-dimensioned matrix equation:

$$A H = B \tag{12.2.23}$$

with

$$A_{k,k-1} = \frac{1}{a^2} \qquad A_{k,k} = -\frac{2}{a^2} - \frac{q}{\epsilon\epsilon_0} \frac{\partial Q_{s,k}}{\partial \Phi}$$

$$A_{k,k+1} = \frac{1}{a^2} \qquad B_k = -\frac{\partial^2 \phi_k}{\partial x^2} + \frac{q}{\epsilon\epsilon_0} Q_{s,k}.$$

Note that the sparse matrix A has only the diagonal and next-to-diagonal elements nonzero. The matrix equation can be solved by inversion as follows (although there exist faster numerical methods for the solution of problems involving sparse matrices):

$$H = A^{-1}B .$$ (12.2.24)

This procedure is done iteratively, replacing after each iteration the potential ψ_k by its new value $\psi_k + h_k$ and rescaling simultaneously the carrier densities and defect-state occupation probabilities under the assumption of constant QFLs as described in Sects. 12.1.3 and 12.2.1 (see (12.2.8) and (12.2.9)). In this way the coupling between Poisson and continuity equations is taken into account.

Continuity Equations
Solving the continuity equations for electrons and holes can in principle be performed in an analogous fashion. Better convergence and less problems with computer precision is reached using the Scharfetter–Gummel approach (Scharfetter and Gummel 1969). In this method an analytical solution of the continuity equation for each grid interval is taken under the assumption of constant current density, constant electric field and constant mobility, as if no charge generation was occuring within the interval. This results in an exponential dependence of the carrier density on the position, which relates the current density in the interval to the carrier densities at the grid points (the endpoints of the interval). The current density is assumed to be representative for the center of the interval and the difference of two neighboring current density center values is related to the excess generated charge carriers in the region between the two centers.

In the following this method is demonstrated for electron and hole densities. The current density J is given by

$$\frac{J_n}{q} = D_n \frac{\partial n(x)}{\partial x} + q\mathcal{E}\mu_n n(x)$$ (12.2.25)

$$\frac{J_p}{q} = -D_p \frac{\partial p(x)}{\partial x} + q\mathcal{E}\mu_p p(x) ,$$ (12.2.26)

which, using Einstein's relation connecting diffusion and mobility

$$D = \frac{kT}{q}\mu \equiv V_T\mu$$ (12.2.27)

can be rewritten as

$$\frac{\partial n(x)}{\partial x} + \frac{\mathcal{E}}{V_T}n(x) = \frac{1}{D_n}\frac{J_n}{q}$$ (12.2.28)

$$-\frac{\partial p(x)}{\partial x} + \frac{\mathcal{E}}{V_T}p(x) = \frac{1}{D_p}\frac{J_p}{q} .$$ (12.2.29)

The general solution to this linear differential equation is

$$n(x) = \frac{V_T}{\mathcal{E}D_n}\frac{J_n}{q} + n_0 e^{-\frac{\mathcal{E}}{V_T}(x-x_0)} \tag{12.2.30}$$

$$p(x) = \frac{V_T}{\mathcal{E}D_p}\frac{J_p}{q} + p_0 e^{\frac{\mathcal{E}}{V_T}(x-x_0)} \tag{12.2.31}$$

with n_0 , p_0 and x_0 arbitrary constants.

Considering the interval x_k to x_{k+1}, choosing $x_0 = x_k$, and setting the electric field as $\mathcal{E} = \frac{\phi_{k+1}-\phi_k}{x_{k+1}-x_k}$, one finds with the notation $x_{k+1} - x_k = \xi$; $\frac{\phi_{k+1}-\phi_k}{V_T} = \eta$ that

$$n_{k+1} = \frac{V_T}{\mathcal{E}D_n}\frac{J_n}{q} + n_0 e^{-\frac{\phi_{k+1}-\phi_k}{V_T}} = \frac{\xi}{\eta D_n}\frac{J_n}{q} + n_0 e^{-\eta}$$

$$n_k = \frac{V_T}{\mathcal{E}D_n}\frac{J_n}{q} + n_0 = \frac{\xi}{\eta D_n}\frac{J_n}{q} + n_0$$

$$n_{k+1} - n_k e^{-\eta} = \frac{\xi}{\eta D_n}\frac{J_n}{q}\left[1 - e^{-\eta}\right] \tag{12.2.32}$$

$$\frac{J_n}{q} = \frac{D_n}{\xi}\left[n_{k+1}\frac{\eta}{1-e^{-\eta}} - n_k \frac{\eta e^{-\eta}}{1-e^{-\eta}}\right] , \tag{12.2.33}$$

and the equivalent notation for holes

$$p_{k+1} = \frac{V_T}{\mathcal{E}D_p}\frac{J_p}{q} + p_0 e^{\frac{\phi_{k+1}-\phi_k}{V_T}} = \frac{\xi}{\eta D_p}\frac{J_p}{q} + p_0 e^{\eta}$$

$$p_k = \frac{V_T}{\mathcal{E}D_p}\frac{J_p}{q} + p_0 = \frac{\xi}{\eta D_p}\frac{J_p}{q} + p_0$$

$$p_{k+1} - p_k e^{\eta} = \frac{\xi}{\eta D_p}\frac{J_p}{q}\left[1 - e^{\eta}\right] \tag{12.2.34}$$

$$\frac{J_p}{q} = \frac{D_p}{\xi}\left[p_{k+1}\frac{\eta}{1-e^{\eta}} - p_k \frac{\eta e^{\eta}}{1-e^{\eta}}\right] . \tag{12.2.35}$$

With the Bernoulli function

$$B(x) = \frac{x}{e^x - 1} , \tag{12.2.36}$$

one has for electrons and equivalently for holes in the interval x_k to x_{k+1}

$$\frac{J_n}{q} = \frac{D_n}{\xi}\left[n_{k+1}B(-\eta) - n_k B(\eta)\right] \tag{12.2.37}$$

$$\frac{J_p}{q} = \frac{D_p}{\xi}\left[-p_{k+1}B(\eta) + p_k B(-\eta)\right] . \tag{12.2.38}$$

Setting equal the difference of current densities between two consecutive intervals with the excess charge generation rate (with negative sign for electrons) one has

$$-\frac{1}{q}(J_n^+ - J_n^-) = (G_n - R_n)\frac{x_{k+1} - x_{k-1}}{2} \qquad (12.2.39)$$

$$\frac{1}{q}(J_p^+ - J_p^-) = (G_p - R_p)\frac{x_{k+1} - x_{k-1}}{2} \ . \qquad (12.2.40)$$

(J_n^+ and J_n^- being the current densities in the intervals $x_k \leftrightarrow x_{k+1}$ and $x_{k-1} \leftrightarrow x_k$ respectively), and then one obtains at grid point k

$$- n_{k-1} D_n B\left(\frac{\phi_k - \phi_{k-1}}{V_T}\right)\frac{1}{x_k - x_{k-1}}$$

$$+ n_k\left[D_n B\left(-\frac{\phi_k - \phi_{k-1}}{V_T}\right)\frac{1}{x_k - x_{k-1}} + D_n B\left(\frac{\phi_{k+1} - \phi_k}{V_T}\right)\frac{1}{x_{k+1} - x_k}\right]$$

$$- n_{k+1} D_n B\left(-\frac{\phi_{k+1} - \phi_k}{V_T}\right)\frac{1}{x_{k+1} - x_k}$$

$$= (G_n - R_n)_k\frac{x_{k+1} - x_{k-1}}{2}$$

$$= \frac{x_{k+1} - x_{k-1}}{2}(n_i^2 - n_k p_k)\sum_j \beta_j \ . \qquad (12.2.41)$$

For holes one finds equivalently

$$- p_{k-1} D_p B\left(-\frac{\phi_k - \phi_{k-1}}{V_T}\right)\frac{1}{x_k - x_{k-1}}$$

$$+ p_k\left[D_p B\left(\frac{\phi_k - \phi_{k-1}}{V_T}\right)\frac{1}{x_k - x_{k-1}} + D_p B\left(-\frac{\phi_{k+1} - \phi_k}{V_T}\right)\frac{1}{x_{k+1} - x_k}\right]$$

$$- p_{k+1} D_p B\left(\frac{\phi_{k+1} - \phi_k}{V_T}\right)\frac{1}{x_{k+1} - x_k}$$

$$= (G_p - R_p)_k\frac{x_{k+1} - x_{k-1}}{2}$$

$$= \frac{x_{k+1} - x_{k-1}}{2}(n_i^2 - p_k n_k)\sum_j \beta_j \ . \qquad (12.2.42)$$

In order to solve (12.2.41) and (12.2.42) using the matrix formalism we write all terms containing the variable to be solved (electron concentration n in equation (12.2.41)) to the left side. We encounter here the problem that the excess generation rate $G_n - R_n$ does not only depend explicitly on the electron concentration n but also implicitly through the dependence of the generation–recombination factor β on n (and p). This dependence can be approximately taken into account by performing a linear expansion of β around the approximate carrier concentrations n_k, p_k. For simplicity and clarity, however, β will be assumed constant during the iteration step in the following.

In order to solve these equations one has to introduce boundary conditions for the first x_0 and last x_n position. Assuming thermal equilibrium, carrier concentration at these positions is equivalent in assuming infinite surface recombination velocity. Electron and hole concentrations at the boundaries are then fixed and (12.2.41) and (12.2.42) are valid for $k = 1$ to $k = n - 1$. For electrons the matrix formulation of (12.2.41) for the carrier density $C = (n_1, n_2, ...n_{n-1})$ becomes

$$A\,C = B \qquad\qquad (12.2.43)$$

with

$$A_{k-1,k} = -D_n B \left(\frac{\phi_k - \phi_{k-1}}{V_T} \right) \frac{1}{x_k - x_{k-1}}$$

$$A_{k,k} = \left\{ D_n B \left(-\frac{\phi_k - \phi_{k-1}}{V_T} \right) \frac{1}{x_k - x_{k-1}} + D_n B \left(\frac{\phi_{k+1} - \phi_k}{V_T} \right) \frac{1}{x_{k+1} - x_k} \right.$$

$$\left. + p_k \sum_j \beta_k^j \frac{x_{k+1} - x_{k-1}}{2} \right\}$$

$$A_{k,k+1} = -D_n B \left(-\frac{\phi_{k+1} - \phi_k}{V_T} \right) \frac{1}{x_{k+1} - x_k}$$

$$B_1 = n_i^2 \sum_j \beta_1^j \frac{x_2 - x_0}{2} + n_0 D_n B \left(\frac{\phi_1 - \phi_0}{V_T} \right) \frac{1}{x_1 - x_0}$$

$$B_k = n_i^2 \sum_j \beta_k^j \frac{x_{k+1} - x_{k-1}}{2}$$

$$B_{n-1} = n_i^2 \sum_j \beta_{n-1}^j \frac{x_n - x_{n-2}}{2} + n_n D_n B \left(-\frac{\phi_n - \phi_{n-1}}{V_T} \right) \frac{1}{x_n - x_{n-1}} \ .$$

The special form for B_1 and B_{n-1} is due to the boundary conditions chosen. For holes a similar formulation can be found from (12.2.42).

12.2.3 Example of a Stationary Problem

The example to be presented has been selected to demonstrate the principal possibilities of simulations when including high concentrations of deep-level defects. It does not correspond to a realistic case, as only one type of radiation-induced defect has been assumed whose properties and doping profiles are chosen arbitrarily. However, the chosen example is able to explain some prominent properties observed in irradiated detectors.

The effect of high-concentration deep-level acceptors in a diode is investigated by comparison with an identical ("unirradiated") device without these deep-level acceptors. The original wafer is n-type silicon with bulk doping of $N_D = 1 \times 10^{12}\,\mathrm{cm}^{-3}$ and a minority carrier lifetime of $\tau_g = 1\,\mathrm{ms}$. Thin highly doped p^+ (at $x = 0$) and n^+ regions are located at the two surfaces of the $250\,\mu\mathrm{m}$-thick device. The exact doping profiles of these layers and the corresponding effects near the wafer surfaces will not be discussed in the following. The discussion will concentrate on the situation within the bulk instead.

A uniform introduction of deep-level slightly-above-midgap ($E_t - E_i = 50\,\mathrm{meV}$) acceptor-type defects of significantly higher concentration ($N_t =$

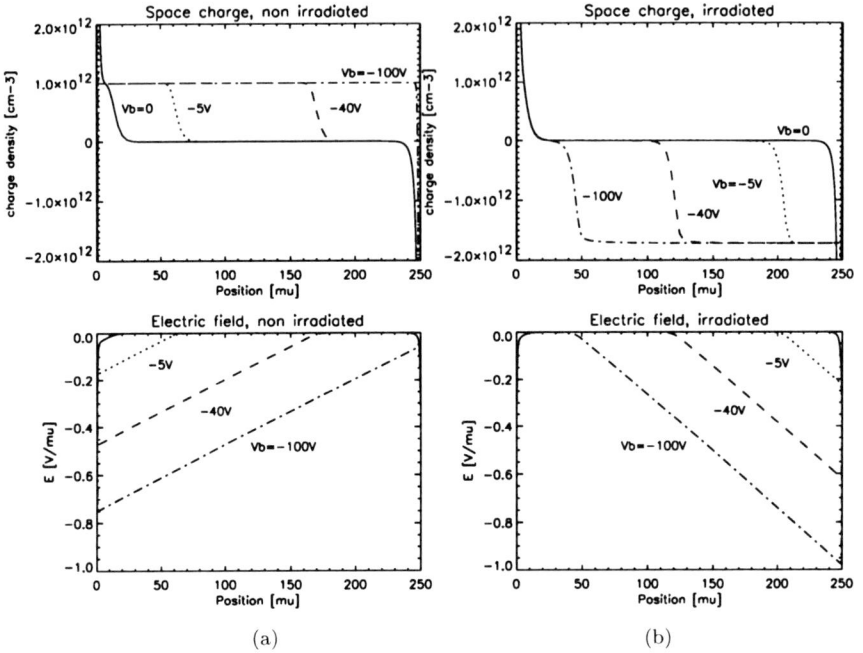

Fig. 12.1a,b. Space-charge densities and electric field for an unirradiated (a) and an irradiated type inverted (b) p^+-n-n^+ detector without and with application of a reverse bias V_b

$3 \times 10^{13}\,\mathrm{cm}^{-3}$) than the original shallow donors is assumed to be caused by irradiation. Electron and hole capture cross-sections $\sigma_n = 1 \times 10^{15}\,\mathrm{cm}^2$, $\sigma_p = 5 \times 10^{15}\,\mathrm{cm}^2$ are assumed for the simple deep-level acceptor.

Figure 12.1 compares the space-charge densities and electric fields of the "unirradiated" and "irradiated" devices. At zero bias one notices in both devices the space-charge regions in the vicinity of the surfaces caused by the doping variations close to the surface. Before irradiation these are the p^+-n junction on the left and the $n-n^+$ junction on the right side of Fig. 12.1a. Applying a reverse-bias voltage to the unirradiated detector, the space-charge region on the left side grows into the bulk, yielding a positive space-charge density equal to the bulk doping density. For the irradiated detector the growth starts from the right side and the space-charge density Q_s has the opposite sign, its magnitude being much lower than the defect density $N_t = 3 \times 10^{13}\,\mathrm{cm}^{-3}$.

The space charge is composed of the electron and hole densities, the fully ionized shallow dopants and, for the irradiated sample, by the partially ionized deep-level acceptors.

Charge carrier densities are shown in Fig. 12.2 and the ionization fraction of the deep-level acceptor in Fig. 12.3. For the unbiased devices electron n and hole p densities are constant throughout the bulk, their product being equal to the square of the intrinsic concentration n_i^2. While the bulk of the unirradiated

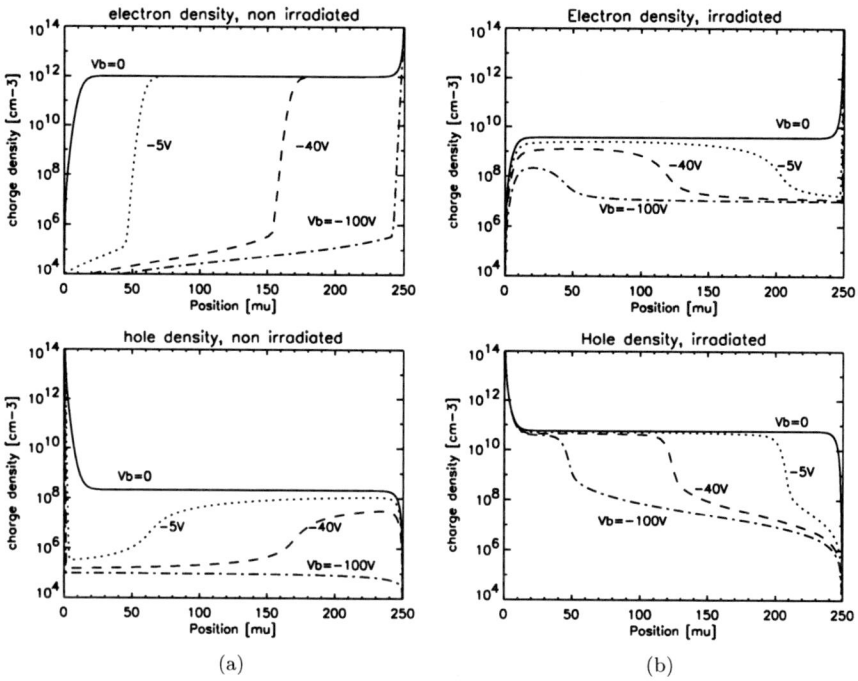

(a)   (b)

Fig. 12.2a,b. Electron and hole densities in an unirradiated (a) and an irradiated type-inverted (b) p^+–n–n^+ detector without and with reverse bias V_b

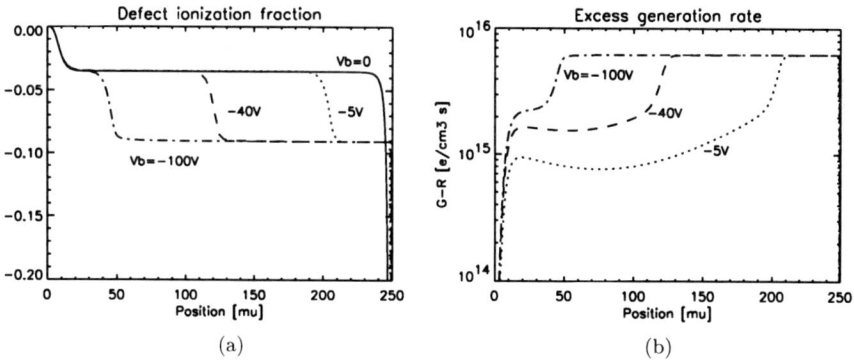

(a)   (b)

Fig. 12.3a,b. Fractional ionization factor of deep-level acceptors (a) and excess generation rate (b) in the irradiated type-inverted p^+–n–n^+ detector for several reverse bias voltages

device is of n-type with $n = N_D$, in the irradiated device the hole concentration exceeds the electron concentration, its magnitude being two orders of magnitude below the deep-level acceptor-type defect density N_t. Slightly more than

3% of the deep-level acceptors are ionized already in the unbiased situation (Fig. 12.3).

After reverse-biasing the unirradiated device, both types of charge carriers are removed in the expanding space-charge region near the p^+–n junction, while in the neutral region the majority carriers remain unchanged and the minority carrier density is reduced due to diffusion into the space-charge region. For the irradiated device the very strong reduction of both carrier types is seen in the space-charge region on the right, but is also noticeable in the left-hand neutral region. It is accompanied by an increase in the fraction of ionization in the space-charge region. As the space charge in the depleted region is negative, one speaks of effective p-type doping and type inversion.[38]

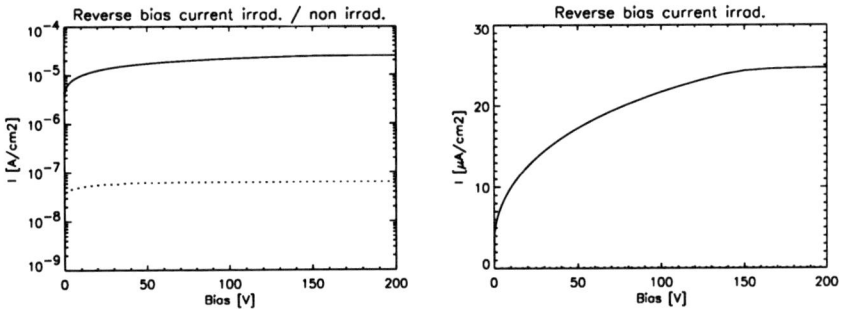

Fig. 12.4. Reverse-bias current as a function of the applied voltage for the unirradiated (*dotted line*) and the irradiated type inverted (*continuous line*) p^+–n–n^+ detector in logarithmic and linear representation

The reverse-bias current–voltage characteristics of the irradiated and unirradiated devices are shown in Fig. 12.4. The reverse-bias current of the irradiated detector is several orders of magnitude higher due to deep-level defects acting as generation/recombination centers. Generation is largest in the space-charge region, as can be seen in the right-hand side of Fig. 12.3. Notice, however, that significant generation also occurs in the neutral region on the left-hand side due to the reduction of carrier concentrations in this region, as seen in Fig. 12.2.

12.3 Simulation of Time-Dependent Situations

The approach for the numerical solution of time-dependent situations differs significantly from that described so far. One will take as a starting condition a solution of the stationary problem. One again will have to solve the Poisson equation (12.1.3) and continuity equations for electrons (12.1.1) and holes (12.1.2). The situation will have to be considered at different time points. While

---

[38]The defect density at which type inversion occurs does not necessarily coincide with the one in which majority carriers in the undepleted region change from electrons to holes.

Poisson's equation can be solved separately at each time point, it is still coupled to previous time points through the space charge (see (12.1.4)), which contains the defect charges Q_t and their dependency not only on the electron and hole densities but also on the previous history. This is a significant difference from standard approaches, not taking into account the change of charge states of deep-level defects.

The dependence of the continuity equations on time is seen directly by the time derivative of the carrier concentrations on the left-hand side of (12.1.1) and (12.1.2). In discretizing these equations, these derivatives are replaced by differences between neighboring time points.

We will not go further into this subject here; however, it may be worthwhile pointing out that the method of taking care of the coupling between a Poisson equation and the continuity equations through the introduction of QFLs does not apply in nonstationary situations. While in the stationary case the change in the equilibrium charge-carrier density and defect-occupation state due to the potential change can be anticipated, the space charge in the nonstationary case is obtained from the defect charge states and the solution of the time-dependent continuity equations only, and both of these depend on the values at the previous time point.

Part III

Reference Material

Appendix A
Frequently Used Symbols

Symbols	Meaning	Units
$\mathcal{B}$	magnetic induction	Vs/cm^2
c_n, c_p	$c = \nu_{\text{th}}\sigma$, electron and hole capture coefficients	cm^3/s
C_G	gate capacitance	F
C_{ox}	oxide capacitance per unit area	F/cm^2
d_{ox}	oxide thickness	cm
d_s	depletion layer depth	cm
D_n, D_p	diffusion constants of electrons and holes	cm^2/s
$\mathcal{E}$	electric field	V/cm
E_F	Fermi level	eV
E_F^n, E_F^p	Quasi-Fermi levels of electrons and holes	eV
E_d	defect energy level	eV
E_i	intrinsic energy level	eV
E_C	lower edge of conduction band	eV
E_V	upper edge of valence band	eV
E_G	band gap	eV
$\mathcal{E}_s$	electric field at semiconductor surface	V/cm
E_t	defect energy level	eV
f	frequency	s^{-1}
f	fraction of charge induced in the channel	1
$\mathcal{F}$	force	eV/cm
$F(E)$	Fermi–Dirac occupation probability for electrons	1
$F(E)$	Fermi–Dirac distribution function	1
$F_n(E)$	occupation probability for electrons in the conduction band	1
$F_p(E)$	occupation probability for holes in the valence band	1
F_n	flux density of electrons	$\text{cm}^{-2}\text{s}^{-1}$
F_p	flux density of holes	$\text{cm}^{-2}\text{s}^{-1}$
g	$= \partial I_D/\partial V_D$ transistor conductance	A/V
g_m	$= \partial I_D/\partial V_G$ transistor transconductance	A/V
g_q	$= \partial I_D/\partial Q_{\text{sig}}$ DEPFET charge steilheit	s^{-1}
$g_{m,\text{sat}}$	$= \partial I_{D,\text{sat}}/\partial V_G$ transconductance in the saturation region	A/V
g_l	defect charge state degeneration factor	1

h	Planck's constant	Js
h	depletion depth	cm
h	potential change in iterative solution of Poisson equation	V
I_D	drain current	A
$I_{D,sat}$	drain current at saturation	A
J_n, J_p	electron and hole current densities	A/cm^2
G	generation rate	$cm^{-3}s^{-1}$
G_n, G_p	electron and hole generation rates	$cm^{-3}s^{-1}$
k	Boltzmann's constant	J/K, eV/K
k_τ	radiation damage constant (lifetime)	cm^2/s
K_F	$1/f$ noise parameter	C^2/cm^2
L	transistor channel length	μm
L_n, L_p	diffusion length of electrons and holes	cm, μm
m_n	effective electron mass	kg
m_p	effective hole mass	kg
n	free electron concentration	cm^{-3}
n_i	intrinsic carrier concentration	cm^{-3}
n_n, n_p	electron concentration in n- and p-regions	cm^{-3}
N_A	acceptor concentration	cm^{-3}
N_d	defect concentration	cm^{-3}
N_D	donor concentration	cm^{-3}
N_{eff}	effective doping concentration	cm^{-3}
N_t	defect density	cm^{-3}
N_C	effective density of states in the conduction band	cm^{-3}
N_V	effective density of states in the valence band	cm^{-3}
$N(E_{kin})$	density of states	1
p	readout pitch	μm
p	free hole concentration	cm^{-3}
p_n, p_p	hole concentration in n- and p-region	cm^{-3}
$P_{t,l}$	probability of defect being in charge state l	1
q	elementary charge	As
Q_c	channel surface charge density	C/cm^2
Q_{inv}	inversion layer surface charge density	C/cm^2
Q_{sig}	signal charge	C
Q_s	space–charge density (in units of e)	cm^{-3}
Q_t	defect charge density (in units of e)	cm^{-3}
R	recombination rate	$cm^{-3}s^{-1}$
R_n, R_p	electron and hole capture rates	$cm^{-3}s^{-1}$
T	absolute temperature	K
U	excess recombination rate	$cm^{-3}s^{-1}$
v_n	noise voltage	V
V_{bi}	built-in voltage	V
V_c	channel potential	V
V_D	drain potential	V
$V_{D,sat}$	drain–source saturation voltage	V

V_F	flat-band voltage	V
V_G	gate potential	V
$V_{G,\text{eff}}$	effective gate voltage	V
V_p	pinch-off voltage	V
V_{sub}	substrate potential	V
V_S	source potential	V
V_T	threshold voltage	V
V_T	$= \frac{kT}{q}$ thermal voltage	V
W	transistor channel width	μm
$x(E)$	$=\exp(\frac{E-E_1}{kT})$	1
α	radiation-damage constant (current)	A/cm
α_l	ratio of probabilities of neighboring defect charge states	1
α_T	transistor base-transport factor	1
α_0	transistor common-base current gain	1
β	generation constant	$\text{cm}^{-3}\text{s}^{-1}$
β	transistor current gain	1
γ	transistor emitter efficiency	1
ϵ	dielectric constant	1
ϵ_0	permittivity in a vacuum	F/cm
Θ_n, Θ_p	Hall angle of electrons and holes	1
μ_n, μ_p	mobility of electrons and holes	cm^2/Vs
μ_n^H, μ_p^H	Hall mobility of electrons and holes	cm^2/Vs
ν_n, ν_p	drift velocity of electrons and holes	cm/s
$\nu_{\text{th},n}, \nu_{\text{th},p}$	thermal velocity of electrons and holes	cm/s
ρ	charge density	C/cm^3
σ	root-mean-square deviation	μm
σ_n, σ_p	electron and hole capture cross-sections	cm^2
τ_c	mean free time between collisions	s
τ_r	recombination life-time of minority carriers	s
τ_g	generation life-time of minority carriers	s
Φ	potential	V
Φ	radiation fluence	cm^{-2}
$q\Phi_m$	work function of metal	eV
$q\Phi_s$	work function of semiconductor	eV
$q\Phi_{B_n}, q\Phi_{B_p}$	barrier height of metal–semiconductor contact	eV
Ψ_B	$\frac{1}{q}|E_i - E_F|$, distance intrinsic level to Fermi level	V
Ψ_c	channel potential	V
Ψ_s	potential at the semiconductor surface	V
$q\chi$	electron affinity	eV

Appendix B
Physical Constants

Quantity	Symbol	Value (in one or two units)
Ångström unit	Å	$1\,\text{Å} = 10^{-8}\,\text{cm} = 10^{-10}\,\text{m}$
Electron volt	eV	$1.6022 \times 10^{-19}\,\text{J}$
Speed of light	c	$2.99792 \times 10^{10}\,\text{cm/s}$
Permittivity of free space	ϵ_0	$8.85418 \times 10^{-14}\,\text{F/cm}$
Permeability of free space	μ_0	$1.25663 \times 10^{-8}\,\text{H/cm}$
Planck's constant	h	$6.62617 \times 10^{-34}\,\text{J s}$
Reduced Planck constant	$\hbar$	$1.05458 \times 10^{-34}\,\text{J s}$
Elementary charge	q	$1.60218 \times 10^{-19}\,\text{C}$
Electron rest mass	m	$0.91095 \times 10^{-27}\,\text{g}$
Proton rest mass	M_p	$1.67264 \times 10^{-27}\,\text{kg}$
Ideal gas constant	R	$1.98719\,\text{cal/mole K}$ $= 8.3145\,\text{J/mole K}$
Boltzmann's constant	$k = R/N_A$	$1.3087 \times 10^{-23}\,\text{J/K}$
Avogadro's number	N_A	$6.0221 \times 10^{23}\,\text{mole}^{-1}$
Thermal voltage at 300 K	$V_T = kT/q$	$0.0259\,\text{V}$
Wavelength of 1-eV quantum	λ	$1.23977\,\mu\text{m}$

References

Books and Reviews

Beadle, W.E., Tsai, J.C.C. and Plummer, R.D. (1984): "Quick Reference Manual for Silicon Integrated Circuit Technology" John Wiley and Sons, New York, Chichester, Brisbane, Toronto, Singapore, 1984

Gatti, E. and Manfredi, P.F. (1986): "Processing the signals from solid state detectors in elementary particle physics" Rivista di Nuovo Cimento 9, Ser. 3 (1986) 1-145

Gray, P.R. and Meyer, R.G. (1993): "Analysis and design of analog integrated circuits" 3rd ed., Wiley, New York 1993

Grove, A.S. (1967): "Physics and technology of semiconductor devices" Wiley, New York 1967

Henke, B.L. (1982) et al.: Atomic Data Tables, Vol. 27 (1982) 1

Kittel, C. (1976): "Introduction to solid state physics" Wiley, New York 1976

Knoll, Glenn F. (1989): "Radiation detection and measurement" 2nd ed., Wiley and Sons, New York 1989

Landolt–Börnstein vol. III/17a (1982): "Semiconductors, Physics of Group IV Elements and III-V Compounds" Springer, Berlin Heidelberg New York, 1982

Landolt–Börnstein vol. III/17c (1984): "Semiconductors, Technology of Si, Ge and SiC" Springer, Berlin Heidelberg New York, 1984

Landolt–Börnstein vol. III/22a (1987): "Semiconductors, Intrinsic Properties of Group IV Elements and III-V, II-VI and I-VII Compounds" Springer, Berlin Heidelberg New York, 1987

Leo, W.R. (1994): "Techniques for Nuclear and Particle Physics Experiments" Springer, Berlin Heidelberg New York

Muller–Kamins (1986): "Device electronics for integrated circuits" 2nd ed., John Wiley & Sons, New York

Nicollian, E.H. and J.R. Brews (1982): "MOS Physics and Technology" John Wiley & Sons, New York

Palik, E.D. (1985): "Handbook of optical constants of solids" Academic Press, New York

Shockley, W. (1950): "Electrons and holes in semiconductors" D. van Nostrand Company Inc., New York

Smith, R.A. (1979): "Semiconductors" 2nd ed., Cambridge University Press, London 1979

Spenke, E. (1965): "Elektronische Halbleiter" 2nd ed., Springer Verlag, Berlin, Heidelberg, New York 1965

Sze, S.M. (1981): "Physics of semiconductor devices" 2nd ed. Wiley, New York 1981

Sze, S.M. (1983): "VLSI Technology" McGraw-Hill, New York 1983

Sze, S.M. (1985): "Semiconductor devices, physics and technology" Wiley, New York 1985

Sze, S.M. (1994): "Semiconductor sensors" Wiley, New York 1994

Van Lint, V.A.J., Flanagan, T.M., Leadon, R.E., Naber, J.A., and Rogers, V.C. (1980): "Mechanisms of radiation effects in electronic materials, volume 1" John Wiley and Sons, 1980

Veigele, W.J. (1973): Atomic Data Tables, Vol. 5 No.1 (1973) 51

Wang, S. (1989): "Fundamentals of Semiconductor Theory and Device Physics" Prentice Hall, Englewood Cliffs, New Jersey

Articles

Abe, K., Arodzero,A. (1997) et al: "Design and performance of the SLD vertex detector: a 307 Mpixel tracking system" Nucl. Instr. and Meth. A400 (1997) 287-343

Alig, R.C., Bloom, S. and Struck, C.W. (1980): "Scattering by ionization and phonon emission in semiconductors" Phys. Rev. B, Vol 22, no.12 (1980) 5565-5582; "Scattering by ionization and phonon emission in semiconductors II. Monte Carlo calculations" Phys. Rev. B, Vol 27, no. 2 (1983) 968-977

Anghinolfi, F., Dabrowski, W. (1997) et al.: "SCTA – a rad-hard BiCMOS analogue readout ASIC for the ATLAS semiconductor tracker" IEEE Trans. Nucl. Sci. 44 (1997) 298-302

Bailey, R., Damerell, C.J.S. (1983) et al.: "First Measurements of Efficiency and Precision of CCD Detectors for High Energy Physics" Nucl. Instr. and Meth. 213 (1983) 201-215

Bak, J.F., Burenkov, A. (1987) et al.: "Large departures from Landau distributions for high-energy particles traversing thin Si and Ge targets" Nucl. Phys. B288 (1987) 681-716

Belau, E., Klanner, R. (1983a) et al.: "Charge collection in silicon strip detectors" Nucl. Instr. and Meth. 214 (1983) 253-260

Belau, E., Kemmer, J. (1983b) et al.: "Silicon detectors with $5\,\mu m$ spatial resolution for high energy particles" Nucl. Instr. and Meth. 217 (1983) 224-228

Bethe, H.A. (1930): "Zur Theorie des Durchgangs schneller Korpuskularstrahlen durch Materie" Ann. d. Phys. 5 (1930) 325

Beutenmüller, R.H., Kraner, H.W. (1987) et al.: "Silicon position sensitive detectors for the Helios (NA34) experiment" Nucl. Instr. and Meth. A253 (1987) 500-510

Beuville, E., Borer, K. (1990) et al.: "AMPLEX, a low-noise low-power analog CMOS signal processor for multielement silicon particle detectors" Nucl. Instr. and Meth. A288 (1990) 157-167

Bichsel, H. (1988): "Straggling in thin Silicon detectors" Rev. Mod. Phys. 60 (1988) 663-699

Bloch, F. (1933): "Bremsvermögen von Atomen mit mehreren Elektronen" Z. Phys. 81 (1933) 363

Buskulic, D., Casper, D. (1995) et al.: "Performance of the ALEPH detector at LEP" Nucl. Instr. and Meth. A360 (1995) 481-506

Buttler, W., Lutz, G. (1988) et al.: "Low-noise, low power monolithic multiplexing readout electronics for silicon strip detectors" Nucl. Instr. and Meth. A273 (1988) 778-783

Caccia, M., Evensen, L. (1987) et al.: "A Si Strip Detector with Integrated Coupling Capacitors" Nucl. Instr. & Meth. A260 (1987) 124-131

Castoldi, A., Rehak, P. (1996) et al.: "A new drift detector with reduced lateral diffusion" Nucl. Instr. and Meth A377 (1996) 375-380

Cesura, G., Findeis, N. (1996) et al.: "New pixel detector concepts based on junction field effect transistors on high resistivity silicon" Nucl. Instr. and Meth A377, 521-528 (1996)

Chen, W., Kraner, H. (1992) et al.: "Large area cylindrical silicon drift detector" IEEE Trans.Nucl.Sci. 39 (1992) 619-628

Cottini, C., Gatti, E., Gianelli, G. and Rozzi, G. (1956): "Minimum noise preamplifiers for fast ionization chamber" Nuovo Cimento (1956) 473-483;

Damerell, C.J.S., English, R.L. (1987) et al.: "CCDs for Vertex Detection in High Energy Physics" Nucl. Instr. and Meth. A253 (1987) 478-481;

Damerell, C.J.S., English, R.L. (1990) et al.: "A CCD based vertex detector for SLD" Nucl. Instrum. Methods A288 (1990) 236-239

Decamp, D., Deschizeaux, B. (1990) et al.: "A detector for electron-positron annihilations at LEP" Nucl. Instrum. Methods A294 (1990) 121-178

Dentan, M., Abbon, P. (1996) et al.: "DMILL, a mixed analog-digital radiation-hard BICMOS technology for High Energy Physics electronics" IEEE Trans. Nucl. Sci. 43 (1996) 1763-1767

Di Maria, D.J., Arnold, D. and Cartier, E. (1993): "Impact ionization and degradation in silicon dioxide films on silicon", in: C.R. Helms and B.E. Deal (eds.), The Physics and Chemistry of SiO_2 and the Si–SiO_2 Interface 2 (Plenum Press, New York, 1993)

Fano, U. (1947): "Ionization yield of radiations II: The fluctuations of the number of ions" Phys. Rev. 72 (1947) 26-29

Farell, R., Vanderpuye, K. (1994) et al.: "Radiation detection performance of very high gain avalanche photodiodes" Nucl. Instr. and Meth. A353(1994) 176-179

Feick, H., Fretwurst, E. (1996) et al.: "Long term damage studies using silicon detectors fabricated from different starting materials and irradiated with neutrons, protons and pions" Nucl. Instr. and Meth. A377 (1996) 217-223

Fraser, G.W., Abbey, A.F. (1974) et al.: "The X-ray energy response of silicon" Nucl. Instr. and Meth. A350 (1994) 368-378

Fretwurst, E., Herdan, H. (1990) et al.: "Silicon detector developments for calorimetry: technology and radiation damage" Nucl. Instr. & Meth. A288, (1990) 1-12

Fretwurst, E., Feick, H. (1994) et al.: "Reverse annealing of the effective impurity concentration and long term operational scenario for silicon detectors

detectors in future collider experiments" Nucl. Instr. & Meth. A342, (1994) 119-125

Gadomski, S., Hall, G. (1992) et al.: "The deconvolution method of fast pulse shaping at hadron colliders" Nucl. Instr. & Meth. A320, (1992) 217-227

Gajewski, H., Heinemann, B., Langmach, H., Telschow, G., and Zacharias, K. (1992): "TOSCA – Two dimensional Semiconductor Analysis Package", Handbuch, Karl-Weierstraß Institut, Berlin 1992

Gatti, E. and Rehak, P. (1984a): "Semiconductor Drift Chamber - An Application of a Novel Charge Transport Scheme" Nucl. Instr. and Meth. 225 (1984) 608-614;

Gatti, E., Rehak, P. (1984b) et al.: "Silicon Drift Chambers - First results and optimum processing of signals" Nucl. Instr. and Meth. 226 (1984) 129-141;

Gatti, E., Rehak, P. (1985) et al.: "Semiconductor Drift Chambers" IEEE Trans.Nucl.Sci. 32 (1985) 1204-1208;

Hall, R.N. (1952): "Electron-hole recombination in germanium" Phys. Rev. 87 (1952) 387

Hartmann, R., Strüder, L. (1997) et al.: "Ultrathin entrance windows for silicon drift detectors" Nucl. Instr. and Meth. A387 (1997) 250-254

Heijne, E.H.M., Hubbeling, L. (1980) et al.: "A silicon surface barrier microstrip detector designed for high energy physics" Nucl. Instr. and Meth. 178 (1980) 331-341

Heijne, E.H.M., Antinori, F. (1994) et al.: "First operation of a 72 k element hybrid silicon micropattern pixel detektor array" Nucl. Instr. and Meth. A349 (1994) 138-155

Heijne, E.H.M., Antinori, F. (1996) et al.: "LHC1: a semiconductor pixel detector readout chip with internal, tunable delay providing a binary pattern of selected events" Nucl. Instr. and Meth. A383 (1996) 55-63

Hofmann R., Lutz, G. (1984) et al.: "Development of readout electronics for monolithic integration with diode strip detectors" Nucl. Instr. and Meth. 226 (1984) 196-199

Kandiah, K., Deighton, M.O. and Whiting, F.B. (1981): "Low frequency noise mechanisms in field effect transistors" Proc. 6th Conf. on noise in physical systems, NBS Publication 614, U.S. Department of Commerce,1981

Kandiah, K. (1983): "Energy levels of bulk defects responsible for L.F. noise in Si JFETs" Noise in Physical Systems and $1/f$ Noise, pg 287-290, Elsevier Science Publishers B.V., 1983

Kandiah, K. (1986): "Low frequency noise mechanisms in field effect transistors" Noise in Physical Systems and $1/f$ Noise, pg 19-25, Elsevier Science Publishers B.V., 1986

Kandiah, K., Deighton, M.O. (1989) et al.: "A physical model for random telegraph signal currents in semiconductor devices" J. Appl. Phys. 66, (1989) 937-948

Kemmer, J. (1980): "Fabrication of low noise silicon radiation detectors by the planar process" Nucl. Instr. & Meth. A169 (1980) 499-502

Kemmer, J. and Lutz, G. (1987): "New semiconductor detector concepts" Nucl. Instr. & Meth. A253 (1987) 356-377

Kemmer, J., Lutz, G. (1987) et al.: "Low capacitive drift diode" Nucl. Instr. & Meth. A253 (1987) 378-381

Kemmer, J. and Lutz, G. (1988): "New structures for position sensitive semiconductor detectors" Nucl. Instr. & Meth. A273 (1988) 588-598

Kemmer, J., Lutz, G. (1990) et al.: "Experimental confirmation of a new semiconductor detector principle" Nucl. Instr. & Meth. A288 (1990) 92-98

Kemmer J. and Lutz, G. (1993): "Concepts for simplification of strip detector design and production" Nucl. Instr. and Meth. A326 (1993) 209-213

Kleinfelder, S.A., Carithers, W.C. jr. (1988) et al.: "A flexible 128 channel silicon strip detector instrumentation integrated circuit with sparse readout" IEEE Trans. Nucl. Sci. 35 (1988) 171-175

Landau, L. (1944): J. Phys. (USSR) 8 (1944) 201

Lechner, P., Hartmann, R., Soltau, H. and Strüder, L. (1996): "Pair creation energy and Fano factor of silicon in the energy range of soft x-rays" Nucl. Instr. and Meth. A377 (1996) 206-208

Li, Z. (1994): "Modelling and simulation of neutron induced changes and temperature annealing of N_{eff} and changes in resistivity in high resistivity silicon detectors" Nucl. Instr. and Meth. A342 (1994) 105-118

Longoni, A., Gatti, E. and Sacco, R. (1995): "Trapping noise in semiconductor devices: a method for determining the noise spectrum as a function of the trap position" Journal of Applied Physics 78 (1995) 6283-6297

Longoni, A. (1990), Sampietro, M. and Strüder, L.: "Instability of the breakdown of high resistivity silicon detectors due to the presence of oxide charges" Nucl. Instr. and Meth. A288 (1990) 35-43

Lumb, D. (1997) et al: "X-ray Multi-Mirror Mission – an overview" SPIE 2808 (1997) 326

Lutz, G. (1986): "Present and Future Semiconductor Tracking Detectors" Vertex Detectors, Plenum Press, New York, 1988, pp. 195-224

Lutz, G. (1991): "Correlated noise in silicon strip detector readout" Nucl. Instr. and Meth. A309 (1991) 545-551

Lutz, G. (1994): "A simplistic model for reverse annealing in irradiated silicon" Nucl. Instr. and Meth. B95 (1995) 41-49

Lutz, G. (1996): "Effects of deep level defects in semiconductor detectors" Nucl. Instr. and Meth. A377 (1996) 234-243

Matheson, J. (1996), M. Robbins and S. Watts: "The effect of radiation induced defects on the performance of high resistivity silicon diodes", Nucl. Instr. and Meth. A377 (1996) 224-227

Mours, B., Boudreau, J. (1996) et al.: "The design, construction and performance of the ALEPH silicon vertex detector" Nucl. Instr. and Meth. A379 (1996) 101-115

Nyquist, H. (1928): "Thermal agitation of electrical charge in conductors" Phys. Rev. 32 (1928) 110-113

Parker, S.I., Kenney, C.J. (1994) et al.: "A prototype monolithic pixel detector"

Pinotti, E., Bräuninger, H. (1993) et al: "The pn-CCD on-chip electronics" Nucl. Instr. and Meth. A326 (1993) 85-91

Pitzl, D., Cartiglia, N. (1992) et al: "Type inversion in silicon detectors" Nucl. Instr. and Meth. A311 (1992) 98-104

Radeka, V., Rehak, P. (1989) et al.: "Implanted silicon JFET on completely depleted high resistivity devices" IEEE Electron device letters 10 (1989) 91-95

Rehak, P., Gatti, E. (1985) et al.: "Semiconductor drift chambers for position and energy measurements" Nucl. Instr. & Meth. A235 (1985) 224-234

Rehak, P., Walton, J. (1986) et al.: "Progress in semiconductor drift detectors" Nucl. Instr. & Meth. A248 (1986) 367-378

Rehak, P., Gatti, E. (1989) et al.: "Spiral silicon drift detectors" IEE Trans. Nucl. Sci. 36 (1989) 203-209

Rehak, P., Rescia, S. (1990) et al.: "Feedback charge amplifier integrated on detector wafer" Nucl. Instr. and Meth. A288 (1990) 168-175

Richter, G. (1996) et al.: "ABRIXAS, A Broadband Imaging X-ray All-sky Survej"; L. Bassani, G.di Cocco (eds.): Imaging in High Energy Astronomy. Experim. Astron., (1996) 159

Richter, R.H., Andricek, L. (1996) et al.: "Strip detector design for ATLAS and HERA-B using two dimensional device simulation" Nucl. Instr. and Meth. A377 (1996) 412-421

Sansen, W. (1987): "Integrated low noise amplifiers in CMOS technology" Nucl. Instr. and Meth. A253 (1987) 427-433

Scharfetter, D.L. and Gummel, H.K. (1969): "Large-signal analysis of a silicon Read diode oscillator", IEEE Trans. ED-16 (1969) 64-77

Shapiro, S.L., Dunwoodie, W.M. (1989) et al.: "Silicon PIN diode array hybrids for charged particle detection" Nucl. Instr. and Meth. A275 (1989) 580-586

Sedlmeir, J. (1985): "Untersuchung über einseitig und zweiseitig auslesbare Siliziumstreifendetektoren" Diplomarbeit, Technische Universität München, 15. August 1985

Shockley, W. and Read, W.T. (1952): "Statistics of the recombination of holes and electrons" Phys. Rev. 87 (1952) 835-842

Soltau, H., Holl, P. (1996) et al.: "Performance of the pn-CCD detector system designed for the XMM satellite mission" Nucl. Instr. and Meth. A377 (1996) 340-345

Strüder, L., Bräuninger, H. (1990) et al: "The MPI/AIT x-ray imager (MAXI) - high speed pn-CCDs for x-ray detection" Nucl. Instr. and Meth. A288 (1990) 227-235

Strüder, L., Bräuninger, H. (1993) et al: "First results with the pn-CCD detector system for the XMM satellite mission" Nucl. Instr. and Meth. A326 (1993) 129-135

Strüder, L., Bräuninger, H. (1997) et al: "A 36 cm^2 large monolythic pn-charge coupled device x-ray detector for the European XMM satellite mission" Rev.Sci.Instrum. 68 (1997) 4271-4274

Tapan, I., Duell, A.R. (1997) et al.: "Avalanche photodiodes as proportional particle detectors" Nucl. Instr. and Meth. A388 (1997) 79-90

Toker, O., Masciocchi, S. (1994) et al.: "VIKING, a CMOS low noise monolithic 128 channel frontend for Si-strip detector readout" Nucl. Instr. and Meth. A340 (1994) 572-579

Vasilescu, A. (1998) et al.: "Fluence normalisation based on the NIEL scaling hypothesis", 3rd ROSE Workshop, DESY Hamburg, 12-14 February 1998, DESY-Proc. 1998-2

Walker, J.T., Parker, S. (1984) et al.: "Development of high density readout for silicon strip detectors" Nucl. Instr. and Meth. 226 (1984) 200-203

Wunstorf, R. (1992a): "Systematische Untersuchungen zur Strahlenresistenz von Silizium-Detektoren für die Verwendung in Hochenergiephysik-Experimenten" Ph.D. thesis, Universität Hamburg, 1992

Wunstorf, R., Benkert, M. (1992b) et al.: "Results on radiation hardness of silicon detectors up to neutron fluences of $10^{15} \, n/cm^2$" Nucl. Instr. and Meth. A315 (1992) 149-155

Index

Lightning Source UK Ltd.
Milton Keynes UK
18 May 2010

154364UK00002B/62/P